A TEXT BOOK OF

WASTE WATER TREATMENTS

(Elective - III)

FOR
SEMESTER – VI

THIRD YEAR (T.Y) B. TECH COURSE IN CIVIL ENGINEERING

Strictly According to New Revised Credit System Syllabus of Babasaheb Ambedkar Technological University (BATU), Lonere, (Dist. Raigad) Maharashtra,
(w.e.f. June 2019-20)

Dr. S. S. JAHAGIRDAR
M.E. ((Envir.), Ph.D. (Civil, NITK, Surathkal)
Professor in Civil & Engg. Deptt.
Dean R & D
N. K. Orchid College of Engineering & Technology
SOLAPUR.

Dr. V. S. RAJAMANYA
M.E. (Civil), Ph.D.
Head of Civil Engg. Deptt.
Mahatma Basveshwar Education Society's
College of Engineering
Ambajogai, Dist. BEED.

Dr. R. K. LAD
Ph.D. (Engg.)
Campus Director,
JSPM Narhe Technical Campus,
Narhe, PUNE.

Dr. M. R. GIDDE
Ph.D.
Professor, Dept. of Civil Engineering,
BVDU's College of Engineering,
Dhankawadi, PUNE.

N1127

WASTE WATER TREATMENTS (CIVIL, BATU) ISBN : 978-93-89825-86-2

First Edition	:	**January 2020**
©	:	**Authors**

Published By :

NIRALI PRAKASHAN

Abhyudaya Pragati, 1312, Shivaji Nagar,
Off J.M. Road, Pune – 411005
Tel - (020) 25512336/37/39, Fax - (020) 25511379
Email : niralipune@pragationline.com

➢ DISTRIBUTION CENTRES

Pune

Nirali Prakashan (Local) : 119 Budhwar Peth, Jogeshwari Mandir Lane, Pune 411002, Maharashtra
Tel : (020) 2445 2044, Mobile : 9657703145, Email : niralilocal@pragationline.com

Nirali Prakashan (Outstation) : S. No. 28/27 Dhayari, Near Asian College, Dhayari, Pune 411041, Maharashtra
Tel : (020) 2469 0204, Fax : (020) 2469 0316, Mobile : 9657703143
Email : bookorder@pragationline.com

MUMBAI

Nirali Prakashan : 385 S.V.P. Road, Rasadhara Co-op. Hsg. Society Ltd., Girgaum, Mumbai 400004, Maharashtra
Tel : (022) 2385 6339 / 2386 9976, Fax : (022) 2386 9976, Mobile : 9320129587
Email : niralimumbai@pragationline.com

➢ DISTRIBUTION BRANCHES

JALGAON

Nirali Prakashan : 34 V. V. Golani Market, Navi Peth, Jalgaon 425001, Maharashtra
Tel : (0257) 222 0395, Mob : 94234 91860, **Email** : niralijalgaon@pragationline.com

KOLHAPUR

Nirali Prakashan : New Mahadvar Road, Kedar Plaza 1st Floor, Opp. IDBI Bank
Kolhapur 416012, Maharashtra. Mobile : 9850046155
Email : niralikolhapur@pragationline.com

NAGPUR

Nirali Prakashan : Above Maratha Mandir, Shop No 3, Second Floor,
Rani Jhanshi Square, Sitabuldi, Nagpur 440012, Maharashtra
Tel : (0712) 254 7129, Email : niralinagpur@pragationline.com

Nirali Prakashan : **DELHI**
4593/15 Basement, Agarwal Lane, Ansari Road, Daryaganj
Near Times of India Building, New DelhiV 110002 Mobile : 8505972553
Email : niralidelhi@pragationline.com

BENGALURU

Nirali Prakashan : Maitri Ground Floor, Jaya Apartments, No. 99, 6th Cross, 6th Main,
Malleswaram, Bengaluru 560003, Karnataka
Mobile : 9449043034, **Email** : niralibangalore@pragationline.com

niralipune@pragationline.com | www.pragationline.com
Also find us on www.facebook.com/niralibooks

Dedicated to . . . .

Our Beloved Students

...Authors

PREFACE

It gives us great pleasure to present the book **"Waste Water Treatments" (Elective III)** for the students of **Semester VI Third Year (T.Y.) B. Tech. Course Civil Engineering of Dr. Babasaheb Ambedkar Technological University (BATU), Lonere, Dist. Raigad (Maharashtra)**. This book is strictly as per the new revised syllabus 2019-20 Pattern, effective from the Academic Year July 2019-20.

In New Revised Syllabus, there will be Internal Assessment (20 Marks), Mid Sem. Exam. (20 Marks) and End Sem. Exam. (60 Marks). End Sem. Exam. will be based on all six units and each unit will carry 12 Marks. The Theory Course will have 3 Credits.

The basic objective of this book is to bridge the gap between the vast contents of the reference books, written by the renowned International Authors and the concise requirements of Undergraduate Students. This book has been written in a comprehensive manner using Simple and Lucid language, keeping in mind students' requirements. The main emphasis has been given on exploring the basic concepts rather than merely the Information. Solved Examples and Exercises have been provided throughout the book and at the end of the Unit. Also, we have given **Model Question Papers** for practice at the end of book.

Our special thanks to our family members, students and all those who directly or indirectly supported us in this project.

We take this opportunity to express our thanks to Shri. Dineshbhai Furia, Shri. Jignesh Furia, Mrs. Nirali Verma, Shri. M. P. Munde and entire team of Nirali Prakashan, namely Mrs. Deepali Lachake (Co-ordinator), and her colleagues who really have taken keen interest and untiring efforts in publishing this text.

Suggestions of our esteemed readers are most welcome and will be highly appreciated.

Pune   **Authors**

SYLLABUS

Module 1: Waste Water Treatment   **(5 Lectures)**

Introduction of wastewater, its types and various sources, Concept of sewage, sullageand storm water, Necessity of treatment of waste water

Preliminary Treatment: screening and grit removal units, oil and grease removal, Primary treatment,

Secondary Treatment: Activated sludge process, trickling filter, sludge digestion, drying bed. Stabilization pond, septic tank, soakage system, Imhoff Tank, recent trends and advanced wastewater treatment: nutrient removal, solids removal

Module 2: Low Cost Wastewater Treatment Methods   **(7 Lectures)**

Principles of waste stabilization pond, Design and operation of oxidation pond, aerobic & anaerobic Lagoons, Aerated Lagoon, Oxidation ditch, Septic tank. Concept of recycling of sewage Disposal of waste water-stream pollution, Self Purification, DO sag curve, Streeter Phelp's Equation, Stream classification, disposal on land, effluents standards for stream

and land disposals

Module 3: Industrial Waste Water Treatment Management   **(6 Lectures)**

Sources of Pollution: Physical, Chemical, Organic and Biological properties of Industrial Wastes – Differences between industrial and municipal waste waters –Effects of industrial effluents on sewers and treatment plants, Prevention vs Control of Industrial Pollution

Pre and Primary Treatment: Equalization, Proportioning, Neutralization, Oil Separation by Floatation, Prevention v/s Control of Industrial Pollution

Module 4: Waste Water Treatment Methods   **(7 Lectures)**

Nitrification and De-nitrification – Phosphorous removal – Heavy metal removal – Membrane Separation Process–Reverse osmosis– Chemical Oxidation–Ion Exchange – Air Stripping and Absorption Processes – Special Treatment Methods – Disposal of Treated Waste

Common Effluent Treatment Plants (CETPs): Need, Planning, Design, Operation & Maintenance Problems

Module 5: Environmental Sanitation   **(6 Lectures)**

Communicable diseases, Methods of communication, Diseases communicated by discharges of intestines, nose and throat, other communicable diseases and their control

Insects and Rodent Control - Mosquitoes, life cycles, factors of diseases control methods - natural &chemical, Fly control methods and fly breeding prevention, Rodents and public health, plague control methods, engineering and bio-control methods

Module 6: Rural Sanitation   **(5 Lectures)**

Rural areas, Population habits and environmental conditions, problems of water supply and sanitation aspects, low cost excreta disposal systems, Rural sanitation improvement schemes

CONTENTS

WASTE WATER TREATMENT

1.1 INTRODUCTION

- Wastewater engineering is that branch of environmental engineering in which the basic principles of science and engineering are applied to the problems of water pollution control.
- Related to sources of generation, wastewater may be defined as a combination of the liquid or water carried wastes removed from residences, institutions and commercial and industrial establishments, together with such ground water, surface water and storm water as may be present.
- Wastewater or sewage is the byproduct of many uses of water. There are the household uses such as showering, dishwashing, laundry and, of course, flushing the toilet. Additionally, companies use water for many purposes including processes, products, and cleaning or rinsing of parts. After the water has been used, it enters the wastewater stream, and it flows to the wastewater treatment plant. When people visit a treatment plant for the first time, often it is not what they perceived it would be. These wastewater plants are complex facilities and provide high quality end product.

1.2 SOURCES OF SEWAGE OR WASTEWATER FLOWS   (May 17, Aug. 17)

The following are the sources of wastewater flows, which are dependent on the type of collection system:

1. Domestic or sanitary wastewater.
2. Industrial wastewater.
3. Infiltration / Inflow.
4. Storm water.

1. **Domestic Wastewater :** The principal sources of domestic wastewater are residential and commercial buildings, other important sources include institutional and recreational facilities.

   (i) Residential : For small residential areas, wastewater flows are commonly determined on the basis of the average per capita contribution of wastewater and population density.

   For large residential areas, wastewater flows are determined by considering 75% to 80% of the accounted water supplied from the water works. Residential sources of wastewaters are apartments, hotels, individual dwellings, etc.

   (ii) Commercial : Commercial wastewater flows are generally based on existing or anticipated future development or comparative data.

   Commercial sources of wastewater are airports, automobile service stations, hotels, industrial buildings (excluding industry and cafeteria), laundries (self-service), motels, motels with kitchen, offices, restaurants, shopping centers, etc.

   (iii) Institutional Facilities : Institutional sources of wastewaters are hospitals (medical), hospitals (mental), prisons, schools, etc.

   (iv) Recreational Facilities : Recreational sources of wastewaters are apartments (resorts), cafeteria, coffee shops, country clubs, day camps (no meals), dining halls, hotels (resort), swimming pools, theaters, stores (resort), etc.

2. **Industrial Wastewater :** Industrial wastewater flow rates vary with the type and size of the industry, the supervision of the industry, the degree of water reuse, etc.

3. **Infiltration / Inflow :** The total quantity from both infiltration and inflow.

   (i) Infiltration : round water or subsoil water may enter into the sewers through leaky joints.

   (ii) Inflow : The water discharged into a sewer system, including service connections.

4. **Storm Water :** Storm is generated due to rain water flowing over the ground surface, pavements, roofs, etc.

1.3 NECESSITY OF WASTEWATER TREATMENT

- It's a matter of caring for our environment and for our own health. There are a lot of good reasons why keeping our water clean is an important priority:

- **Fisheries:** Clean water is critical to plants and animals that live in water. This is important to the fishing industry, sport fishing enthusiasts, and future generations.

- **Wildlife Habitats:** Our rivers and ocean waters teem with life that depends on shoreline, beaches and marshes. They are critical habitats for hundreds of species of fish and other aquatic life. Migratory water birds use the areas for resting and feeding.

- **Recreation and Quality of Life:** Water is a great playground_ for us all. The scenic and recreational values of our waters are reasons many people choose to live where they do. Visitors are drawn to water activities such as swimming, fishing, boating and picnicking.

- **Health Concerns :** If it is not properly cleaned, water can carry disease. Since we live, work and play so close to water, harmful bacteria have to be removed to make water safe.

- If wastewater is not properly treated, then the environment and human health can be negatively impacted. These impacts can include harm to fish and wildlife populations, oxygen depletion, beach closures and other restrictions on recreational water use, restrictions on fish and shellfish harvesting and contamination of drinking water.

- Some examples of pollutants that can be found in wastewater and the potentially harmful effects these substances can have on ecosystems and human health are as follows:

  ➢ Decaying organic matter and debris can use up the dissolved oxygen in a lake so fish and other aquatic biota cannot survive;

  ➢ Excessive nutrients, such as phosphorus and nitrogen (including ammonia), can cause eutrophication, or over-fertilization of receiving waters, which can be toxic to aquatic organisms, promote excessive plant growth, reduce available oxygen, harm spawning grounds, alter habitat and lead to a decline in certain species;

  ➢ Chlorine compounds and inorganic chloramines can be toxic to aquatic invertebrates, algae and fish;

  ➢ Bacteria, viruses and disease-causing pathogens can pollute beaches and contaminate shellfish populations, leading to restrictions on human recreation, drinking water consumption and shellfish consumption;

  ➢ Metals, such as mercury, lead, cadmium, chromium and arsenic can have acute and chronic toxic effects on species.

  ➢ Other substances such as some pharmaceutical and personal care products, primarily entering the environment in wastewater effluents, may also pose threats to human health, aquatic life and wildlife.

1.4 IMPORTANT TERMS AND DEFINITIONS

- **Sullage :** Sullage is a term used to indicate the wastewater from bathrooms, kitchens, wash basins etc. As organic matter in it is either absent or is of negligible amount, so it does not create bad smell.

- **Sewage :** It includes sullage, discharge from latrines, urinals, stables, industrial waste and also the ground surface and storm water that may be admitted into the sewer.

- **Domestic Sewage :** It consists of liquid wastes originating from urinals, latrines, bathrooms, kitchen sinks, wash basins etc. of the residential, commercial or institutional buildings. Since, it contains human excreta and urine, it is extremely foul in nature.

- **Industrial Sewage :** It consists of liquid wastes originating from the industrial processes of various industries, such as distillery, dairy, paper mill, textile, brewing etc.

  The quality of the industrial sewage depends upon the type of industry. It may contain objectionable organic compounds and thus may require extensive treatment before being disposed of in public sewers.

- **Sanitary Sewage :** It is the sum total of domestic and industrial sewage.

- **Storm Sewage :** The run-off resulting from the rain storms is called storm sewage or storm drainage or simply drainage.

- **Night Soil :** It is a term used to indicate the human and animal excreta.

- **Sewer :** It is an underground conduit through which sewage is carried to a point of disposal.

  Sewers can be Classified as Follows :

  ➢ **Separate Sewers :** These type of sewers are carrying the household and industrial wastes only.

  ➢ **Storm Water Sewers :** These type of sewers are carrying rain water.

  ➢ **Combined Sewers :** These type of sewers are carrying both sewage and storm water.

  ➢ **House Sewer :** It is a pipe to carry sewage from a building to street sewer.

  ➢ **Lateral Sewer :** This type of sewer collects sewage from the houses.

  ➢ **Branch Sewer (Submain Sewer) :** This is a sewer which receives sewage from laterals and discharge into a main sewer.

  ➢ **Main Sewer (Trunk Sewer) :** This is a sewer which receives sewage from tributary branches and sewers.

- ➢ **Depressed Sewer :** This type of sewer is at lower level than adjacent sewers. This sewer runs full under the force of gravity and at more than atmospheric pressure.
- ➢ **Intercepting Sewer :** This type of sewer flowing parallel to a natural drainage channel, into which a number of main sewers discharge.
- ➢ **Outfall Sewer :** This type of sewer receives the sewage from the collecting system. It's location is at final discharge.
- **Sewarage :** This term is applied to the art of collecting, treating and finally disposing of the sewage.
- **Oxygen Deficit :**
  Oxygen deficit = [Saturation dissolved oxygen – Actual dissolved oxygen]
- **Bio-Chemical Oxygen Demand (BOD) :** The amount of oxygen consumed by aerobic bacteria for the decomposition of wastewater (ofcourse upto oxidation is completed) is called BOD.
- **Chemical Oxygen Demand (COD) :** A known quantity of wastewater is mixed with a known quantity of standard solution of potassium dichromate and the mixture is heated. The organic matter is oxidised in presence of H_2SO_4 by potassium dichromate. The resulting solution of potassium dichromate is titrated. The oxygen used in oxidising the wastewater is known as COD.
- **Total Organic Carbon (TOC) :** TOC test is the method for measuring the small concentrations of organic matter.
- **ThOD :** Theoretical oxygen demand.
- **Treatability Index (T.I.) :** $\text{T.I.} = \dfrac{BOD}{COD - BOD}$.
- **Dilution Factor :**

  $$\text{Dilution factor} = \frac{\text{Volume of the diluted sample}}{\text{Volume of the undiluted sewage sample}}$$

- **Population Equivalent :**

  $$\text{Population equivalent} = \frac{\text{Standard } BOD_5 \text{ of the industrial wastewater in kg/day}}{\text{Standard } BOD_5 \text{ of domestic sewage per kg per person per day}}$$

  Generally, the BOD_5 of domestic sewage is 0.08 kg/day/person.

- **Relative Stability :**

  $$\text{Relative stability} = \frac{\text{Oxygen available in the effluent}}{\text{Total oxygen required to satisfy its first stage BOD demand}}$$

- **Detention Period :**

  $$\text{Detention period} = \frac{\text{Volume of the tank}}{\text{Rate of flow in the tank}}$$

- **Volumetric BOD loading / Organic Loading :**

  $$\text{Volumetric BOD loading/Organic loading} = \frac{\text{Mass of BOD applied per day in gm}}{\text{Volume of the aeration tank in } m^3}$$

- **Food to Micro-Organisms Ratio (F/M) :**

  $$\text{F/M ratio} = \frac{\text{BOD load applied per day to the aerator system in gm}}{\text{Total microbial mass in the system in gm}}$$

- **MLSS :** Mixed liquor suspended solids.
- **MLVSS :** Mixed liquor volatile suspended solids.
- **Sludge Age :**

  $$\text{Sludge age} = \frac{\text{Mass of suspended solids in the system}}{\text{Mass of solids leaving the system per day}}$$

1.5 PRELIMINARY TREATMENT

The Preliminary Treatment Consists of :

- Screening – for removal of floating matter.
- Grit chamber – for removal of sand and grit.
- Skimming tank – for removal of oil and grease.

The Primary Treatment Consists of :

- Settling tanks – for removal of suspended solids.

The possible arrangements of these units are shown in Fig. 1.1.

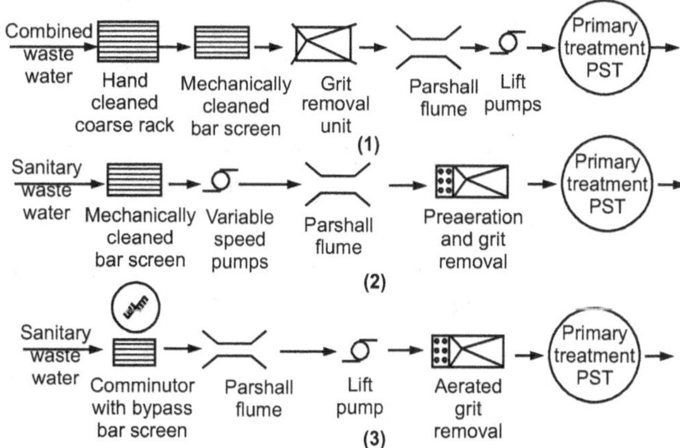

Fig. 1.1 : Possible arrangements of preliminary and primary treatment units in wastewater processes

1.6 SCREENING OF SEWAGE (Aug. 15)

- Screening of sewage normally removes large floating solids from the sewage. Screening of sewage has to be done in the sewerage system also, e.g., in a pumping station, coarse screens are normally provided before the pumps.

- Screens are provided as a common unit in sewage treatment plants to avoid the following troubles that the floating solids may cause :

  - They may clog the pumps.

  - They may interfere with the secondary treatment by clogging the trickling filter media or by obstructing aeration in activated sludge process.

- Sometimes, when the sewage is to be disposed off without any treatment, the gross solids have still to be removed to avoid their floating on the receiving water, thus preventing the reaeration and giving an ugly appearance.

1.6.1 Classification of Screens

The screens are normally classified in number of ways as follows :

- Based on the size of opening – such as coarse, medium and fine screens.

- Depending on the shape – such as disc, drum, band etc.

- The mode of settling such as fixed, movable or moving types.

- The method of cleaning such as manual or mechanical types.

Classification based on Sizes of Opening Normally divides the Screens as follows :

- **Coarse Screens** when the size of opening is above 80 mm, **Medium Screens** when the size of opening is between 20 mm –50 mm and **Fine Screens** when the size of opening is less than 20 mm.

- The disc, band and cage screens are very commonly used in the case of industrial waste treatment and in chemical engineering operations.

- In normal practice, medium bar screens or racks are used in municipal sewage treatment plants. Thus, in a screening chamber, bars, rectangular or circular in shape, are kept at an inclination to the horizontal. Normally, sharp edged rectangular bars (10 mm × 40 mm) set about 25 mm apart are used. The inclination of these bars is normally kept between 30° to 45° for manual cleaned types and 45° to 60° in such cases

where the cleaning is done by using mechanical means. Normal practice in municipal sewage treatment plants is to use fixed screens.

- As the velocity inside the chamber is reduced below the value in the approach channel, the transition should be as far as possible smooth to avoid undue headloss and related problems.

- The velocity through the screen opening is normally kept at 0.6 m per second at 50% clogging of the screens. Thus, on knowing the velocity to be provided, the requisite area of opening can be calculated; on the basis of which the width of screen chamber should be arrived at.

- As far as possible, the depth of flow inside the screening chamber is maintained at the same level as that in the approach channel. As the sewage flows through the screens, head loss occurs and to maintain the continuity of the flowline, normal practice is to depress the invert of the channel on the downstream side of the screen by an equal amount as shown in Fig. 1.2.

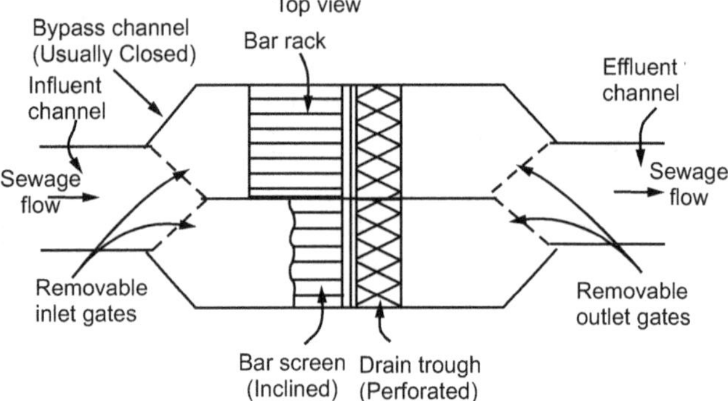

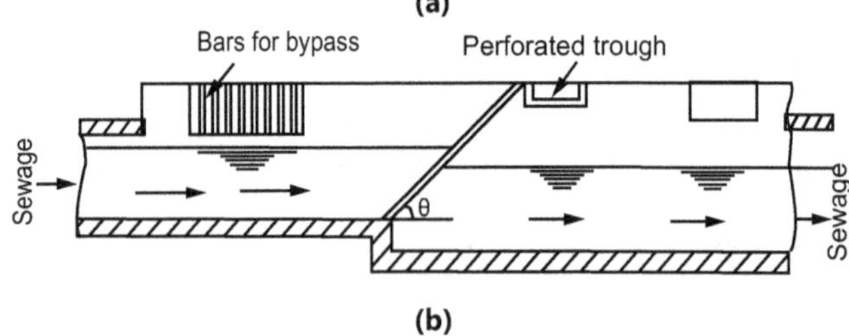

Fig. 1.2 : Screen chamber

- The head loss through the screens can be calculated in the following ways :

(i)
$$h = (V^2 - v^2) \times 0.0729$$

where, h = Head loss in m

V = Velocity through screens in m/sec

v = Velocity before screens in m/sec.

(ii) O' Kirschmer (1926) Formula :

$$h = \beta\left(\frac{W}{b}\right)^{1.33} h_V \sin\theta$$

where, β = 2.42 for sharp edged rectangular bars

= 1.79 for circular bars

$h_V = \dfrac{v^2}{2g}$ where, v = Approach velocity in m/sec

and, θ = Inclination of bars to horizontal.

- As the material goes on getting accumulated on the face of the screens, the liquid level in the chamber starts rising. When the liquid level rises sufficiently high, the mechanical cleaning devices are put into operation. The mechanical cleaning device can be either the time operated switch or float operated switch type.

- However, regulations make it mandatory that a float operated switch must be provided even in such cases where time switch operated cleaning mechanism is used. As can be easily guessed, the float operated switch is sturdier and hence more reliable. The manual cleaning of the screens is normally done once in a day in the case of flows less than 5 mld.

- If the cleaning devices fail and the sewage is likely to overflow the channels, an overflow weir is normally provided, which is equipped with bars set 0.6 m apart, which take the liquid to downstream side of the screen through a channel.

- The material removed from against the screening surface is normally found to contain a high moisture content (70 – 80%) which can be reduced by drying on a draining platform.

- The amount of screenings removed normally depend upon the type of sewage and the size of openings in the screens. In municipal practice, the amount of screenings normally vary between 0.0015 m^3 per million litre with screen sizes of 10 cm and 0.015 m^3 per million litre in case of 2.5 cm size.

- These screenings as they are removed from the screening chamber are found to contain a large moisture content of about 90%. Hence, it is desirable to reduce this moisture content to at least 60% for proper disposal. This is normally done in India, by sun drying. After the moisture content is reduced to 60%, it

can be disposed off by either composting it, burying in pits or by burning them.

- Sometimes when the sewage has to be disposed off without any treatment, fine screening is done which removes most of the floating and lighter solids. The material that is removed, however, also contains a large proportion of organic matter and hence its disposal poses a problem. Also its efficiency is hardly 1/5 of what would be obtained in a sedimentation tank and in addition it entails a high maintenance cost.

- Now-a-days, hence, these are replaced by cutting devices known as **Comminuters or Shredders.** These normally consist of a revolving slotted drum with a number of cutters mounted on its surface which shear the material retained against a brush. The liquid can flow to the downstream side, only by passing through opening of 0.5 to 1 cm size in the drum surface. Thus, the retained particles when reduced to this size escape to the downstream side.

1.6.2   Operation and Maintenance of Screens

- Regular cleaning of hand cleaned screens to prevent backing up of sewage.
- Lubrication of mechanical screens as per instructions of manufacturers.
- Painting of entire mechanism once in a year.
- Visual inspection to check whether screenings are retained between bars or being pushed through them due to more velocity.
- Volume of screening collected should be recorded regularly.
- Prompt and hygienic disposal of screenings.
- Daily record of operation and maintenance should be maintained.

1.6.3   Disposal of Screenings

- Screenings are disposed off by the following methods :
  - ➤ Burial
  - ➤ Incineration
  - ➤ Digestion
  - ➤ Grinding.

SOLVED EXAMPLES ON SCREENS

Example 1.1 : *Design a bar screen for a peak flow 50 million litres per day.*

Solution :

Assume manual cleaning.

Keep the bars at inclination of 45° with horizontal.

Assume the rectangular bars of size 10 mm × 50 mm with 10 mm dimension facing the flow.

Assume clear spacing of 40 mm between the bars.

Assume the velocity through the screen as 0.8 m/s at peak flow.

Maximum rate of flow

$$= 50 \text{ mld} = \frac{50 \times 10^6}{24 \times 60 \times 60 \times 1000}$$

$$= 0.578 \text{ m}^3/\text{sec}$$

$\therefore$ Net area of screen

$$= \frac{0.578}{0.8} = 0.7225 \text{ m}^2$$

$\therefore$ Gross area of screen

$$= 0.7225 \times \frac{50}{40} = 0.903 \text{ m}^2$$

(For 40 mm opening, net area is 0.7225 m². Therefore, for 50 mm, gross area is $0.7225 \times \dfrac{50}{40}$)

As the screen is inclined at 45° with the horizontal,

Gross area of screen

$$= \frac{0.903}{\sin 45°} = 1.277 \text{ m}^2$$

Also velocity of flow above screen $= 0.8 \times \dfrac{40}{50} = 0.64$ m/s

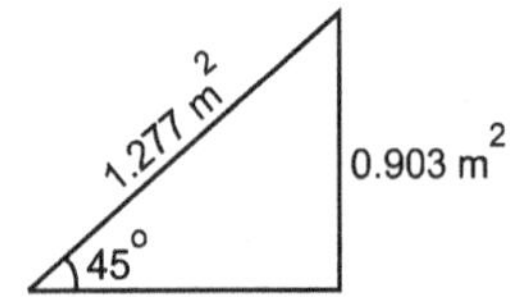

Fig. 1.3

Thus, we have $V = 0.8$ m/s, $v = 0.64$ m/s

$\therefore \quad h_L = 0.0729 \ (V^2 - v^2)$

$$= 0.0729 \ [(0.8)^2 - (0.64)^2]$$

$$= 0.017 \text{ m} = 1.7 \text{ cm}$$

This will be the head loss when the screen is clean.

If, however, the screen is half clogged,

$$V = 2 \times 0.8 = 1.6 \text{ m/s} \qquad \text{...Ans.}$$

$\therefore \quad h_L = 0.0729 \ [(1.6)^2 - (0.64)^2]$

$$= 0.157 \text{ m}$$

$$= 15.7 \text{ cm} \qquad \text{...Ans.}$$

Example 1.2 : *Design the screen chamber of a ETP to treat a peak flow of 60 mld of sewage.*

Solution :

Assume manual cleaning.

Keep the bars at inclination of 60° with horizontal.

Assume the rectangular bars of size 10 mm × 70 mm; 10 mm dimension facing the flow.

Assume clear spacing between bars as 50 mm.

Assume the velocity through the screen as 0.8 m/s at peak flow.

Maximum rate of flow

$$= 60 \text{ mld} = 60,000 \text{ m}^3/\text{day}$$

$$= 0.694 \text{ m}^3/\text{sec}$$

$\therefore$ The net area of screen openings required

$$= \frac{0.694}{0.8}$$

$$= 0.87 \text{ m}^2$$

$\therefore$ Gross area of screen

$$= 0.87 \times \frac{60}{50} = 1.044 \text{ m}^2$$

As the screen is inclined at 60° with the horizontal,

Gross area of screen

$$= \frac{1.044}{\sin 60°} = 1.205 \text{ m}^2$$

Also velocity of flow above screen $= 0.8 \times \dfrac{50}{60} = 0.67$ m/s

Thus, we have $V = 0.8$ m/s, $v = 0.67$ m/s

$\therefore \quad$ Head loss,

$$h_L = 0.0729 \ (V^2 - v^2)$$

$$= 0.0729 \ [(0.8)^2 - (0.67)^2]$$

$$= 0.014 \text{ m}$$

$$= 1.4 \text{ cm}$$

Thus, provide the screen of area $= 1.205 \text{ m}^2$ **...Ans.**

Dimensions of bars $= 10 \text{ mm} \times 70 \text{ mm}$

Clear spacing between bars $= 50 \text{ mm}$ **...Ans.**

Example 1.3 : *A screen consisting of 10 mm diameter bars, at a clear spacing of 40 mm, treats a maximum hourly flow of 1275 m³, velocity of flow through the screen chamber = 75 cm/sec. Work out :*

(i) Length and number of bars.

(ii) Head loss in the chamber

Solution :

Maximum flow $= 1275 \text{ m}^3/\text{hr}$

$$= \frac{1275}{60 \times 60} = 0.35 \text{ m}^3/\text{sec}$$

$$V = 75 \text{ cm/sec}$$

$$= 0.75 \text{ m/sec}$$

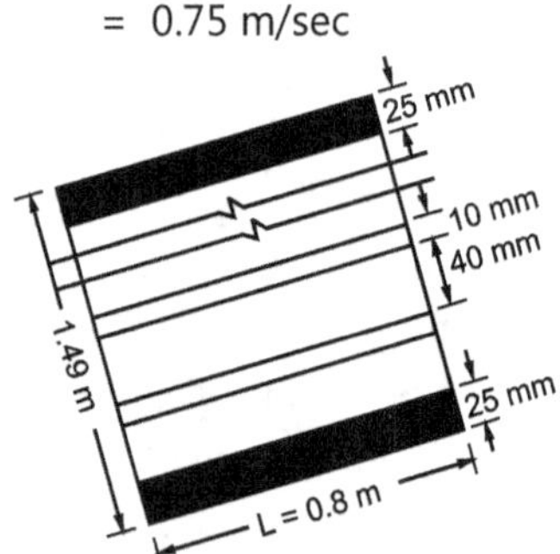

Fig. 1.4

$$Q_{max} = A \times V$$
$$0.35 = A \times 0.75$$
$$\therefore \quad A = 0.466 \text{ m}^2$$

Keeping 100% as excess openings,

$$\therefore \quad A = 2 \times 0.466 = 0.933 \text{ m}^2$$

Assuming length of screen 800 mm,

$\therefore$ Effective width of the screen

$$= \frac{0.933}{0.800}$$
$$= 1.17 \text{ m}$$

Size of opening is 40 mm,

$\therefore$ Number of openings

$$= \frac{1.17}{0.04}$$
$$= 29$$

$\therefore$ Number of bars $= 28$

Let the angle thickness at both ends be 25 mm (diameter of bars is 10 mm).

$\therefore$ Total width of screen

$$= (29 \times 0.04) + (28 \times 0.01) + (2 \times 0.025)$$
$$= (1.16) + (0.28) + (0.05)$$
$$= 1.49 \text{ m} \qquad \textbf{...Ans.}$$

For calculation of head loss please refer example 1.1.

Example 1.4 : *Design the screen chamber of an STP to treat a peak flow of 100 MLD of sewage. Assume inclination of bars 45° with horizontal, Size of bars: 10 mm x 70 mm, 10 mm dimension facing the flow, clear spacing between bars as 50 mm and the velocity through the screen as 0.8 m/sec at peak flow.*   ***(Nov. 15, 5M)***

Solution : Assume, inclination of bars 45° with horizontal size or bars 10 mm $\times$ 70 mm, 10 mm dimension facing the flow. Clear spacing between bars as 50 mm. The velocity through the screen as 0.8 m/s at peak flow.

$\therefore$ Maximum rate of flow

$$= 100 \text{ MLD} = 100000 \text{ m}^3/\text{day}$$
$$= \frac{100 \times 10^6}{24 \times 60 \times 60 \times 1000} = 1.157 \text{ m}^3/\text{sec}$$

$\therefore$ The net area of screen openings required

$$= \frac{1.157}{0.8} = 1.446 \text{ m}^2$$

Gross required of screen $= 1.446 \times \dfrac{100}{50} = 2.892 \text{ m}^2$

As the screen is inclined at 45° with horizontal,

$\therefore$ Gross area of screen $= \dfrac{2.892}{\sin 45°} = 4.089 \text{ m}^2$

Also velocity of flow above screen

$$= 0.8 \times \frac{50}{100} = 0.4 \text{ m/s}$$

Thus, we have V = 0.8 m/s, v = 0.4 m/s

Head loss, $h_L = 0.0729 \ (V^2 - v^2)$

$$= 0.0729 \ [(0.8)^2 - (0.4)^2]$$
$$= 0.035 \text{m} = 3.5 \text{ cm}$$

Thus, provide the area of screen $= 4.089 \text{ m}^2$.

    Dimensions of bars $= 10 \text{ mm} \times 70 \text{ mm}$ **...Ans.**

Clear spacing between bars $= 50$ mm   **...Ans.**

Example 1.5 : *Design a bar screen for a peak flow of 25 MLD.*   ***(May 16, 5 M)***

Solution : Assume manual cleaning keep the bars at inclination of 45° with horizontal.

Assume the rectangular bars of size 10 mm $\times$ 50 mm with 10 mm dimension. Assume clear spacing of 40 mm between the bars.

Assume velocity of the screen as 0.8 m/s at peak flow.

    Maximum rate of flow

$$= 25 \text{ MLD}$$
$$= \frac{25 \times 10^6}{24 \times 3600 \times 1000}$$
$$= 0.289 \text{ m}^3/\text{sec}$$

    Net area of screen

$$= \frac{0.289}{0.8} = 0.361 \text{ m}^2$$

    Gross area of screen

$$= 0.361 \times \frac{25}{40} = 0.225 \text{ m}^2$$

For 40 mm opening, net area is 0.361 m^2

Therefore for 50 mm, gross area is

$$0.361 \times \frac{25}{40} = 0.225$$

As the screen is inclined at 45° with the horizontal,

Gross area of screen $= \dfrac{0.225}{\sin 45} = 0.318 \text{m}^2$

Also velocity of flow above screen $= 0.8 \times \dfrac{40}{25} = 1.28 \text{ m/s}$

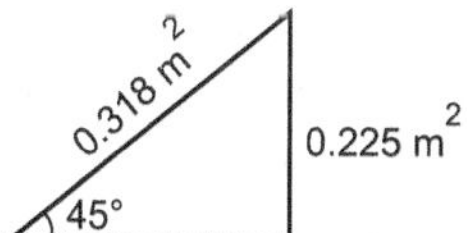

Fig. 1.5

Thus, we have, $V_0 = 0.8$ m/s

$$V = 1.28 \text{ m/s}$$

$$V_2 = 0.0729\ (V_0^2 - V^2)$$
$$= 0.0729\ [(0.8)^2 - (1.28)^2]$$
$$= 7.2\ \text{cm} \qquad \text{...Ans.}$$

This will be the head loss when the screen is clean

If however the screen is half clogged,

$$V = 2 \times 0.8 = 1.6\ \text{m/s}$$
$$V_2 = 0.0729\ [(1.6)^2 - (1.28)^2]$$
$$= 0.067\ \text{m} = 6.7\ \text{cm} \qquad \text{...Ans.}$$

Example 1.6 : *Design a bar screen for a peak discharge average flow 45 million liters day.*

Given data :

(i) Size of bar 9 mm × 50 mm, with 9 mm dimensions facing the flow.

(ii) Clear spacing between the bars 36 mm

(iii) Bars are kept an inclination of 45° with vertical.

(iv) Velocity through screen is 0.8 m/sec. at peak flow.

(Aug. 2016, 5M)

Solution : Maximum peak discharge

$$= 45 \times 10^6\ l/d = \frac{45 \times 10^6}{(24 \times 60 \times 60)}$$
$$= 0.520\ \text{m}^3/\text{sec}$$

Net area of screen $= \dfrac{0.520}{0.80} = 0.65\ \text{m}^2$

Gross area of screen $= 0.651 \times \dfrac{45}{36} = 0.813\ \text{m}^2$

Gross area of screen $= 0.651 \times \dfrac{45}{36} = 0.813\ \text{m}^2$

Gross area of screen needed for bars of 45°

$$= \frac{0.813}{\sin 45} = 1.1508\ \text{m}^2$$

Velocity of flow above screen

$$= 0.8 \times \frac{36}{45} = 0.64\ \text{m/s}$$

Head loss $(H_L) = 0.729\ (V^2 - V^2)$

$$= 0.0729\ (0.8^2 - 0.64^2) = 0.0167\ \text{m}$$

1.7 GRIT CHAMBER (Aug. 15, May 10, 16)

- Grit includes sand, ash, cinder, egg shells, silt, clay, glass pieces, broken crockery pieces, etc. The diameter of grit is less than 0.2 mm. The specific gravity of grit is 2.0 to 2.6. Grit is non-putrescible and has more settling velocity than organic solids.

- This property of difference in settling is utilised in removing non-putrescible grit of more specific gravity and putrescible organic solids in two separate tanks by differential sedimentation.

Sand and Grit Enters into Sewerage System :

- Through the open manhole covers in the case of separate sewerage system.

- Due to habit of Indian housewives of washing utensils by sand, grit, brickbats and such other materials.

- Through unauthorised connection of courtyards of the houses to the sewerage system.

Grit Causes Large Number of Problems such as :

- Erosion of pump impellers.

- Settling in conveying conduits.

- It settles along with the sludge, which when digested does not undergo any stabilisation with the result that the final product is of a poorer quality.

- If sewage is being disposed off without any treatment, it tends to settle out in the receiving water resulting in sludge bank formation with consequent problems.

Objectives of the Grit Removal are :

- Protection of pumps, valves, piping etc.

- Minimising chances of chocking of pipes and conduits with grit.

- Preventing the grit from occupying the volume in biological treatment units.

- It is desirable to remove this grit in the case of municipal sewage treatment plants before further treatment is given. The removal of this material is normally done by using channels known as grit chambers which are based on the **'Differential Sedimentation'** principle.

- The channels which are normally 0.9 m – 1.2 m deep are designed in such a manner that particles of settling velocity equal to or greater than 0.25 – 0.3 m/sec are only removed and those particles which are having a lesser settling velocity are carried ahead by the scouring velocity of the sewage flow.

- This horizontal velocity ensures the removal of sand and grit with diameter 0.2 mm or above specific gravity of 2.65.

1.7.1 Types of Grit Chambers

1. Horizontal Flow Grit Chambers :

- These are designed to maintain a velocity as to 0.3 m/s as practical. Such a velocity will carry most organic particles through the chamber and will tend to resuspend any particle that settle; but will permit heavier particle to settle out. Fig. 1.6 shows the grit chamber, consisting of long narrow open channel with a liquid depth between 0.9 to 1.2 m.

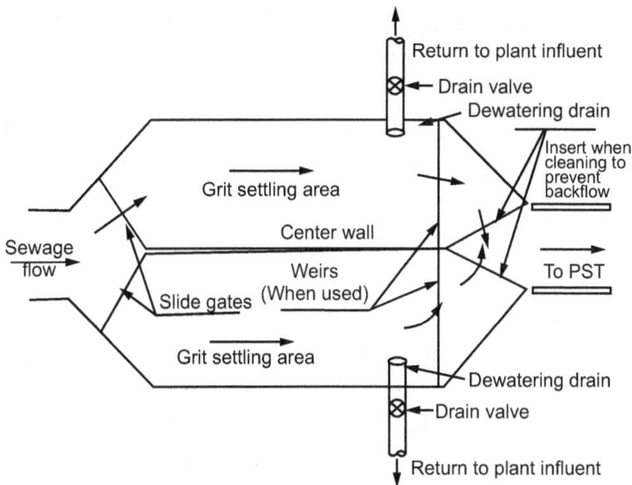

Fig. 1.6 : Grit chamber

2. Aerated Grit Chambers :

- These are designed to provide detention periods of about 3 min at the maximum rate of flow. The cross-section of the tank is similar to that provided for oxidation ditch; except that a grit hopper of about 0.9 m deep with steeply sloping sides is located along one side of the tank under the air diffusers. The diffusers are located about 0.45 m or 0.6 m above the normal plane of the bottom.

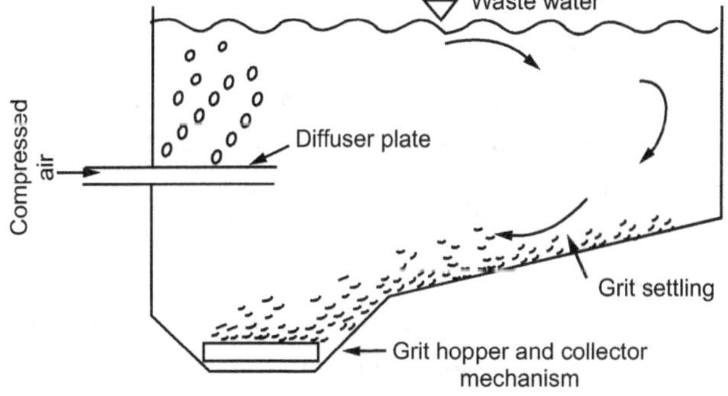

Fig. 1.7 : Aerated grit chamber

1.7.2 Design of Grit Chamber   (May 16)

- Grit chamber can be designed on a rational basis by considering it as a sedimentation basin having discrete settling.

- The settling velocity is given by the Stoke's law, applicable for the particles of 0.1 mm diameter and less.

Settling velocity,

$$V_S = \frac{g}{18\,m}\,(r_S - r)\,d^2$$

$$V_S = \frac{g}{18\,n}\,(S_S - 1)\,d^2$$

where, V_S = Settling velocity (cm/sec)

d = Diameter of particle (cm)

μ = Absolute viscosity (centipoise)

$v = \mu/\rho$ = Kinematic viscosity (centistokes)

ρ_S = Density of particle (gm/cm^3)

ρ = Density of water (gm/cm^3)

S_S = Specific gravity of particle

g = Gravitational acceleration (cm/sec^2)

For particles of diameter more than 1 mm,

$$V_S = \sqrt{3.33\,g\left(\frac{r_S - r}{r}\right)d}$$

Hazen's modified equation to determine settling velocity is

$$V_S = 60.6\,(S_S - 1)\,d\left[\frac{3t + 70}{100}\right]$$

where, t = Temperature of liquid in °C

By putting the value of S_S = 2.65 for grit, we get

$$V_S = d\,(3t + 70)$$

Table 1.1 : Settling Velocities and Overflow Rates for Grit Chambers at 10°C

Diameter of Particles in mm	Settling Velocity cm/sec		Overflow Rate in an Ideal Grit Chamber, m³/d/m²	
	S_S = 2.65	S_S = 1.2	S_S = 2.65	S_S = 1.2
0.2	2.5	0.54	2160	467
0.15	1.8	0.39	1555	337

(Ref. : Manual on Sewerage and Sewage Treatment, 2nd Edition)

1. Design Criteria for Horizontal Flow Grit Chamber

- Detention time → 45 – 90 sec (Typical 60 sec)

- Horizontal velocity → 0.25 – 0.4 m/s (0.3 m/sec)

- Settling velocity for 65 mesh material → 1 – 1.3 m/min (1.15 m/min).

- Head loss in control section as % depth in channel → 30 – 40% (36%).

- Allowance for inlet and outlet turbulence → 2 × M_{a}x depth in channel.

2. Design Information for Aerated Grit Chamber

- Dimensions :

    Depth = 2 – 5 m

    Length = 7.5 – 20 m

    Width = 2.5 – 7.0 m

- Width/depth ratio → 1 : 1 → 5 : 2 (2 : 1).

- Detention time at peak flow → 2 – 5 min (3 min)

- Air supply, m³/m/min of length → 0.5 – 0.45 (0.3).

1.7.3 Velocity Control

- Theoretically, for obtaining a constant velocity of flow irrespective of fluctuation in sewage flow, a parabolic channel having an equation $w = \dfrac{3}{2}\, Q/nv$ has to be constructed. However, in normal practice, it is rather difficult to construct such channels of parabolic shape.
- Hence, a common practice is to use a rectangular channel and to ensure a proper control over the horizontal velocity of flow with the aid of outlet control devices such as :
  1. A proportional flow weir or a Sutro weir. [Fig. 1.8 (a) and (b)].
  2. Parabolic grit chamber. [Fig. 1.8 (c)]
  3. A Parshall flume. [Fig. 1.8 (d)].

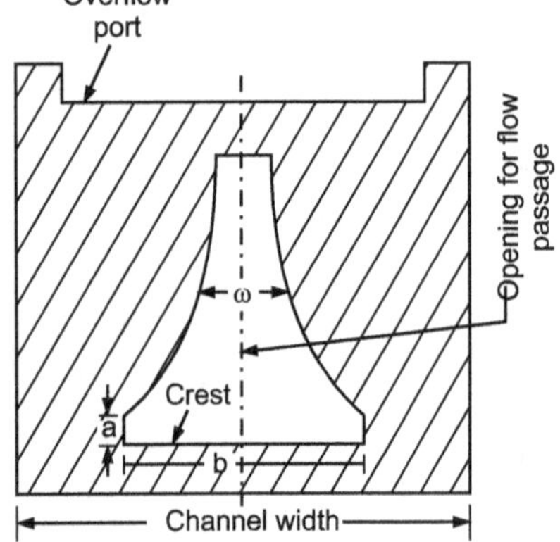

(a) Proportional flow weir (front view)

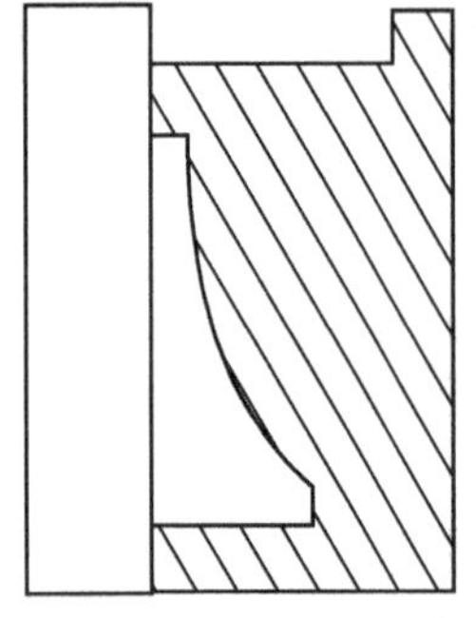

(b) Sutro weir (front view)

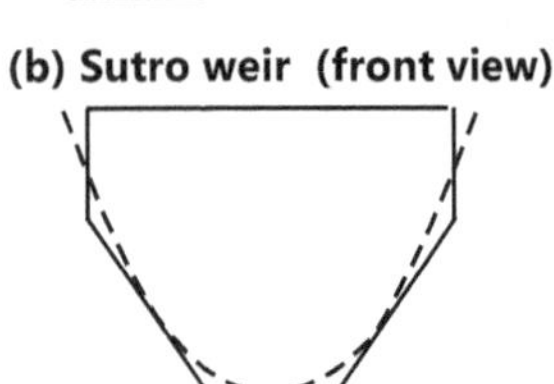

(c) Parabolic grit chamber (front view)

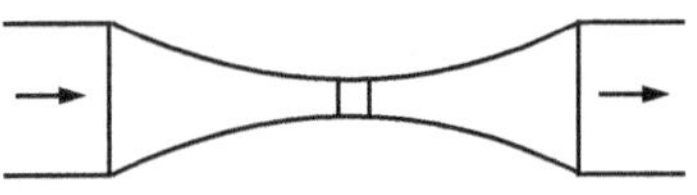

(d) Parshall flume (plan)

Fig. 1.8 : Velocity control devices

- Thus, a normal grit chamber consists of a channel or a number of channels in parallel designed on the following basis :
  - It will have such a cross-sectional area, as it will give a horizontal velocity of flow of 0.25 to 0.3 m per second.
  - It's surface area gives a surface loading 22.50 m³/day/sq.m.
- The design is thus based on the above two criteria and checked for the detention time which should fall between 45 to 90 seconds. The outlet control devices are then designed by following specific methods for each case. In normal practice, the invert of the proportional weir is provided at least 10 cm and normally about 30 cm above the invert of the channel.
- In order to avoid or bring down to a minimum value, the turbulence caused in the grit chamber due to momentum of incoming liquid and by drag of the outgoing liquid, smooth transition must be provided both at the inlet as well as at the outlet of the channel.
- Infact in normal practice, the actual length of the grit chamber is about 50% more than theoretically necessary, to compensate for the above factors.
- The grit collected at the bottom of these grit chambers is normally allowed to flow by gravity to a specific location from where it is then removed under hydraulic head. In such cases, where manual cleaning is practiced, the units have to be always in duplicate so that while one unit is being cleaned, the other can be used.
- The storage space for the grit at the bottom of the channels should be adequate for one day's storage for grit.
- The head loss, that will occur in the grit chambers, mostly depends upon the type of the outlet control device used and varies between 0.06 m to 0.6 m. The head losses are maximum in the case of the weir controlled sections and one lesser in cases where parshall flume is used.
- In case parshall flume is used as a control device, it is necessary to use a parabolic cross-section of the grit chamber. However, in normal practice, it is difficult to construct such units and hence commonly a rectangular channel with chamfered sides are used. It will easily be appreciated, such compromises naturally result in sub-standard performance of the grit chamber.

1.7.4 Washing of Grit

- As known, it is very difficult to obtain a completely ideal performance of the grit chambers with the result that some of the lighter organic solids also settle to its bottom. Also some organic solids deposited on inorganic particles will settle at a faster rate and hence will be removed. Thus, it has been observed that in normal practice, grit contains a minimum of 5% of organic solids.

- It is hence necessary to remove these organic solids before the grit can be properly disposed off. The most common mechanism for doing this is by the use of grit washing equipment. In these devices, by some means, either mechanical or by supply of diffused air, sufficient shear is created between the interface of the organic and inorganic fractions so that the organic portion gets sheared off and is removed. This lighter material is then carried away by the flow of either water or the sewage flowing through the chamber.

- The amount of grit removed is normally about 0.025 – 0.075 m³ per million litre of sewage flow and is normally used in the treatment plant itself for filling low lying areas or for laying minor roads.

- Thus, the grit washing equipment is normally provided alongwith the grit chamber and hence some authorities feel that if inspite of providing the velocity control devices, if a washing arrangement is required, the cost can be brought down and also construction is simplified if no special control device is used. Such tanks which are normally referred to as Detritus tanks do not have any velocity control device and the only control that is exercised is to ensure that the grit particles which might have settled to the bottom are not scoured off.

- The net result is that a larger percentage of organic solids settle to the bottom, but then the design and construction becomes much easier. These tanks are normally designed for a detention time of 3 to 4 minutes and for an average horizontal velocity of flow of 0.3 m per second. These tanks are normally rectangular or square in cross-section and are provided with scrapers at the bottom for removal of collected matter and the grit washing equipment is also provided as in the case of grit chambers.

- Sometimes, grit chambers are also provided with compressed air supply from bottom which enables a better control of grit removal, obviates need of separate grit washing, gives good efficiency even with improperly designed outlet control devices, and helps to remove oily materials here only. The air supply also helps to freshen the flowing sewage.

1.7.5 Proportional Flow Weir (May 10, 11)

- For the efficient removal of grit, the velocity through the grit chamber should be constant inspite of change in flow. The proportional flow weir is most satisfactory type of automatic velocity control device. This is provided at the outer end of chamber. It consists of a rectangular plate, with an opening with curved sides for flow to pass through. This helps in maintaining the constant velocity in grit chamber by varying the cross-sectional area of flow through the weir so that the depth is proportional to the flow.

- The shape of the opening between the plates of the proportional flow weir is made in such a way that the chamber depth will vary directly as the discharge, as a result of which the chamber velocity will remain constant for all flow conditions. The sides are so curved that the area decreases as the three half power of the increasing depth (h) of the flow over the weir.

- The discharge Q in litres/sec over the weir –

$$Q = 1570 \, c \, \sqrt{2g} \, (Wh^{1/2}) \, h$$

where, h = Depth of flow (m) above crest

W = Width of opening (m) at height h

g = Gravitational acceleration (m/sec²)

c = Discharge coefficient

If we take $c = 0.6$ for sharp edged weir, we get

$$Q = 4170 \, (Wh^{1/2}) \, h$$

- This shows that if $Wh^{1/2}$ is made constant, then the depth h will vary directly with the discharge. For different values of h, the corresponding values of width W can be determined and hence the parabolic curvature of the sides of weir could be worked out.

1.7.6 Disposal of Grit

- After separating grit from the sewage, it is washed to remove organic content from it. Washed grit may resemble particles of sand and gravel, which is total inorganic in nature. This may be disposed off by dumping or burying or by sanitary landfill.

- Unwashed grit, when mixed with soil, is valuable as soil conditioner and will give good yield of garden crops like tomatoes, cucumbers, etc.

SOLVED EXAMPLES ON DESIGN OF GRIT CHAMBER

Example 1.7 : *Design a grit chamber for the following data :*

Maximum flow : 20 mld.

Diameter of particle to be removed : 0.2 mm and more.

Specific gravity of particle : 2.65.

Average temperature : 20°C.

Solution :

Assume a grit chamber of rectangular section and also assume that proportional flow weir is provided as velocity control device. Settling velocity as per Hazen's modified equation :

$$V_s = 60.6 \,(S_s - 1)\, \frac{3t + 70}{100} \cdot d$$

$$= 60.6 \,(2.65 - 1)\, \frac{3 \times 20 + 70}{100} \times \frac{0.2}{10}$$

$$= 2.6 \text{ cm/sec}$$

Assume flow through velocity (V_h) as 0.23 m/sec

Hence, Cross sectional area

$$= \frac{\text{Flow (m}^3\text{/sec)}}{\text{Flow through velocity (m/sec)}}$$

$$= \frac{20 \times 10^6}{24 \times 3600 \times 10^3 \times 0.23}$$

$$= 1.006 \text{ m}^2$$

Providing width of 1.2 m, liquid depth required

$$= \frac{1.006}{1.2}$$

$$= 0.838 \text{ m}$$

Provide freeboard of 0.3 m and a space of 0.25 m for sludge accumulation.

∴　Total depth,

$$H = 1.388 \text{ m} \quad \text{say } 1.4 \text{ m}$$

Now, $\quad \dfrac{H}{L} = \dfrac{V_s}{V_h} = \dfrac{2.6}{23} = \dfrac{1}{8.84}$

∴　　　$L = 8.84\,H$　∴　$L = 7.4$ m　　**...Ans.**

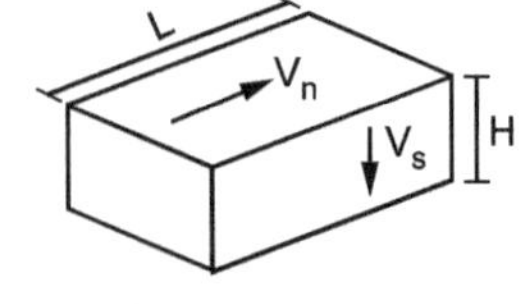

Fig. 1.9 (a)　　　　　**Fig. 1.9 (b)**

Example 1.8 : *Design a grit chamber for the following data :*

Flow : 15,000 m³/day.

Settling velocity of particle : 0.016 to 0.022 m/sec.

Flow through velocity : 0.3 m/sec.

Solution :

Let us provide a rectangular channel with a proportional flow weir.

Now horizontal velocity, $V_h = 0.3$ m/sec.

Assuming settling velocity as 0.02 m/sec (between 0.016 – 0.022 m/sec)

Now, $\quad Q = 15000 \text{ m}^3\text{/day}$

$$= \frac{15000}{24 \times 3600} \text{ m}^3\text{/sec}$$

$$= 0.1736 \text{ m}^3\text{/sec}$$

∴　Cross sectional area

$$= \frac{\text{Flow}}{\text{Velocity}} = \frac{0.1736}{0.3} = 0.578 \text{ m}^2$$

Assuming a depth of 1 m, we have width of basin,

$$W = \frac{0.578 \text{ m}^2}{1 \text{ m}} = 0.578 \text{ m say } 0.6 \text{ m}$$

Now settling velocity,

$$V_s = 0.02 \text{ m/sec}$$

∴　Detention time

$$= \frac{\text{Depth}}{V_s} = \frac{1 \text{ m}}{0.02 \text{ m/sec}} = 50 \text{ sec} \text{ ...\textbf{Ans.}}$$

∴　Length of tank

$$= V_h \times \text{Detention time}$$

$$= 0.3 \times 50 = 15 \text{ m} \qquad \text{...\textbf{Ans.}}$$

Hence, provide a grit chamber of size L = 15 m, W = 0.6 m, H = 1 m.

Example 1.9 : *Design a proportional flow weir for a flow of 0.99 m³/sec.*

Solution :

$$Q_{max} = 0.99 \text{ m}^3\text{/sec} = 990 \text{ lit/sec.}$$

Assume, $\quad h = 1.56$ m

Formula :

$$Q_{max} = 4170 \, bh^{1/2} \cdot h$$

$$990 = 4170 \,(b)\,(1.56)^{3/2}$$

∴　　　$b = 0.12$ m

　　　$b = $ Width of weir at crest level = 12 cm

If flow through velocity is to be constant for changing discharges, Q/a is to maintain constant.

(a = cross sectional area of channel)

$$\frac{Q}{a} = \frac{Q}{bh} \quad (h = \text{depth of flow})$$

From this, we can have $\dfrac{Q}{h} = $ constant.

∴　　　$\dfrac{Q}{h} = 4170 \, bh^{1/2} \cdot h = 4170 \, bh^{1/2} = \text{constant}$

This constant is found from peak flow values.

i.e.   b = 0.12 m and h = 1.56

$\therefore$ $bh^{1/2}$ = $0.12 \times (1.56)^{1/2}$

$\qquad\qquad$ = 0.15 cm

Thus, widths of weir for different depths of flow are calculated.

For h_1 = 0.035 m, $b = \dfrac{0.15}{(0.035)^{1/2}}$ = 80 cm

For h_2 = 0.5 m,

For h_3 = 0.75 m, $b = \dfrac{0.15}{(0.5)^{1/2}}$ = 21.2 cm

For h_4 = 1.0 m,

For h_5 = 1.25 m, $b = \dfrac{0.15}{(0.75)^{1/2}}$ = 17 cm

$\qquad\qquad\qquad b = \dfrac{0.15}{(1.0)^{1/2}}$ = 15 cm

$\qquad\qquad\qquad b = \dfrac{0.15}{(1.25)^{1/2}}$ = 13.4 cm

Now assume sill of weir at 30 cm from bottom.

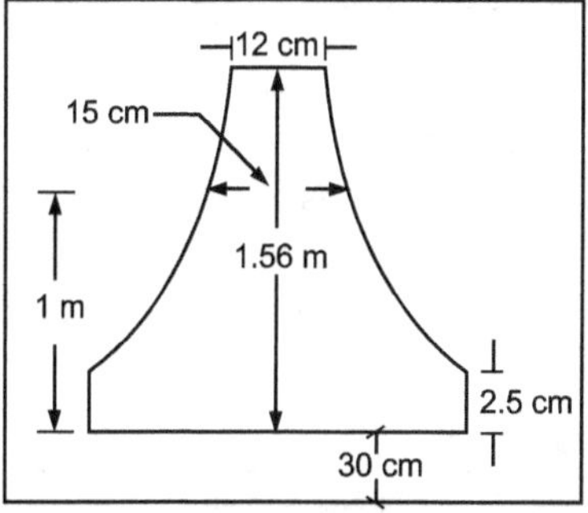

Fig. 1.10 : Proportional flow weir

Example 1.10 *: Design a grit chamber to treat domestic sewage of a town having discharge 1595 m^3/hr. Assume specific gravity of grit as 0.2 mm. Take the temperature of sewage = 10°C. Provide constant velocity of sewage in the chamber of 40 cm/sec.*   **(Aug. 2015, 5M)**

Solution :

Let us provide a rectangular channel with a proportional flow weir.

Now, horizontal velocity, V_h = 0.4 m/sec

Settling velocity as per Hazen's modified equation.

$$V_s = 60.6 \, (S_s - 1) \, \frac{3t + 70}{100} \cdot d$$

$$= 60.6 \, (2.00 - 1) \times \frac{3 \times 10 + 70}{100} \times \frac{0.2}{10}$$

$$= 1.212 \text{ cm/sec.} = 0.012 \text{ m/sec}$$

$$\cong 0.016 \text{ m/sec}$$

$$\text{(between 0.016 - 0.022 m/sec)}$$

Now, $Q = \dfrac{1595}{3600}$ = 0.443 m^3/sec

$$c/s = \frac{flow(m^3/s)}{Flow \text{ through vel (m/s)}}$$

$$= \frac{0.443}{0.4} = 1.1076 \text{ m}^2$$

Assuming, a depth of 1m.

$\therefore$ width of basin, b $= \dfrac{1.1076}{1}$ = 1.1076 m

$$V_s = 0.016 \text{ m/sec}$$

$$\text{Detention Time} = \frac{Depth}{V_s} = \frac{1}{0.016} = 62.50 \text{ sec}$$

Length of lanks = Vh $\times$ D.T.

$$= 0.4 \times 62.50 = 25 \text{ m} \qquad \textbf{...Ans.}$$

Hence, provide a grit chamber of size L = 25m

$$B = 0.6 \text{ m, h} = 1 \text{ m} \qquad \textbf{...Ans.}$$

1.8 OIL AND GREASE REMOVAL

- Removal of free oil and grease from a wastewater stream reduces the potential for equipment problems to occur further downstream. There are three forms of oil encountered in wastewater treatment at a refinery. They are:

1. Free Oil or floating oil is removed by either skimming the surface in the skim tank or by gravity separation in the API separator.

2. Emulsified Oil is comprised of oil droplets in stable suspension within the wastewater. Removal requires chemical addition to lower the pH followed by addition of dissolved oxygen or nitrogen to remove the emulsified oils as they break free from the wastewater.

3. Dissolved Oil is a true molecular solution within the water and can only be removed with biological treatment.

1.8.1 O and G Removal Techniques

API Separators:

- API or American Petroleum Institute Separators are normally the first and most important step in a refineries wastewater treatment. It uses the differences in oil and water's specific gravity to filter out the majority of free oil within the mixture. The lighter oils will remain at the top of the liquid and can be skimmed off, while the heavier oil will settle to the bottom.

- The above figure shows a detail drawing of a typical API separator. A conventional oil-water separator contains a conveyor to assist in the separation of the heavier oils and grease, a scraper/skimmer and a baffle.

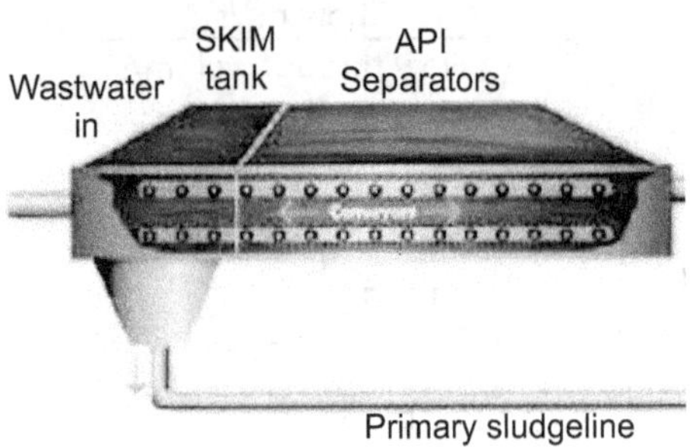

Fig. 1.11

- pH is typically measured at the outlet as an indicator of potential problems, such as flocculation, in the secondary treatment. It is important to note that because emulsified oil is still present in the water mixture, coating will occur causing a sluggish reading.

- At this point in the treatment process the emulsified oil is still in soluble form. The removal of the emulsified oil requires either a chemical addition to lower the pH or use of emulsion breakers.

Dissolved/ Air Flotation:

- Once treated, the wastewater often enters an air flotation unit to remove the emulsified oil as it begins to break free from the wastewater, using dissolved oxygen or nitrogen.

Bioreactors (Activated Sludge Treatment):

- At this phase the wastewater still contains dissolved oil in a true molecular form. The only way to remove the remaining dissolved oil is thru biological treatment. This and the remaining steps are the same as those found in a normal wastewater treatment facility.

- Many components in the process must be in balance in order to obtain complete synergy such as, biomass blends, air, return activated sludge (RAS), waste activated sludge (WAS) and throughput. It is essential to monitor and control all factors which influence the efficiency of the biological conditions in the basin; for example:

Temperature:

- Most biomass achieves optimum efficiency in a temperature range between 10-40°C. Increasing or decreasing the temperature can result in the increasing or decreasing the rate at which the bugs eat and reproduce. All chemical reactions taking place are affected by the process temperature as well.

pH:

- For most systems the pH should be kept between 6.5 to 8.5 pH, when the pH is too high or too low, the biomass loses the ability to convert the food to energy and raw materials. A pH below 6.5 may cause growth of fungi and fungal bulking, and will have to be adjusted using a caustic, lime or magnesium hydroxide.

Low Nutrients:

- If nitrogen and phosphorus are not present in sufficient amounts it can limit the growth rate of the biomass. A sign of nutrient deficiency includes foam on the aeration basin.

Dissolved Oxygen:

- DO is a critical measurement and will be maintained between 1-3 mg/L. The concentration is an indication of the basin environment; whether it is in denitrification (excess nitrate, NO3) or nitrification (excess ammonium, NH4) environment. Essentially the DO measurement is used to minimize ammonium breakthrough. It is not uncommon to see NH4 and DO measurements together.

Septicity/Toxicity:

- Septic wastes contain elevated amounts of sulfides and organic acids (such as acetic acid).Other organic materials and heavy metals are also toxic to the biomass, reducing their efficiency or even destroying them.

1.8.2 Grease Trap

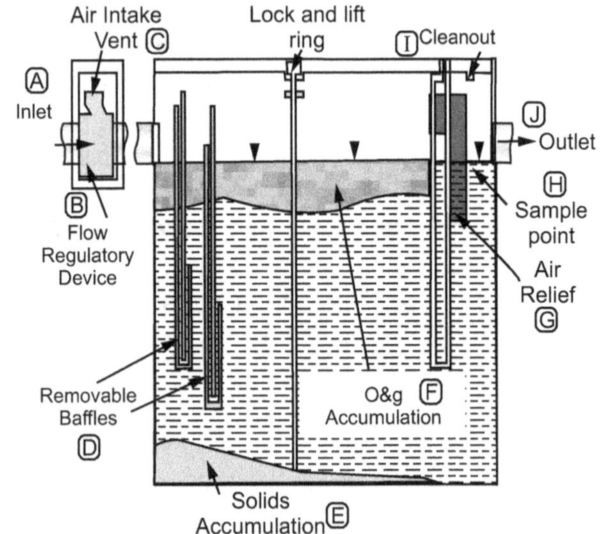

Fig. 1.12

- A Flow from four or fewer kitchen fixtures enters the grease trap.

- B An approved flow control or restricting device is installed to restrict the flow to the grease trap to the rated capacity of the trap.

- C An air intake valve allows air into the open space of the grease trap to prevent siphonage and back-pressure.

- D The baffles help to retain grease toward the upstream end of the grease trap since grease floats and will generally not go under the baffle. This helps to prevent grease from leaving the grease trap and moving further downstream where it can cause blockage problems.

- E Solids in the wastewater that do not float will be deposited on the bottom of the grease trap and will need to be removed during routine grease trap cleaning.

- F Oil and grease floats on the water surface and accumulates behind the baffles. The oil and grease will be removed during routine grease trap cleaning.

- G Air relief is provided to maintain proper air circulation within the grease trap.

- H Some grease traps have a sample point at the outlet end of the trap to sample the quality of the grease trap effluent.

- I A cleanout is provided at the outlet or just downstream of the outlet to provide access into the pipe to remove any blockages.

- J The water exits the grease trap through the outlet pipe and continues on to the grease interceptor or to the sanitary sewer system.

1.8.3 Grease Interceptor

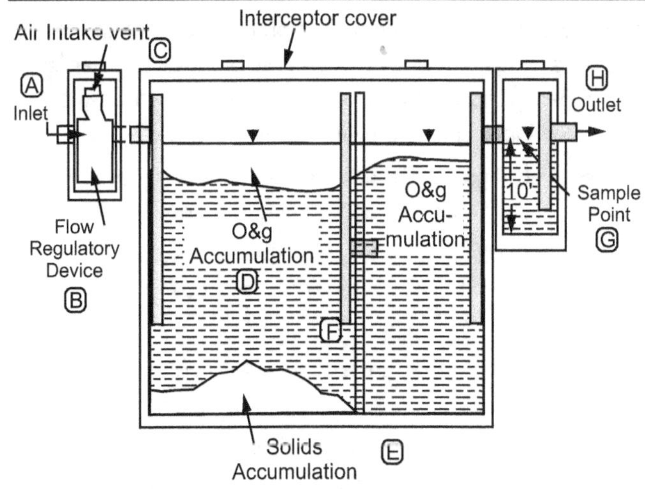

Fig. 1.13

- A Flow from undersink grease traps or directly from plumbing fixtures enters the grease interceptor. The UPC requires that all flow entering the interceptor must enter through the inlet pipe

- B An approved flow control or restricting device is installed to restrict the flow to the grease interceptor to the rated capacity of the interceptor.

- C An air intake valve allows air into the open space of the grease interceptor to prevent siphonage and backpressure.

- D Oil and grease floats on the water surface and accumulates behind the grease retaining fittings and the wall separating the compartments. The oil and grease will be removed during routine grease interceptor cleaning.

- E Solids in the wastewater that do not float will be deposited on the bottom of the grease interceptor and will need to be removed during routine grease interceptor cleaning.

- F Grease retaining fittings extend down into the water to within 12 inches of the bottom of the interceptor. Because grease floats, it generally does not enter the fitting and is not carried into the next compartment. The fittings also extend above the water surface to provide air relief.

- G Some interceptors have a sample box so that inspectors or employees of the establishment can periodically take effluent samples. Having a sample box is recommended by the UPC but not required.

- H Flow exits the interceptor through the outlet pipe and continues on to the sanitary sewer system.

1.8.4 A Skimming Tank

- A skimming tank is a chamber so arranged that the floating matter like oil, fat, grease etc., rise and remain on the surface of the waste water (Sewage) until removed, while the liquid flows out continuously under partitions or baffles.

- It is necessary to remove the floating matter from sewage otherwise it may appear in the form of unsightly scum on the surface of the settling tanks or interfere with the activated sludge process of sewage treatment. It is mostly present in the industrial sewage. In ordinary sanitary sewage, its amount is usually too small.

- The chamber is a long trough shaped structure divided up into two or three lateral compartments by vertical baffle walls having slots for a short distance below the sewage surface and permitting oil and grease to escape into stilling compartments.

- The rise of floating matter is brought about the blowing air into the sewage from diffusers placed in the bottom. Sewage enters the tank from one end, flows longitudinally and leaves out through a narrow inclined duct. A theoretical detention period of 3 minutes is enough. The floating matter can be either hand or mechanically removed.

- Grease traps are in reality small skimming tanks designed with submerged inlet and bottom outlet (Fig. 1.14). The traps must have sufficient capacity to permit the sewage to cool and grease to separate. Frequent cleaning through removable covers is essential for satisfactory operation. Grease traps are commonly employed in case of industries, garages, hotels and hospitals.

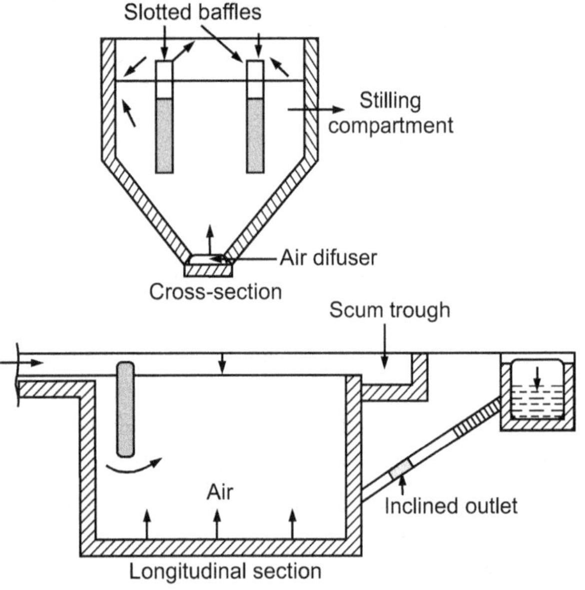

Fig. 1.14

1.9 SEDIMENTATION / SETTLING / CLARIFICATION (Aug. 15, 16)

- Sedimentation is the separation from water, by gravitational settling, of suspended particles that are heavier than water. It is one of the most widely used unit operations in wastewater treatment. (Fig. 1.15).

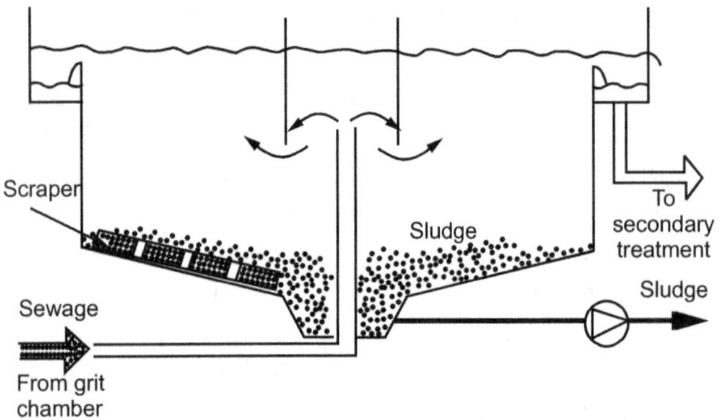

Fig. 1.15 : Schematic sketch of PST

- This is used to remove grit and particulate matter in primary settling basin, to remove biological floc in secondary settling basin of ASP and to remove chemical floc when the chemical coagulation process is used. It is also used for solids concentration in sludge gravity thickeners.

1.9.1 Description

- Depending on the concentration and nature of the suspended solids, discrete type, flocculant type or zone type settling will occur.

- In the discrete type settling, which normally occurs in water supply practice, the solids are of such a nature that they do not change their size, shape and weight while settling. In the case of flocculant type of settling which normally occurs in the case of industrial waste treatment, the solids tend to increase their settling velocity due to coalescence during settling.

- When the concentration of suspended solids becomes too high, the particles are held in fixed position relative to each other by the Van der Waal's force of attraction and the mass settles as a whole. As the interface between the solid blanket and the clear liquid moves down, it encounters resistance from already settled particles and hence a transition zone comes into picture.

- Ultimately, when the solids reach the bottom, they come in direct contact with already settled sludge and the compression zone occurs. Thus in zone settling, the settling velocity decreases with time.

- In the normal sedimentation practice, the Stoke's law is used to calculate the settling velocity of the particles. In designing continuous flow sedimentation basins, the following assumptions are normally made :

  ➤ Settling in a settling tank occurs exactly as it would occur in a quiescent container of liquid.

  ➤ The concentration of suspended solids at right-angles to the direction of flow is constant throughout.

  ➤ A particle that enters the sludge zone stays and is removed.

- The normal sedimentation basins are designed on the basis of surface loading and detention time. The normal values of surface loading are between 25 to 45 $m^3/d/m^2$ and the detention time between 2 to 8 hours. The exact values of the surface loading and the detention times are normally selected depending on the specific purpose of use.

- The normal practice is to use circular sedimentation tanks mainly because of their structural stability and due to their lesser cost for the same volume. They are also found to behave reasonably well from hydraulic point of view.

- Theoretically, a long narrow rectangular tank should give the best performance. Thus, in actual practice, either circular settling tanks having a diameter of 30 to 60 m, a side-water depth of 2.7 to 5 m with a bottom slope of 8% or rectangular tanks having a length to width ratio of 2 to 4, a maximum width of 20 m and a bottom slope of 1%; or square tanks having a bottom slope of 8% and the maximum side length of 20 m are used.

1.9.2 Factors Affecting Sedimentation

- **Characteristics of Solids**
  - ➤ Size of particles.
  - ➤ Specific gravity of settling particles.
  - ➤ Concentration of suspended matter.

- **Characteristics of Liquid**
  - ➤ Temperature
  - ➤ Viscosity
  - ➤ Specific gravity of liquid.

- **Physical Characteristics of Clarifier**
  - ➤ Detention period.
  - ➤ Shape of basins.
  - ➤ Depth of the basin.
  - ➤ Baffling and operation of basin.
  - ➤ Viscosity and length of flow through basin.

1.9.3 Classification of Settling Tanks

- **According to the Shape of Tank**
  - ➤ Rectangular tank. (Fig. 1.16)
  - ➤ Circular tanks. (Fig. 1.17)

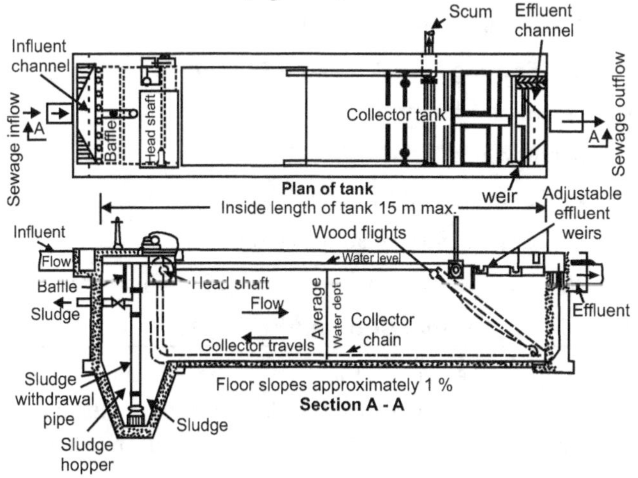

Fig. 1.16 : Rectangular settling tank

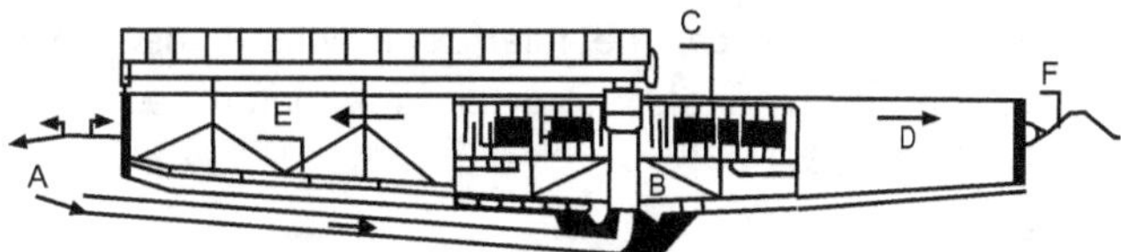

Sedimentation with flocculating compartment.
A = wastewater input, B = flocculation, C = stirrer,
D = sedimentation,
E = sludge scraper, F = outflow

Fig. 1.17 : Circular clarifiers

- **According to Direction of Flow**
  - ➤ Horizontal flow tanks – Longitudinal and radial flow.
  - ➤ Vertical flow tanks.

- **According to Method of Sludge Collection and Removal**
  - ➤ Flat bottom tanks - without scraper.
  - ➤ Hopper bottom tanks.
  - ➤ Flat bottom tanks with mechanical scraper.

- **According to Nature of Working**
  - ➤ Fill and draw type.
  - ➤ Continuous flow type.

- **According to its Location**
  - ➤ Preliminary settling tank (Grit chamber).
  - ➤ Primary settling tank.
  - ➤ Secondary settling tank.
  - ➤ Sludge thickener.

1.9.4 Inlet and Outlet Arrangements

- **Inlet :** The inlet to the tank can be of any of the following types :

  - ➤ **An Over-Flow Weir :** This type of inlet suffers from the drawback that it cannot give a uniform discharge over the whole weir length. But it is simple in construction.

  - ➤ **Baffle Type Inlet :** In this type of inlet, a number of pipes are provided along the width and the momentum of the incoming water is normally broken down by impact on a baffle kept against it. If a solid baffle is used, it is normally taken about 1 m below the water level. However, this sort of inlet is not much preferred now-a-days.

  - ➤ **Slotted Baffle Inlet :** The common practice now-a-days is to go in for the use of slotted baffle infront of the inlet. Thus, two slotted baffles are normally used one after another and the total area of the slots is normally kept from 7 to 20% of the cross-sectional area of the tank.

  - ➤ **Stengel Type of Inlet :** This is very commonly used in European countries.

- In normal practice, the inlet zone, where inefficient settling occurs, is taken to extend for a length equal to depth of tank.

Outlet :

- The outlet invariably consists of a weir. The rate of overflow over unit length of this weir is normally termed as the weir loading. In water supply practice, this weir loading is restricted to a maximum of 250 m^3/d/m. While in the case of settling tanks for the activated sludge process etc., it is kept below 20,000 gpd/ft.

- Due to the flow occurring over the weir, an area in the form of a cylindrical sector having its axis along the outlet weir and a volume of KQ^2/V^2 is ineffective for settling. In this formula, given by the National Research Council of U.S.A., the value of K is normally between 0.35 to 0.55; Q is the weir loading and V is the settling velocity of the design particle. Obviously, the larger the weir loading, the larger will be the ineffective area.

- The bottom slope provided to the tank provides sufficient space for the storage of sludge, which is normal practice, is continuously removed by a system of scrappers which complete one cycle over 15 to 30 minutes time.

1.9.5 Design of Settling Tanks

- Size of the settling tanks is decided based on
  - ➢ Overflow rate, m^3/d/m^2
  - ➢ Detention time, hrs.
  - ➢ Weir loading, m^3/d/m.
  - ➢ Solids loading rate, kg/day/m^2.

$$\therefore \text{ Detention time (hrs)} = \frac{\text{Tank volume in m}^3}{\text{Flow (m}^3/\text{day)}} \times 24 = \frac{V}{Q}$$

$$\left[\begin{array}{l}\text{Surface loading or}\\ \text{overflow rate}\end{array}\right] = \frac{\text{Flow (m}^3/\text{day)}}{\text{Area (m}^2)} = \frac{Q}{A}$$

$$\text{Weir loading rate} = \frac{\text{Flow (m}^3/\text{day)}}{\text{Length of weir (m)}}$$

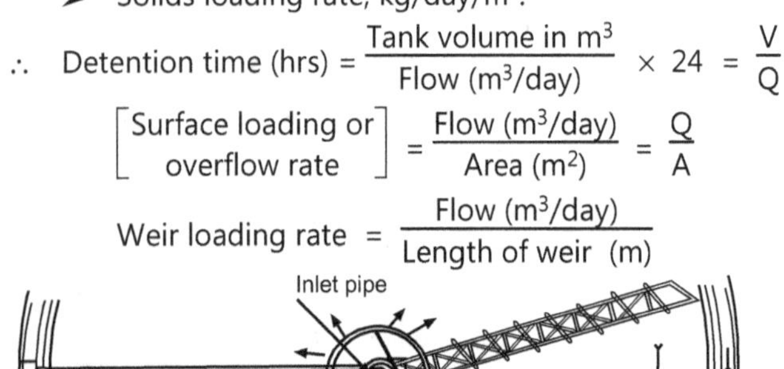

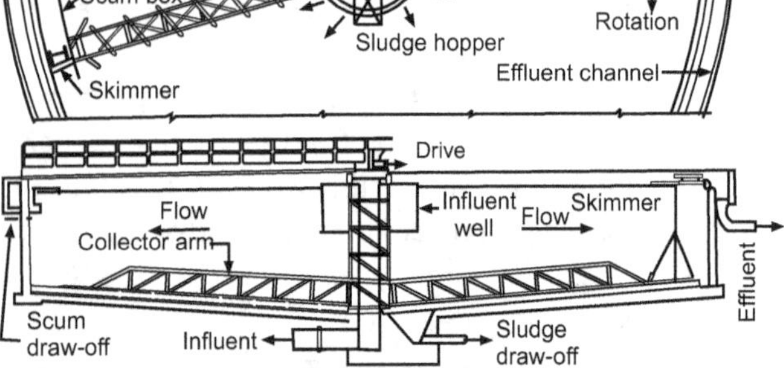

Fig. 1.18 : Circular primary clarifier with a pier-supported center drive and peripheral effluent weir

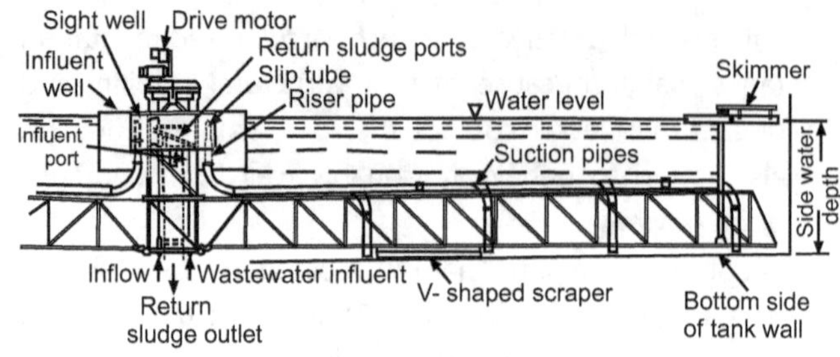

Fig. 1.19 : Final clarifier designed for use with biological aeration

Table 1.2 : Design Information for Settling Tanks

Tank Type		Value	
		Range	Typical
1.　Rectangular			
Depth, m		2.5 – 4.5	3.6
Length, m		15 – 90	25 – 40
Width, m		3 – 21	6 – 10
Flight travel speed, m/min		0.6 –1.2	1.0
2.　Circular			
Depth, m		2.5 – 4.5	4.5
Diameter, m		3.6 – 60.0	12 – 45
Bottom slope, mm/m		60 – 160	80
Flight travel speed, m/min		0.02 – 0.05	0.03

1.9.6　Design Criteria for Various Types of Settling Tanks

Primary Settling Tank Followed by Secondary Treatment :

- Detention time, hrs → 2.0 – 2.5 (2)
- Overflow rate, m^3/d/m^2
  - ➢ For average, flow → 35 – 50
  - ➢ For peak flow → 80 – 120 (100)
- Weir loading, m^3/d/m → 125 – 500 (250).

Primary Settling with Activated Sludge Return :

- Detention time, hrs → 2.0 – 2.5 (2)
- Overflow rate, m^3/d/m^2
  - ➢ For average flow → 25 – 35
  - ➢ For peak flow → 50 – 60 (60)
- Weir loading, m^3/d/m → 125 – 500 (250)

1.9.7 Sludge Collection and Removal

- Sludge collected at the bottom has to be removed from time to time. Now-a-days, sludge collection is invariably carried out with the help of mechanical scrapers.

- The floors of the mechanically scraped tanks should have gentle slopes. It should be around 2% for circular tanks and 8% for rectangular tanks.
- Slope of sludge hoppers without mechanical scraper is 2 to 1 (vertical to horizontal).
- Scraper velocity should be between one revolution in 45 min to one revolution in 80 min.
- Power required for driving scraping mechanism may vary from 0.5 to 1.5 H.P. for tanks of 3 m diameter or length.
- The sludge collected at the bottom is taken out under hydrostatic pressure through an outlet pipe controlled by a sluice valve.

SOLVED EXAMPLES ON SEDIMENTATION TANK

Example 1.11 : *Design a ciruclar PST with following data :*

(i) Average flow = 10 mld.

(ii) Surface loading = 50 $m^3/d/m^2$ of tank area.

(iii) Effluent weir loading = 180 $m^3/d/m$ length of weir (only design of PST and outlet channels).

Solution : Assume detention time = 2 hrs.

Depth of tank = 3 m.

$$Q_{avg} = \frac{10 \times 10^6 \times 10^{-3}}{24} = 416.67 \text{ m}^3/hr$$

Volume of PST = $Q_{avg} \times T$

$$= 416.67 \times 2 = 833.33 \text{ m}^3$$

$$\text{Plan area} = \frac{\text{Volume}}{\text{Depth}} = \frac{833.33}{3} = 277.78 \text{ m}^2$$

$$A = \frac{p}{4} d^2$$

$$277.78 = \frac{p}{4} \times d^2$$

$$\therefore \quad d = 18.80 \text{ m}$$

$\therefore$ Diameter of tank = 18.80 m

$$\text{Surface loading} = \frac{Q}{A} = \frac{416.67}{277.78} \times 24$$

$$= 36 \text{ m}^3/m/day < 50 \text{ m}^3/m/day$$

... Hence O.K.

$$\text{Weir loading} = \frac{416.67 \times 24}{p \times D}$$

$$= 169.3 \text{ m}^3/d/m < 180 \text{ m}^3/m/day$$

... Hence O.K.

Design of Outlet : Total flow in the outlet is

$$10 \text{ mld} = \frac{10 \times 10^6 \times 10^{-3}}{24 \times 60 \times 60} = 0.116 \text{ m}^3/sec$$

$$\text{Discharge is half one} = \frac{0.116}{2} = 0.058 \text{ m}^3/sec$$

Assume width = 0.5 m and v = 1 m/sec

$$\text{Depth} = \frac{\text{Discharge}}{\text{Width} \times \text{Velocity}} = \frac{0.058}{0.5 \times 1} = 0.116$$

$$= 0.12 \text{ m}$$

Slope of channel,

$$R = \frac{A}{P} = \frac{0.116 \times 0.5}{0.74} = 0.078$$

Assume, n = 0.015

$$V = \frac{1}{n} R^{2/3} S^{1/2}$$

$$1 = \frac{1}{0.015} (0.078)^{2/3} S^{1/2}$$

$$\therefore \quad S^{1/2} = 0.081$$

$$\therefore \quad S^{1/2} = \frac{1}{149}$$

Say S = 1 in 160

$$L = \frac{pD}{2} = \frac{p \times 18.8}{2} = 29.53 \text{ m} \qquad \textbf{...Ans.}$$

Difference between levels

$$= L \times \text{Slope} = 29.53 \times \frac{1}{160}$$

$$= 0.185 \text{ m} \qquad \textbf{...Ans.}$$

Example 1.12 : *Design a mechanically cleaned radial flow circular settling tank for treating sewage from a population of 25000 persons.* **(Nov. 2016, 5M)**

Given:

Maximum hourly flow = (1/14) of the daily flow

Volume of sludge = 1.19 l/c/d

Water consumption = 135 l/c/d

Solution : Given data,

Total population = 25000

$$\text{Max hourly flow} = \frac{1}{14} \text{ of daily flows}$$

Volume of sludge = 1.19 l/c/d

Water consumption = 135 l/c/d

(i) Capacity of tank :

$$\text{Tank capacity} = \frac{\text{Daily flow of sewage}}{24} \times \text{D.T.}$$

$$= \frac{135 \times 25000 \times 0.8}{24} \times 2.5 \text{ hours}$$

$$\{ \because \text{Assume DT} = 1.54 \}$$

$$= 281.25 \times 10^3/l = 281.25 \text{ m}^3$$

$\therefore$ Area of tank $= \dfrac{\text{tank capacity}}{\text{effective depth}}$

$= \dfrac{281.25 m^3}{2.5}$ {Let, eff depth = 2.5m}

$A = 112.5\ m^2$

(ii) Dimensions of tank = Surface area of tank

= length $\times$ Width

$A = L \times B$

$A = 4 B \times B$ $\{ \because L = 4 B\}$

$A = 4B^2$

$\therefore$ $112.5 = 4B^2$

$\therefore$ $B = 5.30\ m$

$\therefore$ $L = 4 \times 5.30$

$L = 21.21 m$

Provide '4m' for arrangement at inlet and outlet then,

$L = 21.21 + 4 = 25.21\ m \cong 26\ m$

(iii) Depth of tank :

Let, sludge accumulation depth = 1m

Free board $= 0.5\ m$

Total depth $=$ Eff. Depth + free board

+ sludge accumulation depth

$= 2.5 + 0.5 + 1$

$D = 4\ m$

(iv) Design Parameters :

Provide , Length $= 26\ m$ **...Ans.**

Width $= 5.3\ m$

Depth $= 4\ m$ **...Ans.**

Example 1.13 : *Determine diameter and depth of primary sedimentation tank for sewage flow 10 million liters per day.*

Given data.

(i) Detention time = 2.5 hours

(ii) Surface loading rate = 40000 l/m²/d

(May 2017, 5M)

Solution : Given data,

Detention time = 2.5 hrs

Surface loading rate = 40000 l/m2/d

The quantity of sewage to be treated in 2.5 hrs

$= 10\ M.\ lit \times \dfrac{2.5}{24} = 1.041\ m/lit$

$\therefore$ Capacity of tank $= 1041.67\ m^3$

Now, Surface loading rate

$= \dfrac{Q}{\text{Surface area of tank}}$

$40000 = \dfrac{Q}{p/4 \times d2}$

$40000 = \dfrac{10 \times 106}{40000}$

$D = 17.84\ m$

Say $D = 18\ m$

Now, Effective depth of tank

$= \dfrac{\text{Capacity}}{\text{Area of} \times \text{sation}} = \dfrac{1041}{p/4 \times (18)^2}$

$= 4.166\ m$

Hence, Use a sedimentation tank with 18 m dia and 4.166 m water depth. **...Ans.**

1.10 INTRODUCTION TO ANAEROBIC DIGESTION
(May 10, 17; Nov. 17)

- Anaerobic digestion is by far the most common method of sludge processing with the wastewater sludge containing primary sludge. Primary sludge contains large amounts of readily available organics that would induce a rapid growth of biomass if treated aerobically. Anaerobic decomposition produces considerably less biomass than aerobic processes.

- The principle function of anaerobic digestion, therefore, is to convert as much of the sludge as possible to end products such as liquids and gases, while producing as little residual biomass as possible.

- The bacterial process consists of two successive processes that occur simultaneously in digesting sludge as given below :

1. Breaking down large organic compounds and converting them to organic acids along with gaseous by-products of carbon dioxide, methane and trace amounts of hydrogen sulphide. This is done by variety of facultative bacteria operating in an environment devoid of oxygen.

2. Converting the organic acids to methane and carbon dioxide. This is done by methane formers which are strictly anaerobes.

The reaction is as shown below :

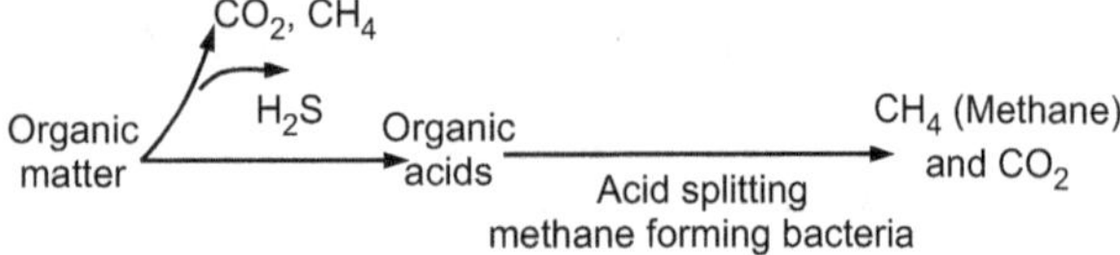

Specific products in the above anaerobic digestion process are shown in Fig. 1.20.

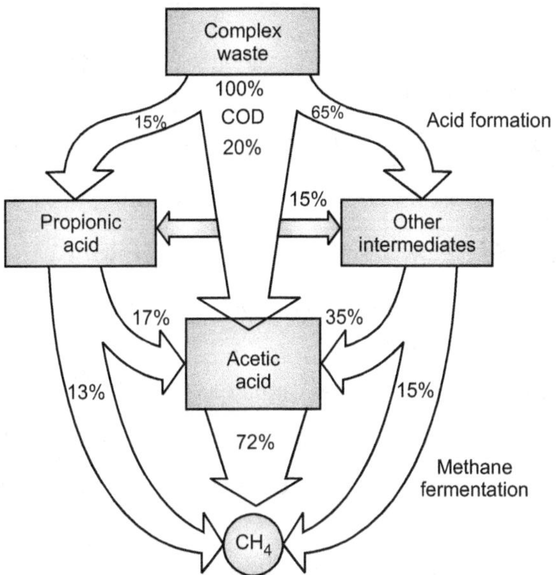

Fig. 1.20 : Pathways and products of anaerobic digestion of wastewater sludge

Typically, about 50 to 60% organics are metabolized, with less than 10% being converted to biomass.

1.11 TYPES OF DIGESTERS

- Reactors for anaerobic digesters consist of closed tanks with airtight covers. Sludge digesters are of two types :
    1. Conventional or low rate digester or standard rate digester.
    2. High rate digesters.

1.11.1 Conventional or Low Rate Digester

(Nov. 15, May 18)

A typical standard rate anaerobic digester consisting of a single stage operation is as shown in Fig. 1.21.

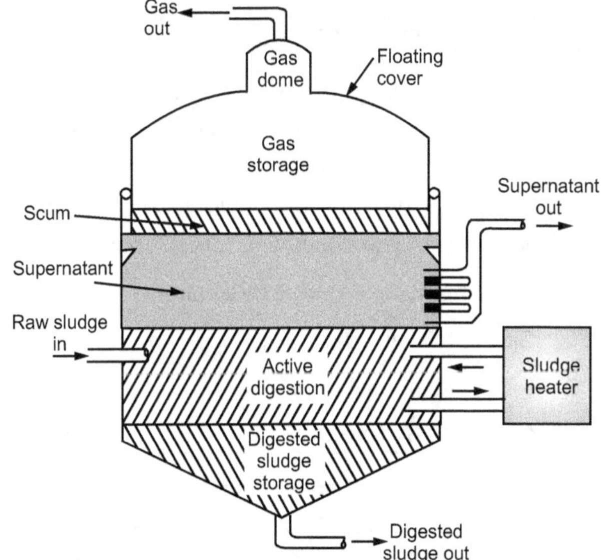

Fig. 1.21 : Diagram of a standard rate anaerobic digester

- The conical bottom facilitates sludge withdrawal while the floating cover accommodates volume changes due to sludge addition and withdrawals.

- The sludge separates in the reactor as shown, although some mixing occurs in the zone of active digestion and in the supernatant because of withdrawal.

- In two stage process, two tanks are provided – first tank meant for digestion and second stage is used for storage, thickening and supernatant formation.

- The high rate digestion process differs from conventional single stage process, in that the solids loading rate is much greater.

- Sludge is fed into the digester on the intermittent basis and the supernatant is withdrawn and returned to the secondary treatment unit.

- The digested sludge accumulates in the bottom, its removal often being determined by subsequent sludge disposal facilities rather than by operational needs of the digester.

Digester Capacity :

- The capacity of the low rate/standard rate digester is determined by loading rates, digestion period, solids reduction and sludge storage. The capacity can be determined by the following rational formula :

$$V = \frac{V_f + V_d}{2} T_1 + V_d T_2$$

- However, since the process of digestion is parabolic, the capacity of the digester is given by the following expression :

$$V = \left[V_f - \frac{2}{3} (V_f - V_d) \right] T_1 + V_d T_2$$

where, V = Volume of digester

V_f = Volume of fresh sludge added per day

V_d = Volume of digested sludge withdrawn per day

T_1 = Digestion time in days

T_2 = Monsoon storage in days

- The checks for digester capacities can be made based on the following parameters :

- **Per Capita Capacity :**
    - ➢ Primary sludge .... 0.05 – 0.075 m³
    - ➢ Combined sludge .... 0.1 – 0.15 m³

- **Loading Factor:**

    Primary or combined sludge ...... 0.3 – 0.75 kg/m³/day

1.11.2 High Rate Digesters (May 18)

- These digesters are more efficient and often require less volume than low rate single stage digesters. The sludge is more or less continuously added and vigorously mixed either mechanically or by recirculating a portion of digestion gases through a compressor.

- Mechanical mixing ensures better contact between the organics and the micro-organisms. The unit is heated to increase the metabolic rate of micro-organisms (thus speeding up the digestion process). Optimum temperature is around 35°C.

- Because of good mixing, there is no stratification and hence loss of capacity does not arise due to supernatant or scum or dead pockets.

- High rate digestion systems normally consist of two tanks operated in series. (Fig. 1.22). The first stage is a complete mixing, heated, floating or fixed cover digester fed as continuously as possible, whose function is anaerobic digestion of volatile solids.

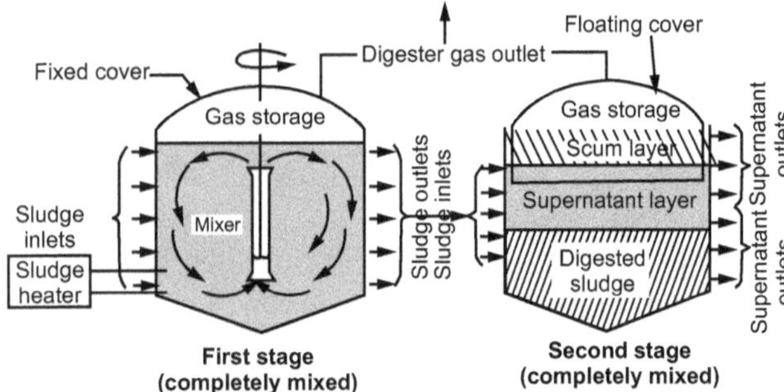

Fig. 1.22 : Diagram of high-rate, two-stage anaerobic sludge digester

In the second stage digester, separation of supernatant accompanied with a reduction in volume of sludge due to gravity thickening takes place. Little gas is generated in the second stage, but the influent is supersaturated with gases that are released in the second stage reactor. Consequently second stage reactor is usually covered and equipped with gas recovery. The second stage reactor is not heated.

Table 1.3 : Design Parameters for Anaerobic Digesters

Parameter	Standard Rate	High Rate
Digestion period, days	30 – 40	10 – 20
Volatile solids loading, kg/m³/day	0.6 – 1.6	1.6 – 6.4
Digested solids concentration, %	4 – 6	4 – 6
Volatile solids reduction, %	35 – 50	45 – 55
Gas production (m³/kg VSS destroyed)	0.5 – 0.9	0.6 – 0.65
Methane content, %	60 – 70	60 – 70

1.12 PERFORMANCE OF DIGESTERS

The following parameters give the idea about the performance of digesters :

- Gas production on per capita basis : 0.025 m³/day

- Gas production per kg of volatile matter added : 0.4 m³

- Gas production per kg of volatile matter destroyed : 0.9 m³

- pH of digesting sludge : 7 to 8

- Methane content of gas produced : 60 to 70%

- Dry solids in digested sludge : 6 to 10%

  ➢ Primary : 10 to 15%

  ➢ Secondary : 6 to 10%

- Volatile acids : 200 to 400 mg/lit

- Colour : Black

- Odour : Inoffensive

- Drainability : Quickly drainable

- Bicarbonate alkalinity : 2000 – 5000 mg/l

- Grease : Practically absent

(Ref. Manual on Sewerage and Sewage Treatment, 2nd Edition)

1.13 SLUDGE THICKENING

- Sludge contains large volumes of water. Thickening of sludge is used to concentrate solids and reduce the volume. Thickened sludge requires less tank capacity and chemical dosage for stabilization and smaller piping and pumping equipment for transport.

- Common methods of sludge thickening :

  1. Gravity thickening

  2. Dissolved air floatation.

1.13.1 Gravity Thickening (May 18)

- This is accomplished in circular sedimentation tank similar to those used for primary and secondary clarification of liquid waste. Solids coming to thickener separate into three distinct zones. The top layer is relatively clear liquid. The next layer is the settling zone, which usually contains a stream of denser sludge moving from influent end toward the thickening zone.

- Water is squeezed out of interstitial spaces and flows upwards to the channels. Arrangement is done to provide gentle stirring of sludge blanket and move the gases and liquid towards the surface. The supernatant from the sludge thickener passes over an effluent weir and is returned at the inlet of plant.

- The thickened sludge is withdrawn from the bottom.

- Gravity thickening is used to concentrate solids in sludge's from the primary clarifier, trickling filter and activated sludge.

- The degree of thickening may vary from 2 to 5 times the concentration of solids in the incoming sludge.

- Maximum solids concentration achieved in gravity thickening is normally less than 10 per cent.

The equipment includes :

- Rotating bottom scraper arm.

- Vertical pickets.

- Rotating scum collecting mechanism with scum baffle plates.

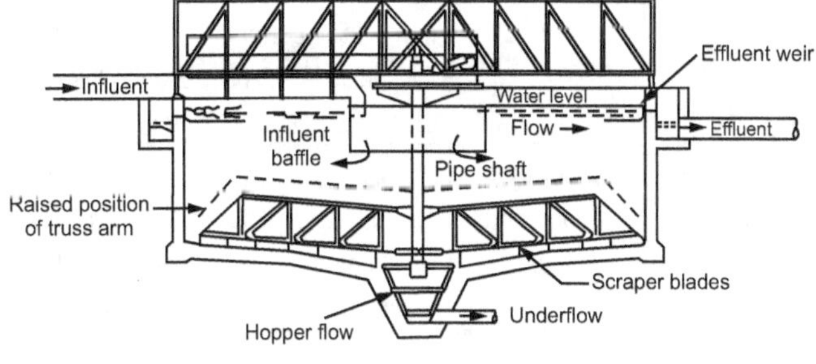

Fig. 1.23 : Gravity sludge thickener

1.13.2 Dissolved Air Floatation

Air floatation is primarily used to thicken the solids in chemical and waste activated sludge. Separation of solids is achieved by introducing fine air bubbles into the liquid. The bubbles attach to the particulate matter which then rise to the surface. In this system, the air is dissolved in the incoming sludge under pressure. The pressurized flow is discharged into a floatation tank that operates at normal pressure.

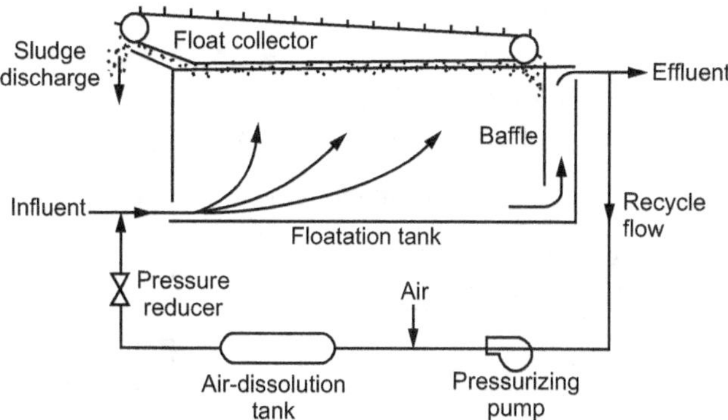

Fig. 1.24 : Schematic diagram of a dissolved air floatation system

Fine air bubbles rise that cause floatation of solids.

The principle advantage of floatation over gravity thickening is the ability to remove more rapidly and completely those particles that settle slowly under gravity. The amount of thickening achieved is 2-8 times the incoming solids. Maximum concentration of solids in the float may reach 4-5 per cent.

The common equipments include:

- Sludge feed pump with air compressors.

- Floatation tank with skimmer.

- Chemical storage and feed system.

- Thickened sludge pump.

1.14 SLUDGE DRYING BEDS (May 16)

- When land is available and the sludge quantity is small, natural dewatering systems are more attractive. These include drying beds and drying lagoons. The mechanical dewatering systems are generally selected where land is not available.

- Sludge drying beds even though oldest method, still used extensively in small to medium size plants to dewater digested sludge. Typical sand bed consists of layer of coarse sand 15-25 cm in depth and supported on a graded gravel bed that incorporates selected tiles or perforated under drains. Sludge is placed on the bed in 20 to 30 cm layers and allowed to dry. The under drained liquid is returned to the plant. The drying period is 10-15 days and moisture content of the cake is 60-70 per cent.

- Depending on climatic conditions and odor control requirements, the drying beds may be open or covered.

Design Criteria :

- Sludge drying bed area :
  - ➢ 0.14 – 0.28 m^2 per capita for uncovered beds.
  - ➢ 0.1 – 0.2 m^2 per capita for covered beds.

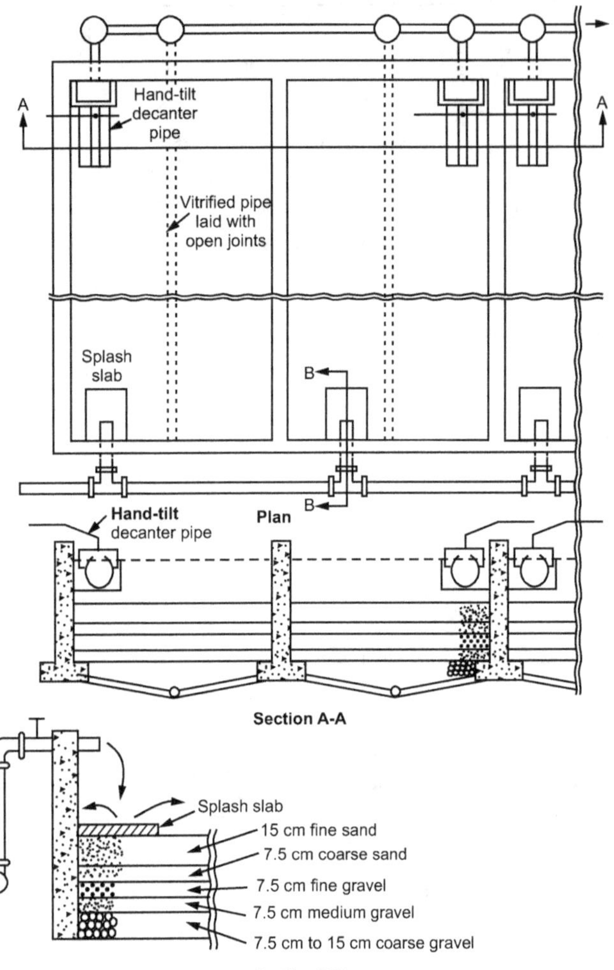

Fig. 1.25 : Details of sludge drying beds

1.15 SLUDGE TREATMENT AND DISPOSAL

A by-product in the treatment of wastewater is the huge volume of sludge that must be disposed off. The problem of disposing sludge is complex because –

- It is largely composed of the substances that make the wastewater objectionable.
- It is composed of organic matter that can decompose making it objectionable.
- The sludge solid itself is only a small portion of the total sludge, resulting in large sludge volume with only little solids to dispose off.
- Sludge contains only from 0.25 to 16% solids. To dispose sludge economically, its volume must therefore be reduced.

1.15.1 Sources of Sludge

The sources of solids in the treatment plant vary according to the type of plant and its method of operation. The principle sources of solids and sludge and the types generated are given in Table 1.4.

Table 1.4 : Sources of Solids and Sludge from a Conventional Wastewater-Treatment Facility

Unit Operation or Process	Type of Solids or Sludge	Remarks
Screening	Coarse solids	Coarse solids are often comminuted and returned to the wastewater for removal in subsequent treatment facilities.
Grit removal	Grit and scum	Scum removal facilities are often omitted on grit removal facilities.
Preaeration	Scum	In some plants, scum removal facilities are not provided in preaeration tanks.
Primary sedimentation	Primary sludge and scum	The quantities of both sludge and scum depend on the nature of the collection system and whether industrial wastes are discharged to the system.
Aeration tank	Suspended solids	Suspended solids are produced from the conversion of BOD. If wasting is from the aeration tank, flotation thickening is normally used to thicken the waste activated sludge.
Secondary sedimentation	Secondary sludge and scum	Provision for scum removal on secondary settling tanks is now a requirement of the U.S. Environmental Protection Agency.
Sludge-processing facilities	Sludge and ashes	The characteristics and moisture content of the sludge and ashes depend on the operations and processes that are used.

1.15.2 Characteristics of Sludge

To treat and dispose off the sludge produced from wastewater treatment plants in the most effective manner, it is important to know the characteristics of the solids and sludge that will be processed. The characteristics vary depending on the origin of the solids and sludge; the amount of aging that has taken place, and the type of processing to which they have been subjected. The physical and chemical characteristics of untreated sludge are given in Tables 1.5 and 1.6 respectively.

Table 1.5 : Characteristics of Sludge Produced during Wastewater Treatment

Solids or Sludge	Description
Screenings	Screenings include all types of organic and inorganic materials large enough to be removed on bar racks. The organic content varies, depending on the nature of the system and the season of the year.
Grit	Grit is usually made up of the heavier inorganic solids that settle with relatively high velocities. Depending on the operating velocities, grit may also contain significant amounts of organic matter, specifically fats and grease.
Scum	Scum consists of the floatable materials skimmed from the surface of primary and secondary settling tanks. It may include grease, vegetable and mineral oils, animal fats, waxes, soaps, food wastes, vegetable skins, hair, paper and cotton, cigarette tips, plastic suppositories, rubber prophylactics, grit particles, and similar materials. The specific gravity of scum is less than 1.0 and usually around 0.95.
Primary sludge	Sludge from primary sedimentation tanks is usually gray and slimy and, in most cases, has an extremely offensive odour. It can be readily digested under suitable conditions of operation.
Chemical-precipitation sludge	Sludge from chemical precipitation tanks is usually dark in colour, though its surface may be red if it contains much iron. Its odour may be objectionable, but not as bad as odour from primary sedimentation sludge. While it is somewhat slimy, the hydrate of iron or aluminium in it makes it gelatinous. If it is left in the tank, it undergoes decomposition like the sludge from primary sedimentation but at a slower rate. It gives off gas in substantial quantities and its density is increased by standing.
Activated sludge	Activated sludge generally has a brown flocculant appearance. If the colour is quite dark, it may be approaching a septic condition. If the colour is lighter than usual, there may have been underaeration with a tendency for the solids to settle slowly. Sludge in good condition has an inoffensive characteristic odour. It tends to become septic rather rapidly and then has a disagreeable odour of putrefaction. It will digest readily alone or mixed with fresh wastewater solids.
Trickling-filter sludge	Trickling-filter humus is brownish, flocculant, and relatively inoffensive when fresh. It generally undergoes decomposition more slowly than other undigested sludges, but when it contains many worms it may become offensive quickly. It is readily digested.
Digested sludge (aerobic)	Aerobically digested sludge is brown to dark brown and has a flocculant appearance. The odour of aerobically digested sludge is not offensive; it is often characterized as musty. Well digested aerobic sludge dewaters easily, and the resulting dry solids are inoffensive.
Digested sludge (anaerobic)	Anaerobically digested sludge is dark brown to black and contains an exceptionally large quantity of gas. When thoroughly digested, it is not offensive, its odour being relatively faint and like that of hot tar, burnt rubber, or sealing wax. When drawn off on porous beds in thin layers, the solids first are carried to the surface by the entrained gases, leaving a sheet of comparatively clear water below them which drains off rapidly and allows the solids to sink down slowly on to the bed. As the sludge dries, the gases escape, leaving a well-cracked surface with an odour resembling that of garden loam.
Septage	Sludge from septic tanks is black. Unless well digested by long storage, it is offensive because of the hydrogen sulphide and other gases it gives off. The sludge can be dried on porous beds if spread out in thin layers, but objectionable odours are to be expected while it is draining unless it has been well digested.

Table 1.6 : Typical Chemical Composition of Untreated Sludge

Item	Untreated Primary Sludge	
	Range	Typical
Total dry solids (TS), %	2.0–8.0	5.0
Volatile solids (% of TS)	60–80	65
Grease and fats (ether-soluble, % of TS)	6.0–30.0	...
Protein (% of TS)	20–30	25
Nitrogen (N, % of TS)	1.5–6.0	4.0
Phosphorus (P_2O_5, % of TS)	0.8–3.0	2.0
Potash (K_2O, % of TS)	0–1.0	0.4

...Conti.

Cellulose (% of TS)	8.0–15.0	10.0
Iron (not as sulphide)	2.0–4.0	2.5
Silica (SiO_2, % of TS)	15.0–20.0	...
pH	5.0–8.0	6.0
Alkalinity (mg/lit as $CaCO_3$)	500–1500	600
Organic acids (mg/lit as HAc)	200–2000	500
Thermal content (MJ/kg)	14–23	16.5

1.15.3 Quantity of Sludge

- The quantity of daily sludge varies depending upon the degree of removal of suspended solids in primary and secondary settling tanks, moisture content and specific gravity of sludge. In case of primary sludge, continuously desludged solids do not generally exceed 4 to 5%, while in secondary settling tanks; they range between 0.5 to 1%.
- When the secondary sludge is combined with the primary sludge, the solids content in the mixed sludge range between 2.5 to 5%. The solids content in the digested sludge is usually in the range between 6 to 13% when reduction in bulk takes place due to separation of supernatant.

Guidelines to Calculate Quantity of Sludge :

- Quantity of suspended solids in raw sewage : 90 gm/day per head.
- Concentration of suspended solids in raw sewage : 200 to 250 mg/lit.
- Solids removed in PST : 60%.
- Solids generated in biological process :
  - ➤ Fixed/Attached growth process : 0.4 to 0.5 kg of BOD_5 applied.
  - ➤ Suspended growth process : 0.2 – 1.0 kg of applied BOD_5.

Typical values of solids are given in Table 1.7.

Table 1.7 : Solids in Sludges (Per Capita Per Day)

Treatment Process	Total (gm)	Volatile (g)	Fixed (g)	Sp. Gr. Of Dry Solids	% Solids in Wet Sludge
(1)	(2)	(3)	(4)	(5)	(6)
Sr. No. **Raw sewage**					
1. Total solids	90	63	27	1.22	
2. Settleable solids (60%)	54	38	16	1.22	
3. Non-settleable solids (40%)	36	25	11	1.22	

		Total	Volatile	Fixed	Sp. Gr.	% Solids
I.	Primary (Pr.) sedimentation and digestion					
1.	Solids removed as fresh sludge	54	38	16	1.22	4–5
2.	Solids digested*	–	–25	+6		
3.	Solids remaining in primary digested sludge	35	13	22	1.6	10–13
II.	Activated sludge Process (AS)					
1.	Non-settleable solids affected	36	25	11	1.22	
2.	Solids digested during activation (10%)	–	–2.5	+0.5		
3.	Solids to be removed	34	22.5	11.5		
4.	Solids removed as fresh activated sludge (90%)	31	20	11	1.25	
5.	Combined primary (I) and secondary excess AS (II)	54 + 31 = 85	38 + 20 = 58	16 + 11 = 27	1.22	2.5 – 4
6.	Solids digested	–	–38	+10		
7.	Solids remaining in digested combined Pr. & Excess AS	57	20	37	1.65	6 – 7
III.	Trickling filter (T.F.) Process					
1.	Non-settleable solids	36	25	11	1.22	
2.	Solids digested during filtration (10%)	–	–2.5	+0.5		
3.	Solids to be removed in sedimentation tank	34	22.5	11.5		
4.	Solids to be removed in SS Tank (40%)	14	9	5		
5.	Combined Pr. (I) and TF humus (III)	68	47	21	1.22	4 – 5
6.	Solids digested in digester	–	–32	+8		
7.	Solids remaining in digested Pr. and T.F. humus	44	15	29	1.66	8 – 10

- The quantity of solids removed in PST can be calculated by the following formula :

$$M_p = \eta \times SS \times Q$$

where, M_p = Mass of primary solids, kg/d

η = Efficiency of PST for solids removal

SS = Total SS in effluent, kg/m^3

Q = Flow rate, m^3/d

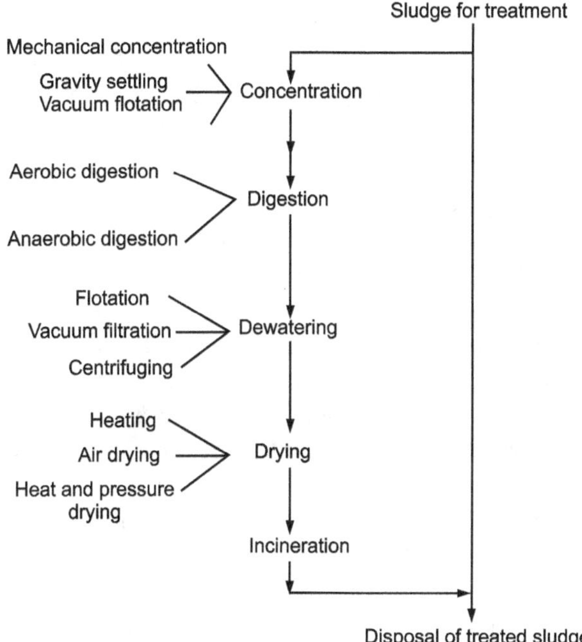

Fig. 1.26 : Flow chart for possible treatment of sludge

- Secondary sludge composed primarily of biological solids, the quantity of which can be estimated by the equation :

$$M_S = Y' \times BOD_5 \times Q$$

where, M_S = Mass of secondary solids, kg/d

Y' = Biomass conversion factor

Q = Flow rate, m^3/day

- To calculate volume of sludge, the following equation can be used :

1. If sludge contains organic (volatile), inorganic (fixed) solids :

$$\frac{W_S}{S_S} = \frac{W_f}{S_f} + \frac{W_v}{S_v}$$

or

$$\frac{100}{S_S} = \frac{\% \text{ of mineral matter}}{\text{Sp. gr. of mineral matter}}$$

$$+ \frac{\% \text{ of volatile matter}}{\text{Sp. gr. of oganic matter}}$$

where, W_S = Wt. of solids

S_S = Sp. gr. of solids

W_f = Wt. of fixed solids

S_f = Sp. gr. of fixed solids (2.4 to 2.65)

W_v = Wt. of volatile matter

S_v = Sp. gr. of volatile matter (1.0 to 1.2)

2. Specific gravity of Wet Sludge (S_{sl}) :

$$\frac{100}{S_{sl}} = \frac{\% \text{ moisture}}{\text{Sp. gr. of water}} + \frac{\% \text{ of solids}}{S_S}$$

3. Once the sp. gr. of wet sludge and % solids in wet sludge are known, volume of wet sludge can be determined, using the relationship :

$$V_{sl} = \frac{W_S}{r_W \, S_{sl} \, P_S}$$

where, V_{sl} = Volume of sludge

W_S = Wt. of dry solids

S_{sl} = Sp. gr. of sludge

P_S = Percent solids expressed as decimal

ρ_W = Density of water

1.16 SEPTIC TANKS

- Septic Tank is a sanitation technology with a clear design which is promoted in areas where a sewerage system is absent. It helps in the primary treatment of liquid waste with solid content. Bureau of Indian Standards (BIS) provides standards for construction of septic tanks and the secondary treatment of fecal sludge matter from the septic tanks.

- However, at the grassroots level construction of septic tanks seldom follow the required standards. Similarly, waste from septic tank needs a secondary treatment, and there is no proper system for emptying the Fecal Sludge Matter which is considerate of the environmental impacts in India currently.

- Unscientific septic tanks and careless disposal of fecal sludge affect the environmental and sanitation goals of the country. The BIS document cautions that 'unsatisfactory design, construction, and maintenance of septic tanks constitute a health hazard'.

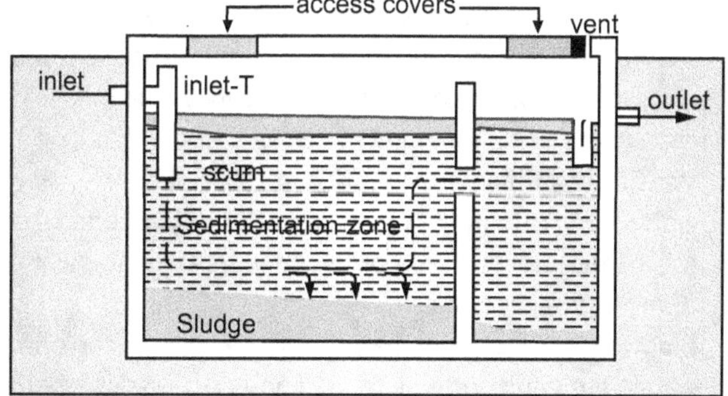

Fig. 1.27 : Section of a Septic Tank

- A septic tank is a rectangular watertight structure used to treat liquid waste with high settleable solids. According to the available literature, a scientifically designed septic tank involves two chambers with an inlet to collect the liquid waste from the pour flush cistern and an outlet to expel the effluent to soakage pit or a sewer. As the black water flows through ST, the solids settle at the bottom of the first chamber and scum moves to the top. The first chamber of the septic tank should be two third of the total length of the septic tank.

- Over time, due to anaerobic digestion, the sludge reduces, and a portion of the solid turns into liquids and gas which rise to the surface in the form of bubbles. At the top of the septic tank, a vent is built to expel the gases produced due to anaerobic digestion of the solid waste

1.16.1 Parameters of Septic Tank Construction (Source: BIS, 1993)

- **Size of the Tank:** The Septic tank shall have a minimum width of 750mm and depth of one metre below the outlet with a liquid capacity of 1000 litres. For rectangular septic tanks, the length of the tank shall be 2 to 4 times the width. For circular tanks, the minimum diameter shall not be less than 1.35 metre, and operating depth shall not be less than 1.0 m.

- **Inlet and Outlet:** For the tanks not more than 1200mm wide, the inlet and outlet are T shaped dip-pipe. The inlet pipe shall be fixed inside the tanks with top limb rising above scum level and the bottom limb extending about 300 mm below the top water level. Outlet pipe should be fixed inside the tank with top limb rising above scum level and the bottom limb extending to about 1/3rd of the liquid depth below top water level.

- **Partitions:** Where the capacity of the septic tank exceeds 2000 litres, the tank may be divided into two chambers using a fixed durable barrier. The partition shall be located so that the capacity of the first chamber is twice that of the second chamber.

- **Ventilating Pipe:** Every septic tank should be provided with a ventilating pipe of at least 50 mm diameter. The top of the pipe shall be provided with a suitable cage of mosquito proof mesh.

- **Floor:** It is essential that the floor of the tank be watertight and adequate strength to resist earth movement and to support the weight of the tank walls and contents. The floor may be of concrete of minimum M15 grade and a minimum slope of 1:10 may be provided towards the sludge outlet to facilitate desludging.

- **Walls:** walls should be thick enough to provide adequate strength and water tightness. Walls built of bricks should not be less than 200 mm thick and should be plastered to a minimum thickness of 12 mm inside and outside.

- **Desludging:** Small domestic tanks for economic reasons may be cleaned at least once in 2 years provided the tank is not overloaded due to use by more than the number for which it is designed. A portion of the sludge not less than 25 mm in depth should be left behind in the tank bottom which acts as the seeding material for the fresh deposits.

- **Commissioning the Tank:** The sewerage system should be complete and ready for operation before connected to the building. The tank should be filled with water to its outlet level before the first time the tank is used. It should preferably be seeded with small quantities of well-digested sludge obtained from the septic tank or sludge digestion tanks.

- **Soak Pit:** Soak pit is a pit through which effluent is allowed to seep into the surrounding soil. It should be 1.5-4 metre in depth and at least 30 m away from any source of drinking water.

1.16.2 Soakage System (Soak Pits) for Disposal of Septic Tank Effluent

Soak Pits

- Soak pits are cheap to construct and are extensively used. They need no media when lined or filled with rubble or brick bats. The pits may be of any regular shape, circular or square being more common. When water table is sufficiently below ground level, soak pits should be preferred only when land is limited or when a porous layer underlies an impervious layer at the top, which permits easier vertical downward flow than horizontal spread out as in the case of dispersion trenches.

- Minimum horizontal dimension of soak pit should be 1 m, the depth below the invert level or inlet pipe being at 1 m. The pit should be covered and the top raised above the adjacent ground to prevent damage by flooding. It is being recommended that these are to be phased out in due course of time.

Dispersion Trenches

- Dispersion trenches consist of relatively narrow and shallow trenches about 0.5 to 1 m deep and 0.3 to 1 m wide excavated to a slight gradient of about 0.25%. Open joined earthenware or concrete pipes of 80 to 100 mm size are laid in the trenches over a bed of 15 cm to 25 cm of washed gravel or crushed stone. The top of pipes shall be covered by coarse gravel and crushed stone to a minimum depth of 15 cm and the balance depth of trench filled with excavated earth and finished with a mound above the ground level to prevent direct flooding of trench during rains.

- The effluent from the septic tank is led into a small distribution box from which several such trenches could radiate out. The total length of trench required shall be calculated from the empirical formula and the number of trenches worked out on the basis of a maximum length of 30 m for each trench and spaced not closer than 2 m apart. Parallel distribution should be such that a distribution box should be provided for 3 to 4 trenches. It is being recommended that these are to be phased out in due course of time.

1.17 IMHOFF TANK

- The imhoff tank is a primary treatment technology for raw wastewater, designed for solid-liquid separation and digestion of the settled sludge. It consists of a V-shaped settling compartment above a tapering sludge digestion chamber with gas vents. In the digestion chamber, the settled solids are anaerobically digested generating biogas.

- The gas is deflected by baffles to the gas vent channels to prevent it from disturbing the settling process. Imhoff tanks are used by small communities and due to the underground construction, land use is very limited. Investment costs are low and operation and maintenance simple. But the treatment efficiency is low and a secondary treatment of the effluent is required. Moreover, the tanks must be desludged regularly.

1.17.1 Principle of Working

A schematic of an Imhoff tank is shown in the Fig. 1.28.

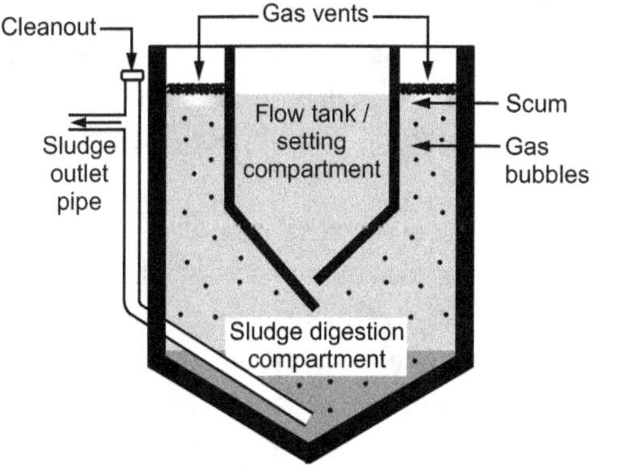

(a) Cross section

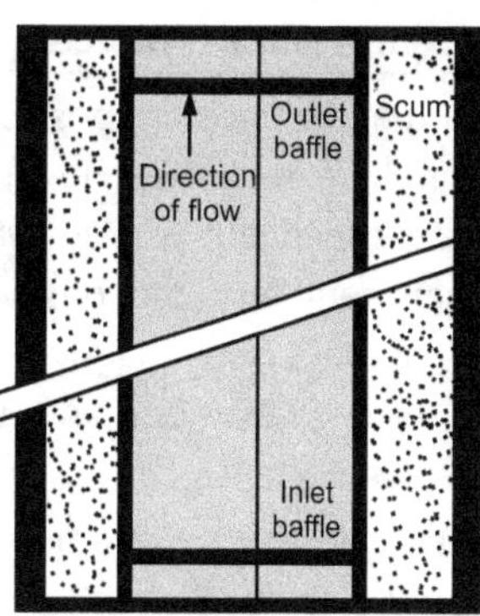

(b) Plan

Fig. 1.28 : Imhoff Tank

- The Imhoff tank works similar to a communal septic tank, is a robust and effective settler that causes a suspended solids reduction of 50 to 70%, COD reduction of 25 to 50%, and leads to potentially good sludgestabilisation – depending on the design and conditions. It is a compact and efficient system for pre-treatment of municipal wastewater.

- The settling compartment has a circular or rectangular shape with V-shaped walls and a slot at the bottom, allowing solids to settle into the digestion compartment, while preventing foul gas from rising up and disturbing the settling process. Gas produced in the digestion chamber rises into the gas vents at the edge of the reactor. It transports sludge particles to the water surface, creating a scum layer. The sludge accumulates in the sludge digestion chamber, and is compacted and partially stabilised through anaerobic digestion.

- The liquid fraction remains only some hours in the tank, while the solids remain for several years in the digestion chamber. There is more biogas production than in septic tanks, but unfortunately the organic load in communal wastewater is usually not high enough for the economical collection and usage (or flaring) of biogas.

- The pre-treated wastewater from the Imhoff tank requires a secondary treatment (e.g. leach field, soak pits, horizontal flow, vertical flow or free-surfaceconstructed wetlands). Also the faecal sludge needs to be correctly disposed and further treated (e.g. small or large scalecomposting, settling - thickening ponds or drying beds). If the sludge is composted either directly or after drying, it can be used as fertiliser to improve the soil quality (see also use of compost or fertilizer from sludge).

1.17.2 Operation and Maintenance

- Operation and maintenance are possible at low cost, if trained personnel are in-charge. Flow paths have to be kept open and cleaned out weekly, while scum in the settling compartment and the gas vents has to be removed daily if necessary. Regular cleaning of the sides of the settling chamber and slot by rake or squeegee is very important.

- Another activity, which should be done twice a month is to reversing the flow (backwash) of water to even up the solids in the digestion chamber. Stabilisedsludge from the bottom of the digestion compartment should be removed according to the design. The sludge can be removed by pumping or hydraulic pressure pipes right from the bottom or by a vacuum truck. A minimum clearance of 50 cm between the sludge blanket and the slot of the settling chamber has to be ensured at all times.

1.17.3 Design Considerations

The Imhoff tank is designed with three compartments:

1. Upper compartment for sedimentation.
2. Lower section for sludge digestion.
3. Gas vent and scum section.

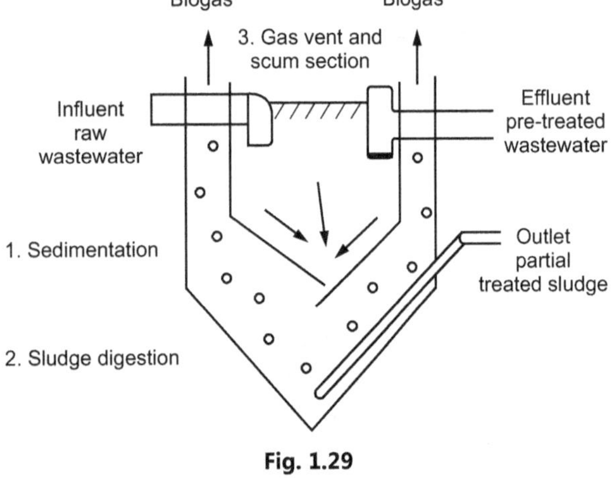

Fig. 1.29

1.17.4 Advantages

- Simple to construct and to operate.
- Imhoff tanks are resistant against shock loads.
- Removes 25 to 50% of COD.
- Solid-liquid separation and sludge stabilisation are combined in one single unit.
- Resistant against organic shock loads.
- Small space requirements.
- The effluent remains fresh (i.e., not septic).
- Low operating costs.
- Suitable for small settlements and house clusters.
- Standardised designs available.
- Simple operation and maintenance.

1.17.5 Disadvantages

- Construction costs are slightly higher than the costs of a septic tank.
- Effluent, sludge and scum require further treatment.
- Very high (or deep) infrastructure; depth may be a problem in case of high groundwater table.
- Requires expert design and construction.
- Low reduction of pathogens.
- Requires desludging.
- Inefficient treatment option if not regularly desludged.
- Odour occurs from escaping gases.
- Less simple than septic tank.

1.17.6 Types of Advanced Wastewater Treatment

- Advanced Wastewater Treatment may be categories as follows on the basis of the type of process adopted:

  1. Tertiary reatment;
  2. Physical-Chemical Treatment; and
  3. Combined Biological-Physical-Chemical Treatment.

- Tertiary treatment may be defined as any treatment process in which unit operations are added to the flow scheme following conventional secondary treatment. Additions to conventional secondary treatment could be as simple as the addition of a filter for suspended solids removal or as complex as the addition of many unit processes for organic, suspended solids, nitrogen and phosphorous removal.

- Physical-chemical treatment is defined as a treatment process in which biological and physical-chemical processes are intermixed to achieve the desired effluent.

- Combined biological-physical-chemical treatment is differentiated from tertiary treatment in that in tertiary treatment any unit processes are added after conventional biological treatment, while in combined treatment, biological and physical-chemical treatment are mixed.

- The main reason for advanced wastewater treatment may be to remove nutrients contained in discharges from secondary treatment plants. The effluents from secondary treatment plants contain both nitrogen (N) and phosphorous (P). N and P are ingredients in all fertilizers. Excessive release to the environment can lead to a buildup of nutrients, called *eutrophication*, which can in turn encourage the overgrowth of weeds, algae, and blue-green algae.

- This may cause an algal bloom, a rapid growth in the population of algae. The algae numbers are unsustainable and eventually most of them die. The decomposition of the algae by bacteria uses up so much of the oxygen in the water that most or all of the animals die, which creates more organic matter for the bacteria to decompose. In addition to causing deoxygenation, some algal species produce toxins that contaminate drinking water supplies.

- Different treatment processes are required to remove nitrogen and phosphorus. In some cases toxic materials, both organic and inorganic are discharged into many sewage collections systems. When these materials are present in sufficient quantities to be toxic to bacteria, it will be necessary to remove them prior to biological treatment.

- In other cases, it is necessary to remove even small amounts of these materials prior to discharge to protect receiving waters or drinking water supplies. Thus, advanced wastewater treatment processes have been used in cases where conventional secondary treatment was not possible due to materials toxic to bacteria entering the plant.

1.18 OXYGEN REQUIREMENT IN AEROBIC PROCESSES

- The BOD removal in a biological reactor is accomplished not entirely by oxidation – a part of the BOD is removed from the system, without being oxidized, in the form of wastage of excess sludge (synthesized biomass) from the system.

- So, the oxygen requirement will be equal to the amount that would have been required, if all the BOD was removed by oxidation only (i.e. the ultimate BOD removed), less, a credit for the fraction of BOD removed by sludge wasting. Therefore one can write :

$\therefore$ Rate of oxygen usage = (Rate of BOD_L removal)– f(Rate of excess sludge wasted)

in which, f is a function that denotes the amount of oxygen "saved" per unit weight of sludge removed from the system. Based on the stiochiometry of cell oxidation, it has been shown that f has a constant value of 1.42 kg of O_2 per kg of cells.

Therefore, an a finite time and mass basis :

O_2 (requirement/day) = Ultimate BOD removed/day

     – 1.42 (Excess sludge wasted per day)

                      ... (1.1)

Now assuming BOD rate constant = 0.23/day at 20°C,

$$\frac{\text{Ultimate BOD}}{\text{5 day 20°C BOD}} = \frac{1}{1 - e^{0.23 \times 5}} = 1.47 \qquad ... (1.2)$$

Using equation (1.2), the equation (1.1) may be modified in the following form :

Weight of oxygen required/day

     = (5 day 20°C BOD removed/day) 1.47 – 1.42 (excess sludge wasted/day)

     = $1.47 \, Q(S_0 - S_1) - 1.42 \, V \, (X/\theta_c)$ ... (1.3)

where,   Q = Rate of inflow of waste

    S_0, S_1 = Substrate concentration at the inlet and outlet of the system respectively, BOD_5 in mg/l

      V = Volume of the reactor

      X = Microorganism concentration in the reactor

and θ_c = Mean cell residence time

The equation (1.3) may be used directly to calculate the theoretical oxygen requirement of the system. It may be noted that, the above expression is not materially different from that given by Eckenfelder and O'Connor, as presented below :

    Oxygen requirement/day = $Q (S_0 - S_1) \, a' + VXb'$

                      ... (1.4)

The equation (1.3) may be rewritten in the following form:

Weight of oxygen required/day

$$= 1.47 \, Q \, (S_0 - S_1) - 1.42 \, V \left(Y \frac{dF}{dt} - k_d X \right)$$

$$= Q \, (S_0 - S_1) \, (1.47 - 1.42 \, Y) + 1.42 \, k_d \, XV \qquad ... (1.5)$$

in which Y is expressed in mg of cell per mg of 5 day BOD. The terms (1.47 – 1.42 Y) and 1.42 k_d of the equation (1.5) are referred to as a' and b' respectively in the equation (1.5).

ACTIVATED SLUDGE PROCESS

1.19 INTRODUCTION TO ACTIVATED SLUDGE PROCESS   (May 11, Nov. 15)

- Aerobic treatment processes utilise a mixed population of micro-organisms. The wastewater is first treated in preliminary and primary treatment units like screen chamber, grit chamber, oil and grease removal tank. Here the floating solids, bigger size material, settleable solids, oil and grease are removed. In these treatment units, around 50 to 60% suspended solids and about 40% of BOD is removed. Then the waste water is treated biologically.

- The objectives of biological treatment of wastewater are to coagulate and remove the non-settleable colloidal solids and to stabilize the organic matter with the help of mixed population of micro-organisms. Many different types of organisms, including bacteria, fungi, protozoa, rotifers, and other higher forms of life can be found in the system. At any particular time the various forms present, bacteria are the most important as they represent the major functional unit of biological system.

- The biological treatment techniques used are of three types :

  1. Attached growth processes.

  2. Suspended growth processes.

  3. Combined processes.

- In the suspended growth process, all the contents of wastewater from the reactor are maintained in suspension with the liquid by employing either natural or mechanical mixing.

- The suspended growth processes are :

  1. Activated sludge process.

  2. Aerated lagoons.

  3. Digestion of sludge.

- In this chapter, the various aspects of activated sludge process are discussed. One should understand that, the micro-organisms convert the soluble organic matter into gaseous form and cell mass and then the insoluble cell mass is removed in the subsequent treatment units.

- The treated effluent obtained from a well operated ASP is of high quality. As compared to trickling filter effluent, the ASP effluent will have low BOD values. BOD removal from 70% to 99.9% can be obtained in this biological treatment process. The advantage of ASP is it requires less land area and quality of sludge is also good.

1.20 PROCESS DESCRIPTION OF ASP

(May 10, Nov. 15, 16)

- The process consists of (a) rapid adsorption of waste substrate by sludge, (b) progressive oxidation of adsorbed organics and synthesis of biomass, (c) separation of biomass and recirculation. Operationally, biological waste treatment with an ASP is typically

accomplished using a typical flow diagram as shown in Fig. 1.30.

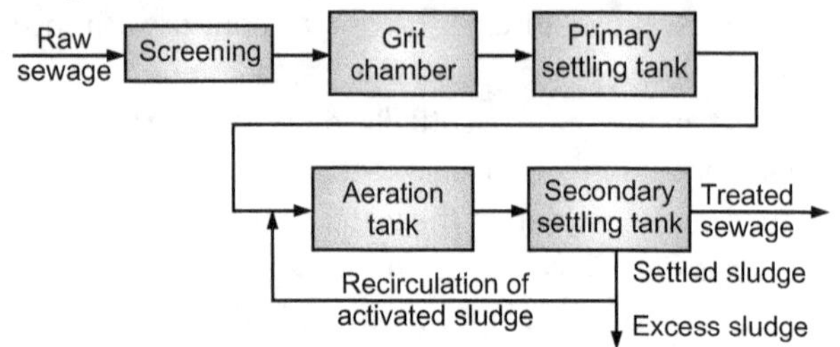

Fig. 1.30 : Flow diagram of ETP with conventional activated sludge process

- In activated sludge process, settled wastewater from primary settling tank is introduced into a reactor (aeration tank) where an aerobic bacterial culture is maintained in suspension. The agitated wastewater is known as mixed liquor.

- In the reactor, the micro-organisms convert soluble organic matter into energy, new cell mass and carbon dioxide gas.

- The aerobic environment into the reactor is maintained with the help of diffused or mechanical aerators.

- These aerators also keep the mixed liquor in suspension, after some specific period of time (hydraulic retention time), the mixture of new cell mass and old cells is transferred into secondary settling tank where the cells (biomass) are separated from treated effluent. A portion of biomass (sludge) is recirculated to maintain the desired concentration of micro-organisms in the reactor, and remaining sludge is either wasted or stabilised.

- In activated sludge process, the bacteria are the most important micro-organisms because they are responsible for decomposition of organic material present in wastewater. In general, the bacteria in ASP are gram negative and include Pseudomonas, Achromobacter, Flavobacterium, Nocardia, Mycobacterium and two nitrifying bacteria Nitrosomonas and Nitrobacter. Also filamentous forms like Thiothrix, Beggiatoa, Geotrichum.

- In ASP, the bacteria degrade the organic waste in the influent, while other micro-organisms like protozoa and rotifers act as effluent polishers. Protozoa consume dispersed bacteria that have not flocculated and rotifers consume small biological floc particles that have not settled. The details of activated sludge process are given in Fig. 1.31.

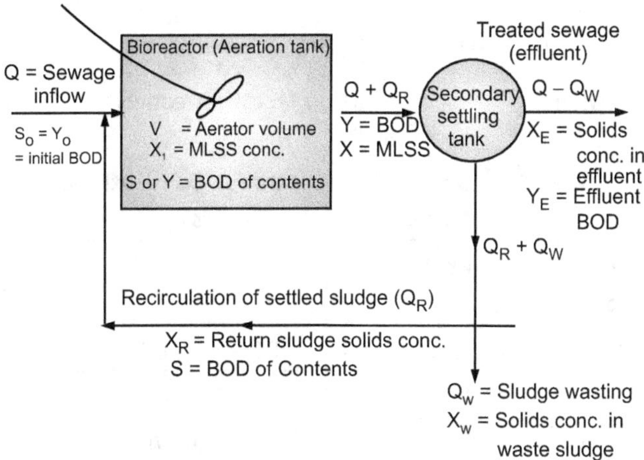

Fig. 1.31 : Flow diagram of conventional activated sludge process

1.20.1 Activated Sludge Process Variables

(Dec. 10; May 10, 16)

The variables of importance in ASP are loading rate, the mixing regime and the flow scheme.

Loading Rate :

- Loading rate is the rate at which wastewater is applied to the aeration tank. The loading parameter here is the hydraulic retention time (HRT) which is expressed as :

    HRT (hours) = $(V \times 24)/(Q)$

    where,   V = Volume of aeration tank, m^3

            Q = Wastewater inflow (sludge recycle excluded), m^3/day

- The organic loadingin ASP should be such that the micro-organisms exhibit flocculant characteristics which will result in efficient separation of sludge from wastewater.

- Volumetric Loading (organic loading) is the kg of BOD applied for unit volume of aeration tank or kg of BOD applied per day.

To calculate kg of BOD to be applied, please see the following example.

e.g. BOD of given wastewater is 800 mg/lit. The daily flow rate is 10,000 m^3/day (10 mld). Calculate daily organic loading in kg.

Organic loading = BOD in mg/lit $\times$ flow lit/day

      = 800 mg/lit $\times$ 10,000 m^3/day $\times$ 1000 lit/m^3

      = 8×10^9 mg/day

      = 8000 kg/day

So,   Volumetric loading

      = (kg of BOD)/m^3/d = $(Q \times L_a \text{ or } S_0)/V$

where, L_a or S_0 = Influent BOD to aeration tank

- This organic loading is also referred as **Food to Micro-Organisms(F/M) Ratio**. This is the ratio of BOD of kg BOD applied per day (representing microbial food) to kg MLSS in aeration tank (representing microbial mass), expressed as under,

$$F/M = \frac{[(Q \times L_a)/V]}{1000 \times X_t} \text{ or } = \frac{Q \cdot S_0}{1000 \times V \times t}$$

where, X_t = MLSS (Mixed liquor suspended solids)

or

      MLVSS (Mixed liquor volatile suspended solids) in mg/lit

also,

$$F/M = \frac{\text{Daily BOD load applied to the aerator system in gm}}{\text{Total microbial mass in the system in gm}}$$

- **Food to Micro-Organisms Ratio** is the main factor controlling BOD removal. F/M can be varied by varying MLSS concentration in the aeration tank.

Solids Retention Time (SRT) :

- This is also called as **Mean Cell Residence Time (MCRT)** or sludge age. It indicates the time period for which the solids are in the system.

- Sludge age

$$\text{or SRT} = \frac{\text{Total weight of solids in biological system (i)}}{\text{Total weight of solids leaving the system per day (i) + (ii)}}$$

- The solids are lost from the reactor through treated effluent and through sludge wasting of SST (Secondary settling tank).

$\therefore$ Total weight of solids in the reactor = Volume of reactor $\times$ MLSS

   (i) Mass of solids removed with wasted sludge per day

        = $Q_W \times X_W$

   (ii) Mass of solids removed with the effluent per day

        = $(Q - Q_W) X_E$

$\therefore$ Total weight of solids leaving the system per day

        = (i) + (ii)

        = $(Q_W \times X_W) + (Q - Q_W) X_E$

$\therefore$ Sludge age or SRT or MCRT,

$$\theta_C = \frac{V \times X_t}{(Q_W \times X_W) + (Q - Q_W) X_E}$$

where, X_t = MLSS in mg/lit

       V = Volume of reactor, m^3

      Q_W = Volume of wasted sludge per day

X_W = Solids concentration in wasted sludge (mg/lit)

Q = Waste water flow per day

X_E = Effluent solids concentration (mg/lit)

Sludge Production and Process Control :

- To design the sludge handling and disposal facilities, it is very important to know the quantity of sludge to be produced per day.

- The quantity of sludge wasted daily can be estimated by the following equation :

Wasted sludge,

$$P_X = \frac{XV}{q_C}$$

where, P_X = Net wasted activated sludge produced each day, as volatile suspended solids, kg/day

Oxygen Requirements :

- To determine the theoretical requirements the following two factors are required :

  (i) BOD_5 of the waste, and

  (ii) The amount of organisms wasted from the system per day.

- The reasoning is as follows. If we will consider that all the BOD_5 were converted to end products, the total oxygen demand would be determined by converting BOD_5 to BOD_L using an appropriate conversion factor.

- The amount of oxygen that must be supplied to the system

  = Total BOD $-$ BOD_L of the wasted cells.

- The following equation shows the BOD_L of a mole of cells.

$$C_5H_7NO_2 + 5O_2 \rightarrow 5CO_2 + 2H_2O + NH_3$$

113 cells   5 (32)

$$\frac{kg\ O_2}{kg\ cells} = \frac{160}{113} = 1.42$$

where, BOD_L = 1.42 (mass of cells, g/m³)

- Therefore, the theoretical oxygen requirements for the removal of carbonaceous organic matter in waste water can be determined by the following equation :

$$kg,\ O_2/day = \left(\begin{array}{c} \text{Total mass} \\ BOD_L\ \text{utilized,} \\ \text{kg day} \end{array} \right) - 1.42 \left(\begin{array}{c} \text{Mass of organisms} \\ \text{wasted, kg/day} \end{array} \right)$$

$$\therefore\ kg,\ O_2/day = \frac{Q(S_O - S)}{f} - 1.42\ (P_X)$$

where, f = Conversion factor for converting BOD_5 to BOD_L

Table 1.8 : Design Parameters for Conventional Activated Sludge Process

Sr. No.	Parameter	Design values
1.	Organic loading rate	0.3 – 0.5 kg BOD per day
2.	MLSS	1500 – 3000 mg/lit
3.	MLVSS	0.8 ·MLSS
4.	F/M ratio	0.3 to 0.4
5.	Hydraulic retention time (HRT) θ	4 to 6 hrs.
6.	SRT or MCRT or θ_C	5 to 8 days
7.	Oxygen requirement	0.8 to 1 kg per kg of BOD removed
8.	Recirculation ratio, Q_R/Q	0.25 to 0.5
9.	BOD removal efficiency	85 – 92%
10.	Air required	40 – 100 m³ per kg of BOD_5
11.	Excess sludge	0.55 to 0.6 kg per kg BOD removed
12.	Surface loading for SST	30 m³/m²/day
13.	Detention time for SST	2 hrs.

(Ref. Manual on Sewerage and Sewage Treatment, 2nd Edition)

1.21 SLUDGE VOLUME INDEX, SLUDGE BULKING, SLUDGE RECYCLE AND RATE OF RETURN SLUDGE   (Nov. 15, Aug. 16)

1. Sludge Volume Index :

- This parameter is used to understand the quality of sludge produced in the aeration tank. The degree of treatment achieved in an aeration process depends directly on settleability of activated sludge in the secondary settling tank. A biological floc that agglomerates and settles by gravity leaves a clear supernatant for discharge.

- Conversely, poorly flocculated particles or buoyant filamentous growths that do not separate by gravity contribute to BOD and suspended solids in the treated effluent. Excessive carry over of floc resulting in inefficient operation is referred to as sludge bulking.

- **Sludge Volume Index** (SVI) is defined as the volume occupied in ml by one gm of dried solids in mixed liquor after settling for 30 minutes. This can be determined in the laboratory. For this the sample is collected from outlet of the aeration tank of ASP. The collected one litre sample is poured in Imhoff cone and allowed to settle for 30 minutes. The settled volume of sludge is recorded. Then the sample is remixed, and is further tested for mixed liquor suspended solids concentration. The SVI is determined by the equation :

$$SVI = \frac{\text{Volume of settled sludge in ml}}{\text{MLSS concentration in mg/lit}} \times 1000$$

- The usual adopted range of SVI is between 50 – 150 ml/gm and such a value indicates good settling sludge. SVI values of 100 – 150 are considered satisfactory in plants operating with MLSS of 1000 – 3500 mg/lit.

2. Sludge Bulking :

Sludge bulking is due to

- Inadequate air supply.
- Low pH.
- Septicity.
- Growth of filamentous organisms.
- **Sludge Bulking** is controlled by eliminating the above causes. In addition to this **Chlorine** or **Hydrogen Peroxide** can be applied to wastewater or return sludge for control of filamentous growth. Limited dissolved oxygen has been noted more frequently than any other cause of bulking.
- The minimum DO should be 2 mg/lit. The food to micro-organisms ratio should be checked to make sure that it is within the range of generally accepted values.

Table 1.9 : Operational Parameters for ASP

SVI, ml/g	Process	MLSS, mg/lit	F/M, d^{-1}	Aeration Time, h	Average Sludge Age, d	BOD removal efficiency, %	Quality of Sludge
150-200	High Rate	500 – 1000	0.5–1.0	3 – 4	5	60 – 75	Poor
50 – (100) –150	Conventional	2000 – 3000	0.2–0.5	6 – 10	10	80 – 90	Good, as indicated by the median value 100 ml/g
25–50	Extended Aeration	4000 – 6000	0.05–0.2	24 – 36	25	90 – 98	Excellent

3. Sludge Recycle and Rate of Return Sludge

Sludge recirculation ratio,

$$\frac{Q_R}{Q} = \frac{X_t}{X_R - X_t}$$

where, Q_R = Sludge recirculation rate in m^3/day

X_t = MLSS in the aeration tank in mg/lit

X_R = MLSS in the returned or wasted sludge in mg/lit

$$X_R = \frac{10^6}{SVI}$$

Now, $$\frac{Q_R}{Q} = \frac{X_t}{X_R - X_t} = \frac{X_t}{\left(\frac{10^6}{SVI} - X_t\right)}$$

The value of return sludge ratios for conventional sludge plant varies between 0.25 to 0.50.

1.22 AERATION SYSTEMS

- Understanding of aeration processes in wastewater treatment needs greater knowledge of biology. The generalised biological process that takes place in the aeration tank is shown in Fig. 1.32.
- Here raw sewage contains organic matter. It enters into aeration tank in the form of food. The micro-organisms present in raw sewage metabolize waste organics, producing new cell mass while taking dissolved oxygen and releasing carbon dioxide. After the addition of large population of micro-organisms, aerating raw wastewater for a few hours removes organic matter from solution by synthesis into microbial cells.

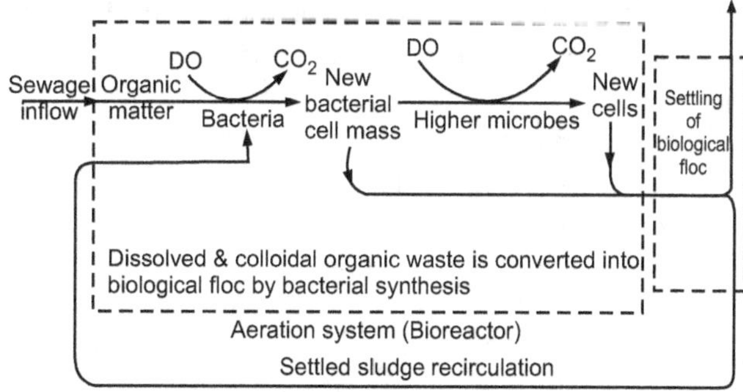

Fig. 1.32 : System biology of ASP

- Mixed liquor is continuously transferred to settling tank for separation of biological floc and discharge of settled effluent. Settled sludge is recirculated into the system as per requirement.
- Oxygen is supplied to the mixed liquor in an aeration tank by dispersing air bubbles through submerged diffusers or by entraining air into the liquid by mechanical aerators. Porous plates, tubes or nozzles

are provided as air diffusers. Compressed air forced through the porous material is released as fine bubbles. Mechanical aerators are horizontal paddle, vertical turbine and vertical turbine draft tube.

1.22.1 Mixing Regime

- The mixing regime provided in the aeration tank may be completely mixed flow or plug flow. Completely mixed flow involves the rapid dispersal of incoming wastewater throughout the tank.
- Plug flow implied that the wastewater moves down progressively along the aeration tank, essentially unmixed with the contents of rest of tank.

1.22.2 Flow Pattern

- This involves the method of wastewater addition and sludge return to the aeration tank and also the method of aeration. Sewage addition may be at the single point at the inlet or it may be at more points along the aeration tank.
- The sludge return may be from settling tank or through sludge reaeration tank. Aeration is done either with mechanical aerators or diffused aerators. Air is applied uniformly with either of the aerators along whole length of tank or it may be tapered from inlet end to outlet end of aeration tank.

1.22.3 Modifications of Activated Sludge Process

(Aug. 15)

The main limitations of conventional ASP are

- Higher aeration tank volume requirement.
- Lack of operational stability.

In order to overcome such difficulties and to meet specific treatment objectives, several modifications in conventional ASP have been suggested.

1. Step Aeration :

- It is a process in which wastewater is introduced at several points in the aeration tank while the return sludge is introduced at the head as shown in Fig. 1.33.

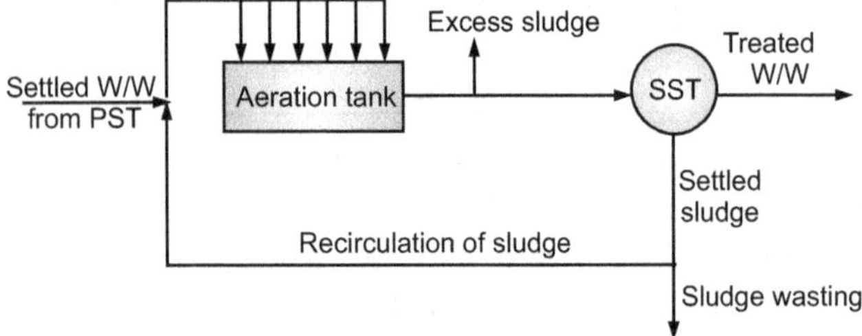

Fig. 1.33 : Flow diagram of step aeration process

- Because of this type of arrangement, there is uniform demand of oxygen along the length of tank. This results in effective use of uniform supply of oxygen in conventional ASP.

Advantages :

- Multiple entry of substrate thereby increasing removal of soluble organics by adsorption and higher BOD loading per unit volume.
- Oxygen demand is more evenly spread over entire length of tank resulting in better utilisation of oxygen supplied.

2. Tapered Aeration :

- Air is supplied at different rates along the length of tank according to need. It ensures higher air supply at inlet and in the initial length of tank, as compared to downstream length. In the aeration tank, as mixed liquor progresses through, its requirement for oxygen goes on decreasing.
- Therefore in this method oxygen is supplied at the higher rates at the inlet and gradually decreased as sewage move towards the outlet end of tank. The aerators are spaced close together at the inlet end to achieve higher oxygenation rate and spaced farther away at outlet end where oxygen demand is less.

Advantages :

- Less air requirement.
- Lower cost.
- Avoidance of over aeration.

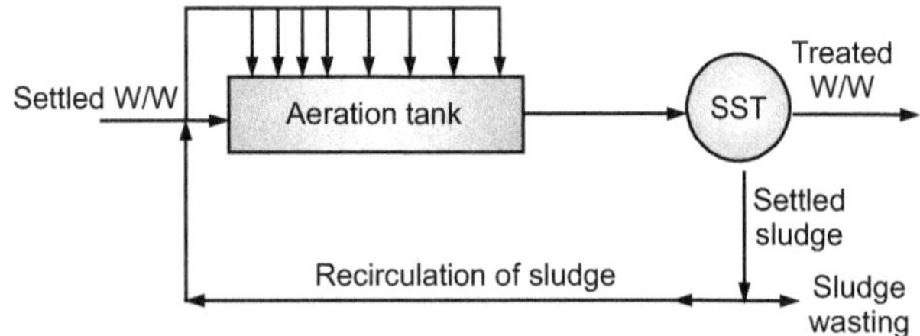

Fig. 1.34 : Flow diagram of tapered aeration process

3. Contact Stabilization :

- This process is also known as **Biosorption**. In this method, the settled wastewater is mixed with recirculated sludge and aerated in a contact tank for 30 to 90 minutes.
- During this period, the organics are absorbed by the sludge floc.
- The sludge is then separated from the treated effluent by sedimentation and the returned sludge is aerated from 3 to 6 hours in a sludge aeration tank.

- During this period the absorbed organics used for energy and production of new cells. A portion of return sludge is wasted prior to recycle, to maintain a constant MLVSS concentration in the tanks.

- It is suitable for wastewaters where a great part of BOD is present in suspended or colloidal form.

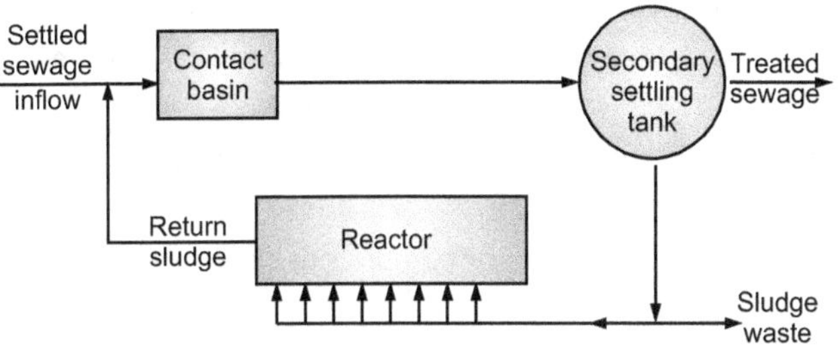

Fig. 1.35 : Flow diagram of contact stabilization process

Advantages :

- It can be loaded higher.

- If toxic conditions occur, the system can be returned to normal condition easily as only a minor fraction of biological organisms is in direct contact with waste flow.

4. Complete Mix :

- In this process, plug flow regime of conventional ASP is replaced by a completely mixed flow regime. The influent wastewater and sludge flow are introduced at several points in the aeration tank from a central channel. The aerated sewage is withdrawn uniformly along the opposite side.

- Mechanical aerators are installed in the centre of the aeration tank. The collected mixed liquor is settled in the secondary clarifier. This process possesses capacity to hold much higher MLSS concentration level in aeration tank. (3000 mg/lit to 6000 mg/lit).

Advantages :

- Higher volumetric loading.
- Low residence time of liquid.
- Complete mixing.

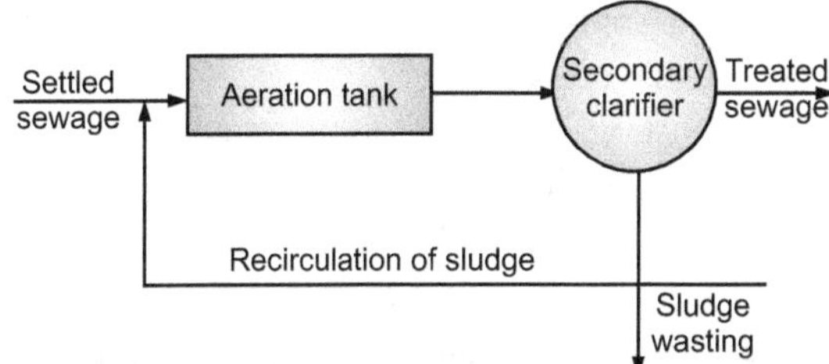

Fig. 1.36 : Flow diagram of complete mix process

Table 1.10 : Characteristics and Design Parameters of Different Activated Sludge Systems

Process Type	F/M	HRT (hr)	Volumetric Loading (kg BOD/m^3)	SRT (days)	$\dfrac{kg\ O_2}{kg\ BOD}$
1. Conventional	0.4–0.2	4–8	0.3–0.7	5–15	0.8–1.1
2. Tapered aeration	0.4–0.2	4–8	0.3–0.8	5–15	0.7–1.0
3. Step aeration	0.4–0.2	3–5	0.7–1.0	5–15	0.7–1.0
4. Contact stabilization	0.5–0.2	3–6	1.0–1.2	5–15	0.7–1.1
5. Complete mix	0.6–0.2	3–5	0.8–2.0	5–15	0.7–1.0
6. Modified aeration	5.0–1.5	1.3–3	1.2–2.4	0.2–0.5	0.4–0.6
7. Extended aeration	0.15–0.05	18–36	0.2–0.4	20–30	1.2–2.0

5. Extended Aeration :

- This process operates in the endogenous respiration phase of the growth curve. The process employs low organic loading, long aeration time, high MLSS concentration and low F/M. Thus, it is generally applicable only to small treatment plants with capacities of less than 4000 m^3/d.

- Continuous complete mixing is either by diffused air or by mechanical aerators and aeration periods are 24 to 36 hours. Because of these conditions, as well as low BOD loading, the biological process is very stable and can accept intermittant loads without upset.

Advantages :

- Excess sludge production is low.
- MLSS undergo endogenous respiration and get well stabilised.
- The excess sludge does not require separate digestion.
- Operation is simple due to elimination of primary settling and separate sludge digestion.

1.22.4 Problems in Activated Sludge Process

(Dec. 2010)

The problems associated with operation of activated sludge process are given in the Table 1.11.

Table 1.11 : Problems and Remedies in ASP

Problem	Probable cause	Remedy
1. Sludge floating at the surface of SST.	(a) Denitrification in SST resulting in nitrogen gas bubbles bouying up sludge. (b) More growth of filamentous organisms.	(a) Increasing rpm of scraper, sludge wasting, return sludge and DO. (b) Chlorine application.

2. Low return sludge concentration.	(a) Filamentous micro-organisms growth. (b) High return sludge rate.	(a) Add chlorine and increase pH and DO. (b) Reduce return sludge rate.
3. Foam	Detergents	Sprinkle water on foam.
4. Dead spots in aeration tank	(a) Less aeration and less DO. (b) Air diffusers plugged.	(a) Increase rate of aeration. (b) Clean or replace air diffusers.
5. Small floc particles in settled effluent.	(a) More disturbance in aeration tank. (b) System devoid of oxygen.	(a) Reduce aeration and increase sludge washing. (b) Increase aeration rate.

Advantages :

- No problem of odours.
- No fly nuisanse.
- Removal of SS and BOD around 90% each.
- Non-putrescible effluent.
- Relatively low cost of installation.
- Area required is small as compared to other conventional treatment methods.
- Excess sludge has good fertilizer value.

Disadvantages :

- No variation in quality of influent is allowed.
- Operation and maintenance cost is high.
- Skilled supervision is required.
- Large quantity of sludge is produced.

1.22.5 Operational Problems and Maintenance in ASP

1. **Sludge Bulking (Filamentous Microrganisms)**
   - **Symptoms :**
     - High TSS content in the secondary clarifier effluent.
     - Filamentous microorganisms in mixed liquor.
   - **Main Causes:**
     - Low nutrient concentration in the incoming wastewater.
     - Toxic compounds in the incoming wastewater.

- Wide pH and temperature oscillations.
- High organic loading rate.
- Insufficient aeration.
- **Investigations and Analyses :**
  - Measure the Sludge Volume Index (SVI).
  - Microscopic observations.
  - Check the C:N:P ratio or BOD5:N:P ratio.
  - Measure temperature, pH and DO (in different sections of the aeration tank).
  - Check F/M, volumetric organic loading rate, SRT.
  - Check toxic compounds in the incoming wastewater.
- **Remedial Actions**
  - Chlorine or oxygen peroxide dosage in the return sludge line (5 – 15 g Cl kg-1SS d-1).
  - Inorganic coagulants (cake, ferric chloride, etc.) dosage.
  - Increase SRT - PH and DO.
  - BOD5:N:P ratio correction in the incoming wastewater.

2. **Foaming**
   - **Symptoms:** Scum presence in the aeration basin.
   - **Main Causes:** High content of foaming agents and/or oils and greases in the incoming wastewater.
   - **Investigations and Analysis:** Evaluate the presence of Nocardia in the mixed liquor and in the foam.
     - Evaluate oil and grease in the influent wastewater.
     - Evaluate temperature oscillation in the aeration basin.
   - **Remedial Actions:** Foam removal through water sprinkling - Chlorine dosage.

3. **Air Diffusers Clogging**
   - **Symptoms:** Air flow rate reduction.
     - Increased headloss in the air line.
   - **Main Causes:** High dust content in the air.
     - Oil content in the air due to air compressor faulty operation.
     - Rust presence in the air pipeline due to the condensing moisture.
     - Organic material growth or solids precipitation over air diffusers.
     - Solid deposition over air diffusers during aeration interruption.

- **Investigations and Analyses**
  - ➢ Check the dust, rust and oil content in the air flowing to the diffusers.
  - ➢ Check breaks presence in the air pipeline through which mixed liquor could enter during aeration interruption.
- **Remedial Actions:**
  - ➢ Keep air compressors as much regularly operated as possible.
  - ➢ Check regularly the air filtration system.
  - ➢ Install oil trap on the air compressor.
  - ➢ Periodical maintenance of the air pipeline. Enhance degritting and screening.

4. **Low DO Value in the AerationBasin**
   - **Symptoms**
     - ➢ Efficiency reduction.
     - ➢ DO reduction in the aeration basin; and temporary bulking.
     - ➢ Mixed liquor dark colour.
   - **Main Causes**
     - ➢ Insufficient aeration.
     - ➢ Wide oscillation of the organic loading rate.
   - **Investigations and Analysis**
     - ➢ Measure DO concentration in the aeration basin specially in the dead zones.
     - ➢ Measure wastewater flow rate and organic concentration during peak time Remedial actions.
     - ➢ Increase the volume of the aeration basin (raising the water level).
     - ➢ Increase the aeration.

1.23 CONCEPT OF SEQUENTIAL BATCH REACTOR (SBR) {ADVANCED WASTEWATER TREATMENT}

- SBRs are used all over the world and have been around since the 1920s. With their growing popularity in Europe and China as well as the United States, they are being used successfully to treat both municipal and industrial wastewaters, particularly in areas characterized by low or varying flow patterns.
- Municipalities, resorts, casinos, and a number of industries, including dairy, pulp and paper, tanneries and textiles, are using SBRs as practical wastewater treatment alternatives. Improvements in equipment and technology, especially in aeration devices and computer control systems, have made SBRs a viable choice over the conventional activated-sludge system.

SBR Plants are Very Practical for a Number of Reasons:

- In areas, where there is a limited amount of space, treatment takes place in a single basin instead of multiple basins, allowing for a smaller footprint. Low total-suspended-solid values of less than 10 milligrams per liter (mg/L) can be achieved consistently through the use of effective decanters that eliminate the need for a separate clarifier.
- The treatment cycle can be adjusted to undergo aerobic, anaerobic, and anoxic conditions in order to achieve biological nutrient removal, including nitrification, denitrification, and some phosphorus removal. Biochemical oxygen demand (BOD) levels of less than 5 mg/L can be achieved consistently.
- Total nitrogen limits of less than 5 mg/L can also be achieved by aerobic conversion of ammonia to nitrates (nitrification) and anoxic conversion of nitrates to nitrogen gas (denitrification) within the same tank. Low phosphorus limits of less than 2 mg/L can be attained by using a combination of biological treatment (anaerobic phosphorus absorbing organisms) and chemical agents (aluminum or iron salts) within the vessel and treatment cycle.
- Older wastewater treatment facilities can be retrofitted to an SBR because the basins are already present.
- Wastewater discharge permits are becoming more stringent and SBRs offer a cost-effective way to achieve lower effluent limits. Note that discharge limits that require a greater degree of treatment may necessitate the addition of a tertiary filtration unit following the SBR treatment phase. This consideration should be an important part of the design process.

Common SBR Characteristics :

- General SBRs are a variation of the activated-sludge process. They differ from activated-sludge plants because they combine all of the treatment steps and processes into a single basin, or tank, whereas conventional facilities rely on multiple basins.
- According to a 1999 U.S. EPA report, an SBR is no more than an activated-sludge plant that operates in time rather than space.

1.23.1 Basic Treatment Process of SBR

The operation of an SBR is based on a fill-and-draw principle, which consists of five steps -

1. Fill
2. React
3. Settle
4. Decant
5. Idle.

These steps can be altered for different operational applications.

1. Fill :

- During the fill phase, the basin receives influent wastewater. The influent brings food to the microbes in the activated sludge, creating an environment for biochemical reactions to take place. Mixing and aeration can be varied during the fill phase to create the following three different scenarios:

(i) Static Fill : Under a static-fill scenario, there is no mixing or aeration while the influent wastewater is entering the tank. Static fill is used during the initial start-up phase of a facility, at plants that do not need to nitrify or denitrify, and during low flow periods to save power. Because the mixers and aerators remain off, this scenario has an energy-savings component.

(ii) Mixed Fill : Under a mixed-fill scenario, mechanical mixers are active, but the aerators remain off. The mixing action produces a uniform blend of influent wastewater and biomass. Because there is no aeration, an anoxic condition is present, which promotes denitrification. Anaerobic conditions can also be achieved during the mixed-fill phase. Under anaerobic conditions the biomass undergoes a release of phosphorous. This release is reabsorbed by the biomass once aerobic conditions are reestablished. This phosphorous release will not happen with anoxic conditions.

(iii) Aerated Fill : Under an aerated-fill scenario, both the aerators and the mechanical mixing unit are activated. The contents of the basin are aerated to convert the anoxic or anaerobic zone over to an aerobic zone. No adjustments to the aerated-fill cycle are needed to reduce organics and achieve nitrification. However, to achieve denitrification, it is necessary to switch the oxygen off to promote anoxic conditions for denitrification. By switching the oxygen on and off during this phase with the blowers, oxic and anoxic conditions are created, allowing for nitrification and denitrification. Dissolved oxygen (DO) should be monitored during this phase so it does not go over 0.2 mg/L. This ensures that an anoxic condition will occur during the idle phase.

2. React :

- This phase allows for further reduction or "polishing" of wastewater parameters. During this phase, no wastewater enters the basin and the mechanical mixing and aeration units are on. Because there are no additional volume and organic loadings, the rate of organic removal increases dramatically.

- Most of the carbonaceous BOD removal occurs in the react phase. Further nitrification occurs by allowing the mixing and aeration to continue the majority of denitrification takes place in the mixed-fill phase. The phosphorus released during mixed fill, plus some additional phosphorus, is taken up during the react phase.

3. Settle :

- During this phase, activated sludge is allowed to settle under quiescent conditions no flow enters the basin and no aeration and mixing takes place. The activated sludge tends to settle as a flocculent mass, forming a distinctive interface with the clear supernatant.

- The sludge mass is called the sludge blanket. This phase is a critical part of the cycle, because if the solids do not settle rapidly, some sludge can be drawn off during the subsequent decant phase and thereby degrade effluent quality.

4. Decant :

- During this phase, a decanter is used to remove the clear supernatant effluent. Once the settle phase is complete, a signal is sent to the decanter to initiate the opening of an effluent-discharge valve. There are floating and fixed-arm decanters. Floating decanters maintain the inlet orifice slightly below the water surface to minimize the removal of solids in the effluent removed during the decant phase.

- Floating decanters offer the operator flexibility to vary fill and draw volumes. Fixed-arm decanters are less expensive and can be designed to allow the operator to lower or raise the level of the decanter. It is optimal that the decanted volume is the same as the volume that enters the basin during the fill phase. It is also important that no surface foam or scum is decanted. The vertical distance from the decanter to the bottom of the tank should be maximized to avoid disturbing the settled biomass.

5. Idle :

- This step occurs between the decant and the fill phases. The time varies, based on the influent flow rate and the operating strategy. During this phase, a small amount of activated sludge at the bottom of the SBR basin is pumped out a process called wasting.

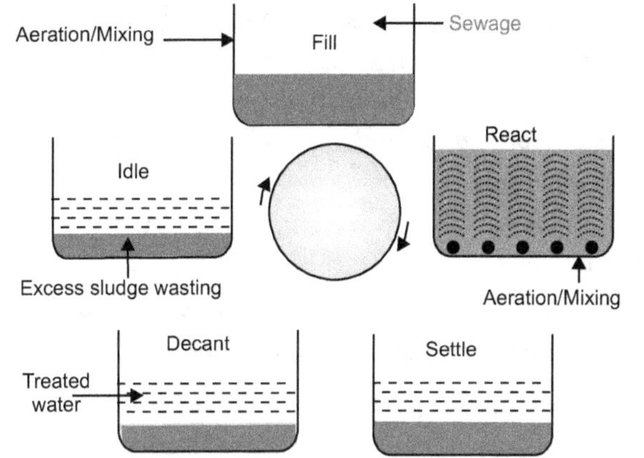

Fig. 1.37

SOLVED EXAMPLES ON DESIGN OF ACTIVATED SLUDGE PROCESS

Example 1.14 : *Data given :*

Municipal waste water flow rate, $Q = 10,000\ m^3/day$

BOD of settled effluent, $S_O = 150\ mg/lit$

BOD of treated effluent, $S_e = 5\ mg/lit$

Yield coefficient, $Y = 0.5\ kg/kg$

Endogenous decay coefficient, $K_d = 0.05\ d^{-1}$

MLVSS concentration, $X = 3000\ mg/lit$

Return sludge solids concentration, $X_r = 10,000\ mg/lit$

Mean cell residence time, $\theta_c = 10\ days.$

Determine :

(i)   *The volume of the reactor.*

(ii) *F/M ratio.*

(iii) *Volumetric loading rate.*

(iv) *Oxygen requirement.*

(v)   *Recycle ratio.*

(vi) *BOD removal efficiency.*

Solution :

(i) The volume of the reactor :

$$V = \frac{YQ\,q_c\,(S_O - S_e)}{X\,(1 + K_d\,q_c)}$$

$$= \frac{0.5 \times 10000 \times 10\,(150 - 5)}{3000\,(1 + 0.05 \times 10)}$$

$$= 1611\ m^3 \qquad \textbf{... Ans.}$$

(ii) F/M ratio :

$$F/M = \frac{Q\,S_O}{V\,X}$$

$$= \frac{10000\ m^3/day \times 150\ mg/lit}{1611\ m^3 \times 3000\ mg/lit}$$

$$= 0.31\ day^{-1} \qquad \textbf{... Ans.}$$

(iii) Volumetric loading rate

$$= \frac{Q\,S_O}{V} = \frac{10000 \times 150}{1611}$$

$$= 931.08\ gm/m^3 \qquad \textbf{... Ans.}$$

(iv) Oxygen requirement :

O_2 demand

$$= 1.47\ Q\,(S_O - S_e) - 1.42\ P_X,\ kg/day$$

$$Q = 10 \times 10^6\ lit/day$$

$$S_O - S_e = 150 - 5 = 145\ mg/lit$$

Wasted sludge,

$$P_X = \frac{XV}{q_c} = \frac{3000 \times 1611 \times 10^3}{10}$$

$$= 4.83 \times 10^8\ mg/day$$

$\therefore$ O_2 demand (kg/day)

$$= [1.47 \times 10 \times 10^6 \times 145 - 1.42$$

$$\times 4.83 \times 10^8]\,10^{-6}$$

$$= 1445.64\ kg/day \qquad \textbf{... Ans.}$$

(v) Recycle ratio,

$$(Q + Q_r)\,X = Q_r\,X_r$$

where, Q = Wastewater flow rate

$$= 10000\ m^3/day$$

X = Aerator MLSS concentration

$$= 3000\ mg/lit$$

X_r = Return sludge solids concentration

$$= 10000\ mg/lit$$

also, $Q_r = \dfrac{Q\,X}{X_r - X} = \dfrac{10000 \times 3000}{10000 - 3000}$

$$= 4286\ m^3/day$$

Recycle ratio,

$$R = \frac{Q_r}{Q} = \frac{4286}{10000}$$

$\therefore$ Recycle ratio $= 0.4286$ **... Ans.**

(vi) BOD removal efficiency

$$= \frac{S_O - S_e}{S_O} \times 100 = \frac{150 - 5}{150} \times 100$$

$$= 96.60\% \qquad \textbf{... Ans.}$$

Example 1.15 : *Given the following data of operating ASP :*

Wastewater flow, Q $= 35000\ m^3/day$

Influent total solids $= 600\ mg/lit$

Influent suspended solids $= 120\ mg/lit$

Influent BOD, S_O $= 175\ mg/lit$

Effluent total solids $= 495\ mg/lit$

Effluent suspended solids $= 22$ *mg/lit*

Effluent BOD, S_o $= 20$ *mg/lit*

MLVSS concentration, X $= 2500$ *mg/lit*

Return sludge solids concentration, X_r $= 9800$ *mg/lit*

Volume of aeration basins $= 10000$ m^3

Determine :

(i) Aeration period.

(ii) BOD load in $kg/m^3/day$.

(iii) F/M ratio.

(iv) Total solids, suspended solids and BOD removal efficiency.

(v) Recirculation ratio.

Solution :

(i) Aeration period $= \dfrac{\text{Volume of reactor, m}^3}{\text{Wastewater flow, m}^3/\text{day}} \times 24$

$$= \dfrac{10000 \text{ m}^3}{35000 \text{ m}^3/\text{day}} \times 24$$

$$= 6.85 \text{ hrs} \qquad \textbf{... Ans.}$$

(ii) BOD load $= \dfrac{Q \cdot S_o}{V}$

$$= \dfrac{35000 \text{ m}^3/\text{day} \times 175 \text{ mg/lit} \times 10^3}{10000 \text{ m}^3 \times 10^6}$$

$$= 0.6125 \text{ kg/m}^3/\text{day} \qquad \textbf{... Ans.}$$

(iii) F/M ratio $= \dfrac{Q \cdot S_o}{V \, X} = \dfrac{35000 \times 175}{10000 \times 2500}$

$$= 0.245 \text{ d}^{-1} \qquad \textbf{... Ans.}$$

(iv) Total solids removal efficiency

$$= \dfrac{600 - 495}{600} \times 100 = 17.5\%$$

Suspended solids removal efficiency

$$= \dfrac{120 - 22}{120} \times 100 = 81.6\% \qquad \textbf{... Ans.}$$

BOD removal efficiency

$$= \dfrac{175 - 20}{175} \times 100 = 88.6\%$$

(v) Recirculation ratio $= R$

$$\therefore \quad Q_r = \dfrac{Q\,X}{X_r - X} = \dfrac{35000 \times 2500}{9800 - 2500}$$

$$= 11986 \text{ m}^3/\text{day}$$

and $\quad R = \dfrac{Q_r}{Q} = \dfrac{11986}{35000} = 0.34 \qquad \textbf{... Ans.}$

Example 1.16 : *Raw sewage at 5000 m^3/day and having a BOD of 450 mg/lit is treated in an activated sludge plant where the aeration tank is 1200 m^3 in volume and the MLVSS is 2500 mg/lit. Find the loading F/M based on solids in tank only.*

Solution : As MLVSS is 2500 mg/lit, the system is conventional.

Assume primary settling tank removes 33% BOD

$\therefore$ Incoming BOD to aeration tank (F)

$$\cong 300 \text{ mg/lit} = \dfrac{300 \times 5000 \times 1000}{1000 \times 1000}$$

$$= 1500 \text{ kg/day}$$

MLVSS in aeration tank (M)

$$= 2500 \text{ mg/lit} = 2500 \text{ kg/1000 m}^3 \text{ vol}$$

$$= 2500 \times \dfrac{1200}{1000} = 3000 \text{ kg}$$

$$\therefore \quad \text{F/M} = \dfrac{1500}{3000} = 0.5 \text{ kg of BOD/kg of MLVSS}$$

$$\textbf{... Ans.}$$

Example 1.17 : *If the above aeration tank was operated as an extended aeration system and primary settling done away with, find the flow of sewage that can be treated at F/M = 0.2 and MLVSS = 5000 mg/lit.*

Solution :

MLVSS in aeration tank (M) $= 5000$ mg/lit

$$= 5000 \text{ kg/1000 m}^3$$

$$= 5000 \times \dfrac{1200}{1000} = 6000 \text{ kg} \qquad \textbf{... Ans.}$$

At $\quad$ F/M $= 0.2$,

$$\dfrac{F}{M} = \dfrac{\text{BOD allowable per day}}{\text{MLSS in aeration tank}}$$

$$0.2 = \dfrac{\text{BOD allowable per day}}{6000}$$

$\therefore$ BOD allowable per day

$$= 0.2 \times 6000 \text{ kg} = 1200 \text{ kg/day}$$

Incoming BOD (without settling)

$$= 450 \text{ mg/lit}$$

$$= 450 \text{ kg/1000 m}^3$$

$$\therefore \quad \text{Permissible flow} = \dfrac{1200}{450} = 2700 \text{ m}^3/\text{day only}$$

$$\textbf{... Ans.}$$

Example 1.18 : *If raw sewage of 300 mg/lit BOD is treated in a conventional activated sludge plant of 5000 m^3/day capacity with an overall efficiency of 90%, find the kg of BOD discharged in treated effluent per day to river.*

Solution :

Incoming BOD $= 300$ mg/lit $= 300$ kg/1000 m³

$$= 300 \times \frac{5000}{1000} = 1500 \text{ kg/day}$$

∴   BOD in effluent $= 0.1 \times 1500 = 150$ kg/day

If the same sewage was treated in an extended aeration plant of 95% efficiency, the BOD in effluent would be

$$= 0.05 \times 1500$$

$$= 75 \text{ kg/day} \qquad \text{... Ans.}$$

Note : Pollution of river is reduced to half for only 5% increase in overall efficiency.

Example 1.19 : *Design a Conventional Activated Sludge plant to treat 5000 m³/day of municipal sewage having BOD of 350 mg/lit. Assume BOD removal in primary settling at a slightly above 40% so that raw settled BOD is 200 mg/lit. Final BOD to be 30 mg/lit.*

Solution :

Settled sewage BOD to aeration tank (F)

$$= 200 \text{ mg/lit} = 200 \text{ kg/1000 m}^3$$

$$= 200 \times \frac{5000}{1000} = 1000 \text{ kg/day}$$

Assume F/M $= 0.5$ kg/kg of MLVSS at 23°C.

$$\frac{1000}{M} = 0.5$$

∴   MLVSS required in the system (M)

$$= \frac{1000}{0.5} - 2000 \text{ kg}$$

Assumethat only 80% of MLVSS are in aeration tank at any given time. The rest being in settling and sludge return system.

∴   MLVSS in aeration $= 0.8 \times 2000$ kg $= 1600$ kg

AssumeMLSS concentration in aeration tank $= 2500$ mg/lit $= 2500$ kg/1000 m³. (Refer table 1.2)

∴   Aeration tank volume required

$$= \frac{1600 \times 1000}{2500} = 640 \text{ m}^3$$

Hence,

Aeration time $= \dfrac{640 \text{ m}^3}{5000 \text{ m}^3/\text{d}} \times 24 \cong 3$ hrs.

Return Sludge :

Assume it is 50% of inflow $= 5000 \text{ m}^3/\text{d} \times \dfrac{50}{100} \times \dfrac{1}{24}$

$= 104.17$ m³/hr.

Hence, provide 3 pumps of 50 m³/hr capacity so that 2 pumps may work and 1 pump may be stand-by.

Aeration Requirements :

For conventional activated sludge plants, it is customary to assume 0.8 kg O_2/kg BOD_5 removed.

BOD to aeration $= 1000$ kg/day as above

BOD in effluent $= 30$ mg/lit

$$= \frac{30 \times 5000 \times 1000}{1000 \times 1000} = 150 \text{ kg/day}$$

∴   BOD removed $= 850$ kg/day

∴   Oxygen required,

$$(N_O) = \frac{850 \times 0.8}{24} = 28 \text{ kg/hr}$$

This is at the actual working conditions. Therefore, it has to be converted to equivalent value in tap water at standard conditions (20°C and zero DO).

∴   Oxygen transfer under field conditions,

$$N = \frac{N_O}{a\left[\dfrac{C_{SW} - C_L}{C_S}\right] \cdot (1.025)^{T-20}}$$

where, N_O = oxygen transfer in water at 20°C and zero DO

Assume : $\alpha = 0.9$ = Oxygen transfer ratio of waste to water

Assume : $C_L = 1.5$ mg/lit = Dissolved oxygen to be maintained in the waste

Assume : $C_S = 9.17$ mg/lit = Oxygen saturation value for tap water at 20°C

Assume : $T = 23$°C = Temperature of waste in tank

Assume : $C_{SW} = 0.9$ to 0.98 of oxygen saturation value for tap water at T°C. (Adjust for barometric pressure also if necessary depending on plant site).

$$= 0.9 \times 8.68 = 7.812 \text{ mg/lit}$$

∴   Substituting

$$N = \frac{28}{0.9\left[\dfrac{7.812 - 1.5}{9.17}\right] \times 1.077}$$

We get, $\qquad = \dfrac{28}{0.9 \times 0.69 \times 1.077} = 42$ kg/hr

Assuming ,

Aerator capacity $= 1.6$ kg/HP-hr ($= 3.5$ *lbs* O_2/HP-hr)

Total HP required $= \dfrac{42}{1.6} = 26$ BHP

Assuming90% efficiency geared drive,

Provide $\dfrac{26}{0.9}$ = 30 HP (say) totally

[**Note :** If no corrections for temperature and nature of waste were done, the HP required would be $\dfrac{28}{1.6}$ = 18 HP only.]

Example 1.20 : *Given the following data of operating ASP :*

Wastewater flow = 20,000 m³/day

Influent BOD₅ = 250 mg/lit

Effluent BOD₅ = 18 mg/lit

Temperature = 25°C

Influent volatile suspended solids to reactor are negligible

Ratio of MLVSS to MLSS = 0.9

Return sludge concentration =12,000 mg/lit of suspended solids

MLVSS = 3500 mg/lit

Mean cell residence time, θ_c = 10 days

Effluent contains 25 mg/lit of biological solids, of which 65 percent is biodegradable

Value of BOD₅ = 0.68 ×BOD_L

Yield coefficient, Y = 0.5 kg/kg

K_d = 0.05 d⁻¹

Determine :

(i) Treatment efficiency.

(ii) Reactor volume.

(ii) Quantity of wasted sludge.

(iv) Oxygen requirements based on BOD_L.

Solution :

Concentration of soluble BOD₅ in the effluent,

Effluent BOD₅ = Influent soluble BOD₅ + BOD₅ of effluent suspended solids

1. BOD₅ of the effluent suspended solids :

Biodegradable portion of effluent biological solids

$$= 0.65 \times 25 = 16.25 \text{ mg/lit}$$

BOD_L of the biodegradable effluent solids

$$= 0.65 \times 25 \times 1.42 = 23.1 \text{ mg/lit}$$

BOD₅ of effluent suspended solids

$$= 23.1 \times 0.68 = 15.71 \text{ mg/lit}$$

2. Influent BOD₅,

$$18 = S_e + 15.71$$

$$S_e = 2.29 \text{ mg/lit}$$

(i) Treatment efficiency,

$$E = \frac{S_O - S_e}{S_O} \times 100$$

1. The efficiency depends on soluble BOD₅

$$E_S = \frac{(250 - 2.29)}{250} \times 100 = 99.10\% \quad \text{... \textbf{Ans.}}$$

2. The overall efficiency,

$$E_{overall} = \frac{(250 - 18)}{250} \times 100 = 92.80\% \quad \text{... \textbf{Ans.}}$$

(ii) Reactor volume,

$$V = \frac{YQq_c\,(S_O - S_e)}{X\,(1 + K_d\,q_c)}$$

$$= \frac{0.5 \times 2000 \times 10\,(250 - 2.29)}{3500\,(1 + 0.05 \times 10)}$$

$$= 471.83 \text{ m}^3 \quad \text{... \textbf{Ans.}}$$

(iii) Quantity of wasted sludge

$$P_X = \frac{XV}{q_c}$$

$$= \frac{3500 \times 471.83}{10} = 1.65 \times 10^8 \text{ mg/day}$$

$$\text{... \textbf{Ans.}}$$

(iv) Oxygen requirements based on BOD_L :

$$Q = 20,000 \text{ m}^3/\text{day} = 20 \times 10^6 \text{ lit/day}$$

$$O_2 \text{ demand} = 1.47\, Q\,(S_O - S_e) - 1.42\, P_X$$

$$= [1.47 \times 20 \times 10^6\,(250 - 2.29)$$

$$- 1.42 \times 1.65 \times 10^8) \times 10^{-6}]$$

$$= 7048.37 \text{ kg/day} \quad \text{... \textbf{Ans.}}$$

Example 1.21 : *An average operating data for conventional activated sludge treatment plant is as follows:* **(Aug. 15, 5M)**

Wastewater flow = 35000 cum/day

Volume of aeration tank = 10900 cum

Influent BOD = 250 mg/L

Effluent BOD = 20 mg/L

MLSS = 2500 mg/L

Effluent SS = 30 mg/L

Waste sludge SS = 9700 mg/L

Quantity of waste sludge 200 cum/day.

Based on information above determine

(i) Aeration period (hrs)

(ii) F/M ratio

(iii) % efficiency of BOD removal

(iv) Sludge age (days)

Solution :

(i) Aeration Period $= \dfrac{\text{Volume of reactor m}^3}{\text{wastewater flow m}^3/\text{day}} \times 24$

$= \dfrac{10900}{35000} \times 24 = 7.47$ hrs   **...Ans.**

(ii) F/M Ratio $= \dfrac{Q.30}{V \times} = \dfrac{35000 \times 250}{10900 \times 2500}$

$= 0.321 \ d^{-1}$ **...Ans.**

(iii) % efficiency of $= \dfrac{250 - 20}{250} \times 100$

BOD removal $= 92 \ \%$ **...Ans.**

(iv) Sludge age (days) $= \dfrac{V.X}{Q_w.\, X_R + (Q - Q_w).\, X_E}$

Where, X = Mixed liquor suspended solids (MLSS)

Q_w = quantity of waste sludge

X_R = waste sludge suspended solids (SS)

Q = wastewater flow

X_E = Effluent suspended solids

$\therefore$ Sludge age (Q_C)

$= \dfrac{(10900 \times 2500)/1000 \ \text{kg}}{\left(\dfrac{200 \times 9700}{100}\right) + (35000 - 200) \times 30/1000}$

$= 9.13$ days   **...Ans.**

Example 1.22 : The mixed liquor suspended solid concentration (MLSS) in an aeration tank is 3000 mg/L and sludge volume after 30 minutes of settling in a 1000 ml graduated cylinder is 135 ml.   **(Aug. 16, 5M)**

Calculate :

(i) SVI

(ii) Required return sludge ratio

(iii) Suspended solids concentration in the recirculated sludge

Solution : (i) SVI $= \dfrac{1000 V_s}{x} = \dfrac{1000 \times 135}{3000}$

$= 45$ ml/gm   **...Ans.**

(ii) Required return sludge ratio

$\dfrac{Q_r}{Q + Q_r} = \dfrac{V_s}{1000} = \dfrac{135}{1000}$

$\therefore$ $\dfrac{Q_v}{Q} = 0.156$

(iii) Suspended solids concentration in the ecirculated sludge

$S_s = \dfrac{10^6}{\text{SVI}} = \dfrac{10^6}{45} = 22222.22$ mg/L   **...Ans.**

Example 1.23 : *An average operating data for conventional activated sludge treatment plant is as follows*

Sewage flow $= 30000 \ m^3/d$

Volume of aeration tank $= 10500 \ m^3$

Influent BOD $= 200 \ mg/l$

Effluent BOD $= 20 \ mg/l$

Mixed liquor suspended solids $= 3000 \ mg/l$

Effluent suspended solids $= 30 \ mg/l$

Waste sludge suspended solids $= 9500 \ mg \ l$

Quantity of waste sludge $= 200 \ m^3/d$

Determine

(i) *Food to microorganism ratio*

(ii) *Sludge age*

(iii) *Percentage of efficiency of BOD removal*

     (May 2017, 6M)

Solution :

(i) Food to micro organism ratio.

F/M Ratio $= \dfrac{Q \times Y_0}{V.X}$

$= \dfrac{30000 \times 200}{10500 \times 3000} = 0.1904 \ d^{-1}$ **...Ans.**

(ii) % efficiency of $= \dfrac{S_o - S}{S_o}$

$= \dfrac{200 - 20}{200} \times 100 = 0.9 \times 100 = 90 \ \%$

   ...Ans.

(iii) Sludge age,

$Q_c = \dfrac{V.X \ t}{Q_w.\, X_R + (Q - Q_w).\, X_E}$

$= \dfrac{10500 \times 3000}{200 \times 95 + (30000 - 200) \times 30}$

$= 11.27$ day $= 12$ days   **...Ans.**

TRICKLING FILTER

1.24 INTRODUCTION TO TRICKLING FILTER

(Nov. 15, May 16)

- The objective of the biological treatment of wastewater is to coagulate and remove the non-settleable colloidal solids and to stabilise the organic matter. In biological treatment system, the living micro-organisms break down the waste organics and use it as food for them. The end product of this process are gases and formation of new cells. Thus, soluble and colloidal organic matter is converted into gases and insoluble biomass.

- The suspended growth systems are discussed in earlier sections. In this section, the attached growth systems will be discussed.

- The attached growth biological systems are those that contact wastewater with microbial growths attached to the surfaces of supporting media. Supporting media is an inert medium, such as rock, slag, redwood or specially designed ceramic or plastic materials. Such processes include : (i) Intermittant sand filters, (ii) Trickling filters, (iii) Rotating biological contactors, (iv) Packed bed reactors.

1.24.1 Biological Process　　(May 10, 11, 16, Nov. 15)

- The trickling filter consists of the filter medium, wastewater distribution system, inlet pipe and wastewater collection system. The settled wastewater from PST is sprayed through the nozzles of distributor. The wastewater sprinkled over the media produces biological slimes that coat the surface. The films consist primarily of bacteria, protozoa and fungi that feed on waste organics.

- Sludge worms, fly larvae, rotifers and other biota are also found. As the wastewater flows over the slime layer, organic matter and dissolved oxygen are extracted and metabolic end products such as CO_2 are released (Fig. 1.38). Dissolved oxygen in the liquid is replenished by absorption from the air in the voids surrounding the filter media. Eventhough the film is very thin, the biological layer is anaerobic at bottom and aerobic at top. In the outer portion of biological film, the organic matter is degraded by aerobic micro-organisms.

- Organisms attached to the media in the upper layer grow rapidly, feeding on the abundent food supply. As the micro-organisms at the outer surface grow, the thickness of the slime layer increases and the diffused oxygen is consumed before it can penetrate the full depth of layer. Hence, the lower (bottom) portion which is near the medium, is in a state of starvation, due to which anaerobic environment is established. As a result of having no external organic source available for cell carbon, the micro-organisms near the media surface enter into endogenous phase of growth and lose their ability to cling to the media surface.

- Eventually there is scouring of slime layer due to flowing liquid and a fresh slime layer begins to grow on the media. This phenomenon of scouring of the slime is called **Sloughing** of the filter. Excess microbial growth sloughing off of the media is removed from the filter effluent by a final clarifier.

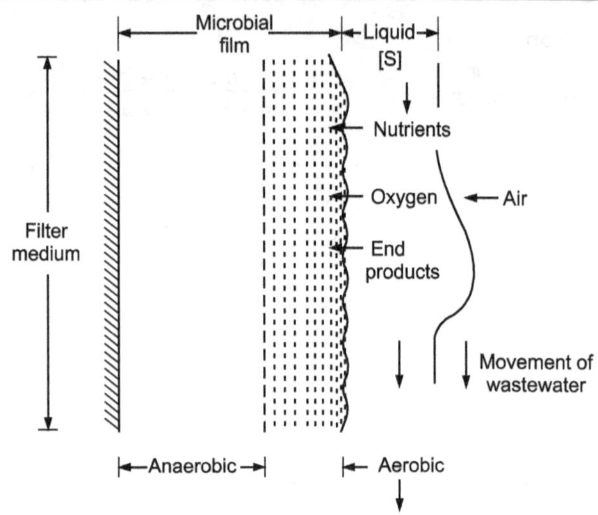

Fig. 1.38 : Schematic representation of exchanges taking place along surfaces biological slimes in trickling filters

1.25 COMPONENTS OF TRICKLING FILTER

It consists of : (See Figs. 1.39 and 1.40)

1. Filter media.
2. Recirculation system.
3. Distribution system.
4. Underdrainage system.
5. Water tight holding tank.
6. Filter floor.
7. Filter walls.

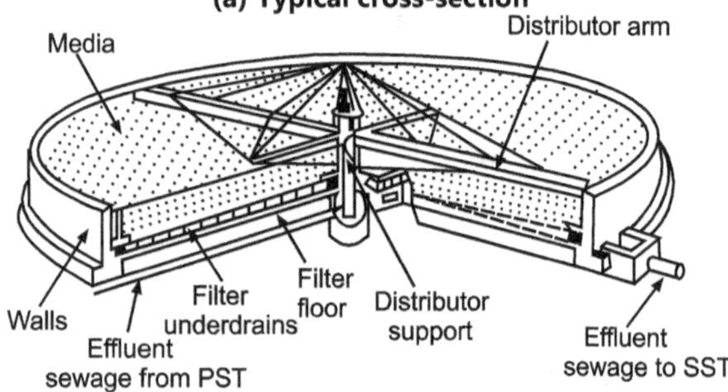

Fig. 1.39 : Trickling filter

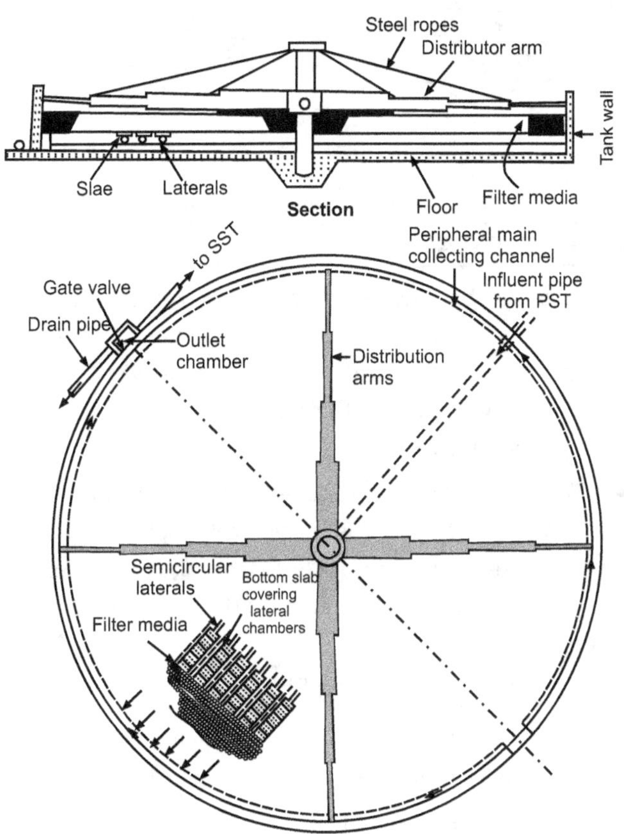

Fig. 1.40 : Typical details of trickling filter

1.25.1 Filter Media

- The ideal filter medium is a material that has a high surface area per unit of volume, is low in cost and has a high durability and does not clog easily. The physical properties of filter media are shown in Table 1.12.

Table 1.12 : Physical Properties of Trickling Filter Media

Medium	Nominal Size (mm)	Mass/unit Bulk Volume (kg/m^3)	Specific Surface Area (m^2/m^3)	Void Space (%)
River rock				
Small	30–50	1200–1500	55–70	40–50
Large	100–130	800–1000	40–50	50–60
Blast furnace slag				
Small	40–80	800–1000	50–80	45–55
Large	80–130	800–1000	50–65	55–65

...Conti.

Plastic				
Conventional	600 × 600 × 1200	30–100	80–100	96–97
High specific surface	600 × 600 × 1200	30–100	100–200	95–97
Redwood	1200 × 1200 × 500	150–175	40–50	75–80

- Plastic media fulfils the requirements except clogging. The most common media in existing filters are crushed rock, slag or field stone that are durable, insoluble and resistant to spalling. The size range preferred for stone media is 21.5 cm to 8 cm.

- If these materials are used, structural problems caused by their weight tend to restrict the bed depth to about 3 m. Although smaller stones provide more surface area for biological growth, the voids tend to plug and limit passage of air and liquid. Bed depths range from 1.5 m to 3 m, greater depths do not materially improve BOD removal efficiency.

- Lightness of plastic media allows much deeper beds. They can be designed to be less prone to plugging by the accumulating slime and a higher rate of BOD removal is possible.

- The popularity of plastic media is increasing. Crushed stone or gravel has a specific surface area of 50 – 100 m^2/m^3 and void volume 30 – 50%. Plastic filter media have a high surface area of greater than 100 m^2/m^3 and high void space of more than 95%.

1.25.2 Recirculation System

- Filter plants return sufficient flow from the final clarifier hopper to the wet well for removal of accumulated settled solids and to prevent stalling of the distributor arm during low wastewater flow.

- Also in plant recirculation of wastewater increases liquid flow through the filter bed to allow greater organic loading without filling the bed voids with biological growths that would inhibit aeration. BOD removal efficiency is enhanced by passing wastewater through a filter more than once.

Advantages of Recirculation :

- High loading rate increases the sloughing of biomass. Thus, thin layer of biomass is maintained.

- The filter influent is freshened due to which odour problems are minimised.

- Self propelled distributors run continuously even at the time of reduced flows.

- The organic loading is reduced because of dilution.

- The applied sewage is seeded with active enzymes.

- Recirculation ratio of 0.5 to 3 usually is maintained. For industrial wastewater, the ratio of more than 5 have been used.

- Recirculation of sewage more than 3 times is not economical because there is less response for BOD removal.

1.25.3 Distribution System

- The rotor distributor used consists of two or more arms mounted in a pivot the center of filter which revolves in horizontal plane. The arms are hollow and contain nozzles through which the wastewater is sprayed over the bed. This can be done either from fixed sprays or moving sprays. In case of fixed sprays, pipe systems, placed evenly over the bed, distribute the wastewater uniformly.

- Among the moving types, the present practice is to provide circular tanks to use only reaction type rotary distributors.

- The reaction type rotary distributor consists of a feed column at the center of the filter, a turn table assembly at the top and two or more hollow radial distributor arms with orifices. The distributor should ensure that the entire surface is wetted and no area is left dry. This type of distributor requires a hydraulic head of 1 to 1.5 m measured from the center line of the distribution arm to the lower water level in the distribution well.

- The rate of rotation may vary from 2 rpm for small distributors to less than 1/3 to 1/2 for large distributors.

1.25.4 Underdrainage System

- This is provided with two objectives : (i) to collect treated wastewater and sloughed biological solids and (ii) to circulate or distribute air through the bed. The underdrains consist of semicircular or equivalent inverts. They are formed of precast vitrified clay or concrete blocks, complete with perforated cover.

- Fig. 1.41 shows the variety of commercially available underdrains. The underdrains have a slope towards the common collecting point or channel. The drains shall be so sized that flow occupies less than 50% of the cross-sectional area with velocity not less than 0.75 m/s at peak instantaneous hydraulic loading.

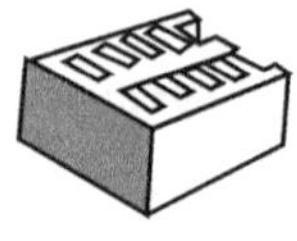  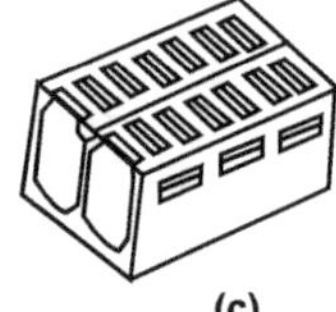

   (a)   (b)   (c)

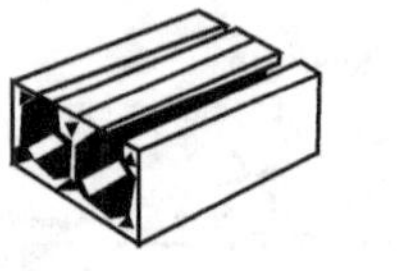 

   (d)   (e)

Fig. 1.41 : Blocks for under-drainage system

1.26 VENTILATION

- This may be natural or forced. For natural ventilation, proper design of underdrains and effluent channels are must. The holes are provided in the walls of filter. Vertical vents along the periphery also improves the natural ventilation.

- Forced ventilation is required in case of deep filters. It consists of forcing the air vertically upwards through the filters by the use of fans or other suitable equipments. The air required is 0.1 to 1 $m^3/min/m^2$ of floor area.

1.26.1 Types of Trickling Filters

- On the basis of hydraulic and organic loading rates, the filters are classified as

  1. Low rate trickling filter.

  2. High rate trickling filter.

- In low rate trickling filters, the hydraulic loading rate is in the range of 1 – 4 $m^3/m^2/d$ and organic loading rate of 80 – 320 gm BOD/m^3/d. In high rate filters, hydraulic loading rate is 10 –30 $m^3/d/m^2$ (including recirculation) and organic loading rate is 1000 to 4000 gm BOD/m^3/day.

- Height is 3 to 6 m, sludge production is 0.4 kg/kg BOD removed and BOD removal efficiency is 40 – 70%. Hydraulic loading rate is the total flow including recirculation, when the organic loading rate is 5 – day 20°C BOD, excluding the BOD of recirculant applied for unit volume per day.

Comparison of Low Rate and High Rate Trickling Filters: **(Dec. 10, May 11)**

The basic difference between high rate trickling filter and slow rate trickling filter is that the rate of filter loading (both hydraulic as well as organic) of the former is several times more than that of the latter. The main drawback of low rate filter is that it has high initial cost; it requires larger area of construction and larger quantity of filter media. Table 1.13 gives the detailed idea about the comparison between low rate and high rate trickling filters.

Table 1.13 : Comparison between Conventional and High Rate Trickling Filters

Sr. No.	Characteristics	Conventional or Low Rate Filter	High Rate Filter
1.	Depth of media	1.8 to 3.0 m	0.9 to 2.5 m
2.	Hydraulic loading ($m^3/d/m^2$)	1 to 4	10 to 40 (including recirculation)
3.	Organic loading as 5 day BOD in $g/d/m^3$	80 to 320	320 to 1000 (excluding recirculation)
4.	Recirculation system	Usually not provided, but can be provided if the hydraulic load does not exceed the limit.	Always provided. Recirculation ratio 0.5 – 3.0
5.	Volume of bed	5 times	1
6.	Interval of dosing	≯ 5 minutes. The sewage is applied at intervals.	≯ 15 seconds. Sewage is thus applied continuously.
7.	Sloughing	Intermittent	Continuous
8.	Cost of operation	More	Less
9.	Land required	More	Less
10.	Characteristics of final effluent	Contains BOD ≤ 20 mg/lit; it is highly nitrified into nitrate stage.	Contains BOD ≥ 30 mg/lit; it is not fully nitrified.
11.	Secondary sludge	Highly oxidized, black colour, having light fine particles.	Not fully oxidized; brownish black colour, containing fine particles.

(Ref. Manual on sewerage and sewage treatment, 2nd edition.)

1.26.2 High Rate Trickling Filter

Because of many drawbacks in low rate filter, many studies were conducted to increase the rate of filtration. Following are the observations :

- The thickness of biofilm reduces with increase in flow rate.

- The film is continuously washed away.

- Thinner biofilm is more active and supplies more nutrients to bacteria.

- The biomass quality collected in secondary settling tank is good.

- Because of less contact period, less degree of treatment and more putrescible organic material reaching to SST.

- The initial cost is less.

- Because of above mentioned favourable reasons, the high rate filters are becoming more popular.

- In high rate filters, better filtering media, with high surface area is provided. The filtering depth is reduced to 1.5 to 2.0 m to provide better aeration. The underdrains are provided of bigger size and steeper slope is given to filter bottom.

1.26.3 Single Stage and Two Stage Plants (Aug. 15)

- In single stage plant, the sewage is passed through single filter, sewage may be recirculated to single stage filters.

- In two stage filters, two filters in series are provided. It consists of primary settling tank and intermediate settling tank. Recirculation is provided to each stage. The treated effluent from first stage filter is applied on second stage filter either after settlement or without settlement.

- The flow diagrams of high rate trickling filter with single stage and two stage are shown in Fig. 1.42.

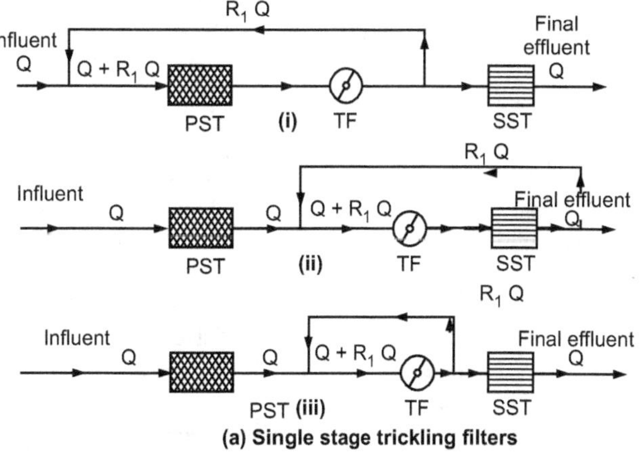

(a) Single stage trickling filters

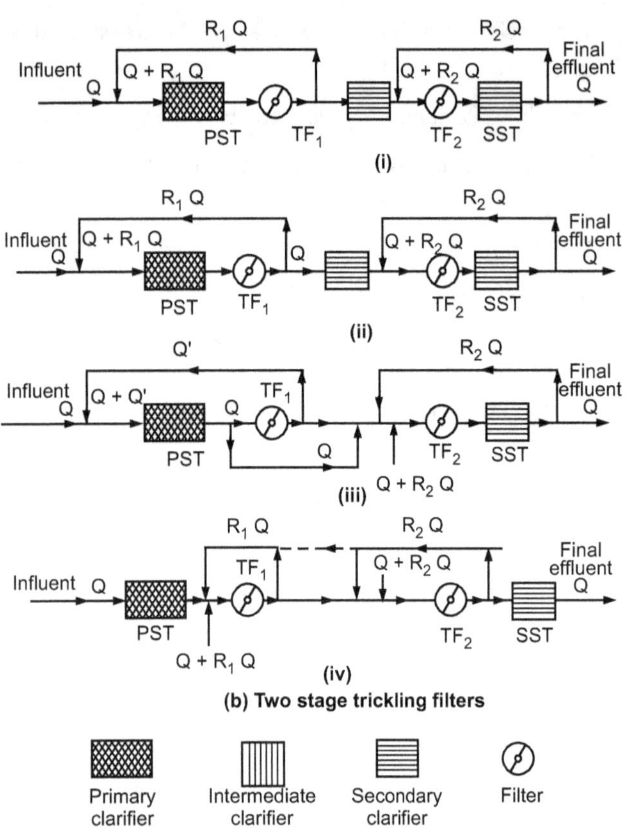

Fig. 1.42 : Flow sheets for high rate trickling filters

1.26.4 Design of Trickling Filters

The important considerations in the design of trickling filters are : (See Table 1.14).

1. Organic loading rate.
2. Recirculation ratio.

Table 1.14 : Typical Design Criteria for Trickling Filters

Item	Low-rate	Intermediate	High-rate	Super-rate
HLR[*] (m³/m²-day)	1 – 4	4 – 10	10 – 40	40 – 200
OLR[*] (kg BOD₅ per m³/day)	0.08 – 0.32	0.2 – 0.5	0.32 – 1.0	0.8 – 6
Depth (m)	1.8 – 3	1 – 3	0.9 – 2.5	4.5 – 12
Recirculation ratio	0	0 – 1	0.5 – 3	1 – 4
Filter media	Rock, slag, gravel etc.	Rock, slag, etc.	Rock, slag, synthetics	Plastic media

...Conti.

Power (kW/ 10³m³)	2 – 5	2 – 5	5 – 10	10 – 20
Filter flies	Many	Intermediate	Few larvae are washed away	Few or none
Sloughing	Intermittent	Intermittent	Continuous	Continuous
Dosing intervals	Not more than 5 min	10 – 50 sec	< 15 sec	Continuous
Effluent	Usually fully nitrified	Partially nitrified	Nitrified at low loadings	Nitrified at low loadings.

(Ref. Manual on sewerage and sewage treatment, 2nd edition).

- *Based on surficial area, HLR – Hydraulic loading rate, OLR – Organic loading rate.

- Once the organic loading rate is selected, the filter volume can be calculated. The depth and surface area are suitably chosen to secure hydraulic loading rates within the prescribed limits.

- A number of equations are available for the determination of plant efficiencies based on organic loading rates and recirculation ratios.

- The National Research Council (NRC) and Rankine have developed the empirical equations for trickling filter performance.

NRC Formula :

- The equations are applicable for both low rate and high rate trickling filters.

- The efficiency of single stage or first stage of two stage filter is given by

$$E = \frac{100}{1 + 0.44\sqrt{\dfrac{W}{VF}}}$$

or

$$E = \frac{100}{1 + 0.44\sqrt{U}}$$

For the second stage of two stage filter, the efficiency is given by

$$E' = \frac{100}{1 + \dfrac{0.44}{1-e}\sqrt{\dfrac{W'}{V'\,F'}}}$$

or $\quad E' = \dfrac{100}{1 + \dfrac{0.44}{1-e}\sqrt{U'}}$

where, $\quad E$ = Percentage BOD removal efficiency

$e = \dfrac{E}{100}$

E' = Percentage BOD removal efficiency for second stage of two stage filter.

W = BOD loading of settled sewage in single stage T.F. (kg/day)

V = Volume of first stage filter (m^3)

F = Recirculation factor

$= \dfrac{1 + R}{(1 + 0.1\,R)^2}$

R = Recirculation ratio, $\dfrac{Q_R}{Q}$

W', V', F' = BOD loading, volume and recirculation factor of second stage of two stage T.F.

$\dfrac{W}{VF} = U$ = Unit organic loading (kg/m^3/day)

1.26.5 Operational Problems in Trickling Filter

Chocking, ponding, fly nuisance, poor efficiency of BOD removal, tilting or stoppage of arms, odour are some of the problems faced. Media deterioration and loss, steel arm support corrosion are other problems.

Fly Nuisance

Filter flies, Psychoda, are nuisance problems near filter during warm weather. They breed in the sheltered area of the media, and on the inside surfaces of the retaining walls. Although wind can carry these small flies considerably long distances, their greatest irritation is to operating personnel.

Remedial Measures :

- Flooding the filter at regular intervals.
- Chlorinating the filter influent at two weeks interval.
- Periodic spraying of the peripheral area and walls of the filter with an insecticide.
- Continuous hydraulic loading to hamper the flies and larvae.
- Sprinkling lime at the site.

Ponding Nuisance

When all the voids of trickling filter are filled up due to chocking by heavy fungus and suspended solids, the problem of ponding of effluent arises. Ponding decreases ventilation, reduces effective volume of filter.

Remedial Measures :

- Flushing of filter with water.
- Reduce strength of filter influent by increasing recirculation.
- Chlorinate the influent.
- Stopping the distributor on ponded area.
- Allow the fungus to dry by keeping them out of operation for 12 to 48 hours.

Example 1.24 : *A trickling filter plant has the following :*

A primary clarifier with 17 m diameter, 2.5 m side water depth, and a single peripheral weir.

A trickling filter 25 m diameter, 2.5 m deep stone filled bed.

Final settling tank with 15 m diameter, 2.5 m side water depth and single peripheral weir.

Normal operating recirculation ratio = 0.5

The daily w/w flow = 6.2 mld

Average BOD of w/w = 180 mg/lit.

Calculate the loading on all units :

Solution :

(i) Primary clarifier :

$A = \dfrac{3.14}{4} \times (17)^2 = 227$ sq.m.

$V = 227 \times 2.5 = 567.5$ m^3

Surface overflow rate

$= \dfrac{6.2 \times 1000 \text{ m}^3/\text{d}}{227 \text{ m}^2}$

$= 27.31$ m^3/d/m^2

Detention time $= \dfrac{\text{Volume}}{\text{Flow}} = \dfrac{567.5 \text{ m}^3}{6200 \text{ m}^3/\text{d}} \times 24$

$= 2.19$ hrs.

Weir loading $= \dfrac{\text{Flow}}{\text{Weir length}} = \dfrac{\text{Flow}}{pD}$

$= \dfrac{6200 \text{ m}^3/\text{d}}{3.14 \times 17} = 116$ m^3/d/m $\qquad$ ... **Ans.**

(ii) Trickling filter :

$A = \dfrac{3.14 \times (25)^2}{4} = 490.6$ sq.m.

$V = 490.6 \times 2.5 = 1226.5$ m^3

Hydraulic loading

$= \dfrac{Q + (R \times Q)}{A}$

$= \dfrac{6200 \text{ m}^3/\text{d} + (0.5 \times 6200) \text{ m}^3/\text{d}}{490.6 \text{ m}^2}$

$= 18.95$ m^3/d/m^2 $\qquad$ ... **Ans.**

BOD / Organic loading

Assume 40% BOD removal by primary settling

∴　Settled wastewater BOD = 0.6×180 = 108 mg/lit

BOD/Organic load

$$= \frac{6.2 \text{ mld} \times 108 \text{ mg/lit}}{1226.5 \text{ m}^3} = \frac{669.6 \text{ kg/day}}{1226.5 \text{ m}^3}$$

$$= 0.546 \text{ kg/day/m}^3$$

$$= 546 \text{ gm/day/m}^3 \qquad \text{... Ans.}$$

(iii) Final settling tank :

$$A = \frac{3.14 \times (15)^2}{4} = 176.6 \text{ m}^2$$

$$V = 176.6 \times 2.5 \text{ m} = 441 \text{ m}^3$$

Surface overflow rate

$$= \frac{6200 \text{ m}^3/\text{d}}{176.6 \text{ m}^2} = 35 \text{ m}^3/\text{d/m}^2$$

$$\text{Detention time} = \frac{\text{Volume}}{\text{Flow}} = \frac{441 \text{ m}^3}{6200 \text{ m}^3/\text{d}} = 1.7 \text{ hrs}$$

$$\text{Weir loading} = \frac{\text{Flow}}{\text{Weir length}}$$

$$= \frac{6200 \text{ m}^3/\text{d}}{3.14 \times 15}$$

$$= 131.6 \text{ m}^3/\text{d/m} \qquad \text{... Ans.}$$

Example 1.25 : *Design a high rate trickling filter plant to treat settled domestic sewage having BOD of 200 mg/lit for an average flow of 20 mldto satisfy an effluent BOD of 10 mg/lit. Adopt peak factor as 2.25.*

Solution :

As the effluent BOD_5 required is less than 30 mg/lit, a two stage filtration plant shall be provided. The filter will be designed for average flow. The distribution arms, underdrainage system and other pipelines etc. shall be for peak flow.

Taking an organic loading of 750 gm/d/m^3,

Volume of first stage filter

$$= 20 \times 10^6 \text{ l/d} \times \frac{200}{10^3} \text{ gm/lit}$$

$$\times \frac{1}{750} \frac{\text{d/m}^3}{\text{gm}}$$

$$= 5333.3 \text{ m}^3$$

Adopting a depth of 1.5 m,

$$\text{filter area required} = \frac{5333.3}{1.5} = 3555.5 \text{ m}^2$$

$$\therefore \quad \text{Diameter of circular filter} = \sqrt{\frac{4 \times 3555.5}{\text{p}}}$$

$$= 67.30 \text{ m}$$

As the diameter is too much high, two units shall be provided.

∴　Area of each unit = 1777.75 m^2

$$\text{Diameter of each unit} = \sqrt{\frac{4 \times 1777.5}{\text{p}}} = 47.60 \text{ m}$$

Assume effluent BOD = 30 mg/lit

Filter dimensions for second stage filter :

Let us choose an organic loading rate of 500 gm/d/m^3.

Volume of second stage filter

$$= \frac{(20 \times 10^6) \ (30 \times 10^{-3})}{500}$$

$$= 1200 \text{ m}^3$$

Let us adopt a depth of 1 m.

$$\therefore \quad \text{Area of filter} = \frac{1200 \text{ m}^3}{1 \text{ m}} = 1200 \text{ m}^2$$

$$\text{and} \quad \text{Filter diameter} = \sqrt{\frac{1200 \times 4}{\text{p}}} = 39 \text{ m} \qquad \text{... Ans.}$$

[Note : This problem can be solved by using NRC equation.]

Example 1.26 : *Design the high rate trickling filter for the following data :*

Sewage flow = 10 mld

Recirculation ratio, R = 1.5

BOD of raw sewage = 250 mg/lit

BOD removal in primary clarifier = 30%

Final effluent BOD desired = 30 mg/lit.

Solution :

Q = 10 mld. BOD concentration = 250 mg/lit.

∴　Total BOD present

$$= (10 \times 10^6) \ (250 \times 10^{-3}) \times 10^{-3}$$

$$= 2500 \text{ kg/day}$$

BOD removal in primary tank = 30%

∴　BOD left in settled sewage

$$= 2500 - 0.3 \times 2500$$

$$= 1750 \text{ kg/day}$$

Desired BOD concentration in effluent

$$= 30 \text{ mg/lit.}$$

∴　Total BOD left in effluent

$$= 10 \times 30 = 300 \text{ kg/day}$$

Hence, BOD removed by filter

$$= 1750 - 300$$

$$= 1450 \text{ kg/day}$$

∴ Efficiency of filter

$$= \frac{\text{BOD removal}}{\text{Total BOD present}} \times 100$$

$$= \frac{1450}{1750} \times 100 = 82.86\%$$

But efficiency, given by NRC equation is

$$E = \frac{100}{1 + 0.44\sqrt{\dfrac{W}{VF}}}$$

where,   W = Total BOD applied to filter

      = 1750 kg/day

    F = Recirculation factor

$$= \frac{1 + R}{(1 + 0.1\,R)} = \frac{1 + 1.5}{1 + (0.1 \times 1.5)^2}$$

$$= 1.89$$

$$82.86 = \frac{100}{1 + 0.44\sqrt{\dfrac{1750}{V \times 1.89}}}$$

Substituting

$$1 + 0.44\sqrt{\frac{1750}{V \times 1.89}} = \frac{100}{82.86} = 1.21$$

$$0.44\sqrt{\frac{1750}{V \times 1.89}} = 1.21 - 1 = 0.21$$

$$\sqrt{\frac{1750}{V \times 1.89}} = \frac{0.21}{0.44} = 0.48$$

$$\frac{1750}{V \times 1.89} = (0.48)^2 = 0.23$$

$$\frac{1750}{0.23} = V \times 1.89$$

$$7608.70 = 1.89 \times V$$

∴ $V = \dfrac{7608.70}{1.89}$

∴ $V = 4026 \ m^3$

Let us assume the depth of filter as 1.8 m.

∴   Surface area of filter $= \dfrac{4026\ m^3}{1.8\ m} = 2236.67 \ m^2$

Provide two units.

∴   Area of each unit = 1118.33 m^2

Hence, diameter of each filter $= \sqrt{\dfrac{1118.33 \times 4}{p}}$

      = 37.74 m ≅ 38 m   ... **Ans.**

Hence, provide two single stage high rate filters of 8.5 m and 1.8 m deep filter media with a recirculation factor of 1.5.

Example 1.27 : *A single stage filter is designed for an organic loading of 10,000 kg of BOD in raw sewage per 10^4 sq.m. per day with a recirculation ratio of 1.2. This filter treats a flow of 10 mld of raw sewage with a BOD of 250 mg/lit. Using NRC equation, determine the strength of effluent.*

Solution :

     Total BOD of raw sewage

$$= (10 \times 10^6) \times (250 \times 10^{-6})$$

$$= 2500 \ \text{kg/day}$$

     Required filter area

$$= \frac{\text{Total BOD of raw sewage, kg/day}}{\text{Permissible BOD loading, kg/m}^2\text{/day}}$$

$$= \frac{2500}{10000} \times 10^4 = 2500 \ m^2$$

∴   Volume of tank

$$= 2500 \times 1.8 = 4500 \ m^3$$

Assume depth of tank is 1.8 m.

Let us assume that primary clarifier removes 40% of BOD.

∴   BOD of influent applied to filter

$$= 0.6 \times 2500$$

$$= 1500 \ \text{kg/day}$$

The efficiency of filter,

$$E = \frac{100}{1 + 0.44\sqrt{\dfrac{W}{V \times F}}}$$

Here,   W = 1500 kg/day

      V = 4500 m^3

$$F = \frac{1 + 1.2}{1 + (0.1 \times 1.2)^2} = 2.169$$

∴   Substituting

$$E = \frac{100}{1 + 0.44\sqrt{\dfrac{1500}{4500 \times 2.169}}}$$

∴   E = 85.29%

∴   Total BOD of effluent

$$= (1 - 0.8529) \times 1500$$

$$= 220.65 \ \text{kg/day}$$

BOD concentration of effluent

$$= \frac{\text{Total BOD}}{\text{Sewage volume}}$$

$$= \frac{220.65 \times 10^6}{10 \times 10^6}$$

$$= 22.07 \ \text{mg/lit} \qquad \text{... } \textbf{Ans.}$$

Example 1.28 : *Determine the size of a high rate trickling filter for the following data:*   **(Aug. 15, 5M)**

   (i) Sewage flow = 4.5 MLD

   (ii) Recirculation ratio = 1.5

   (iii) BOD of raw sewage = 250 mg/L

   (iv) BOD removal in primary tank = 30%

   (v) Final Effluent BOD desired = 30 mg/L

Solution :

$Q = 4.5$ MLD, BOD concentration = 250 mg/L

Total BOD present $= (4.5 \times 10^6) \times (250 \times 10^{-3}) \times 10^{-3}$

$$= 1125 \text{ kg/day}$$

BOD removal in primary tank = 30 %

BOD Left in settled sewage $= 1125 - 0.3 \times 1125$

$$= 787.5 \text{ kg/day}$$

Desired BOD concentration in effluent = 30 mg/l

$\therefore$ Total BOD left in effluent

$$= 4.5 \times 30 = 135 \text{ kg/day}$$

Hence, BOD removed by filter = $787.5 - 135 = 652.5$ kg/ day

Efficiency of filter

$$= \frac{\text{BOD removal}}{\text{Toatl BOD present}} \times 100$$

$$= \frac{652.5}{787.5} \times 100$$

$$E = 82.85 \%$$

But efficiency, given by NRC equation is

$$E = \frac{100}{1 + 0.44 \sqrt{\dfrac{W}{VF}}}$$

$$82.85 = \frac{100}{1 + 0.44 \sqrt{\dfrac{787.5}{V \times F}}}$$

$\therefore$ F = Recirculation factor $= \dfrac{1 + R}{(1 + 0.1 R)}$

$$= \frac{1 + 1.5}{1 + (0.1 \times 1.5)^2} = 2.44$$

$\therefore$ $82.85 = \dfrac{100}{1 + 0.44 \sqrt{\dfrac{787.5}{V \times 1.89}}} = 2.44$

Substituting, $1 + 0.044 \sqrt{\dfrac{787.6}{V \times 2.44}} = \dfrac{100}{82.86}$

By solving above equation

Volume of filter, $V = 1882. \ 56$ m^3

Let us assume, the depth of filter as 1.8 m

Surface area of filter $= \dfrac{1882.56}{1.8} = 10.45.86$ m^2

Provide two units,

   area of each unit = 522.93 m^2 **...Ans.**

Hence, dia. of each filter $= \sqrt{\dfrac{522.93 \times 4}{p}} = 25.80$ m

                                          ...Ans.

Hence,

Provide two single stage high rate filter of 25.80 m and 1.8 m deep filter media with a recirculation factor 1.5.

Example 1.29 : *Determine the size of a high rate trickling filter for the following data.*   **(Aug. 2016, 5M)**

   (i) Sewage flow = 4 MLD

   (ii) Recirculation ratio = 1.5

   (iii) BOD of raw sewage = 230 mg/L

   (iv) BOD removed in primary sedimentation tank = 30%

   (v) Final effluent BOD = 20 mg/L

   (vi) Depth of filter = 2 m

Solution : Total BOD present,

$$= 10 \times 10^6 \times 230$$

$$= 2300 \times 10^6 \text{ mg/d} = 2300 \text{ kg/cl}$$

BOD left after PST in,

$$W = 2300 \times 0.7 = 1610 \text{ kg/d}$$

Total BOD left in effluent

$$= 10 \times 10^6 \times 20 = 200 \times 10^6 \text{mg/d}$$

$$= 200 \text{ kg/d}$$

Hence BOD removed by filter

$$= 1610 - 200 = 1410 \text{ kg/d}$$

$\therefore$ Efficiency of filter

$$= \frac{1410}{1610} \times 100 = 87.57 \%$$

Efficiency by μRC equation,

$$E = \frac{100}{1 + 0.44 \sqrt{\dfrac{W}{VF}}}$$

$$= \frac{100}{1 + 0.44 \sqrt{\dfrac{1610}{V \times 1.89}}} = 87.57$$

$$F = 8185.35 \text{ m}^3$$

$\therefore$ Surface Area of filter

$$= \frac{8185.35}{2} = 4092.67 \text{ m}^2$$ **...Ans.**

Diameter of filter = 72.18 m   **...Ans.**

Example 1.30: *Calculate the effluent BOD of a two stage trickling filter with the following data.*   **(Nov. 16, 6M)**

(i)   Flow $= 2.30\ m^3/min$

(ii) $BOD_5 = 300\ mg/l$

(iii) Volume of filter 1 $= 900\ m^3$

(iv) Volume of filter 2 $= 900\ m^3$

(v)   Filter depth $= 2\ m$

(vi) Recirculation ratio for both the filter $= 1.5$

Use NRC formula.

Solution :(i) Surface area of filter

$$\frac{\text{Volume of filter}}{\text{Depth of filter}} = \text{S.A. of filter}$$

$\therefore$ Volume of filter 1 = Volume of filter 2 $= 900\ m^3$

Depth of both filter $= 2m$

$\therefore$ S.A. for both filters $= \dfrac{900}{2} = 450\ m^2$ **...Ans.**

(ii) Diameter of filter :

$$\text{Surface Area} = \frac{p}{4} \times D^2$$

$\therefore$ $D = 23.936\ m$

$D \approx 24\ m$

$\therefore$ $D_1 = D_2 = D = 24\ m$ **...Ans.**

(iii) Efficiency of filter 1 by NRC formula

$$\eta = \frac{100}{1 + 0.0044\sqrt{\dfrac{Y}{V \times F}}}$$

Y = Total BOD $= 3312 \times 10^3 \times 300$

$= 993.6 \times 10^6\ mgl\text{day}$

$= 993.6\ kg/day$

$V = 900\ m^3 = 0.09\ ham$

$$F = \frac{HR}{(1 + 0.1R)^2} = \frac{1 + 5}{[1 + 0.1\,(1.5)]^2}$$

$F = 1.89$

$$\eta = \frac{100}{1 + 0.0044\sqrt{\dfrac{993.6}{0.09 \times 1.89}}}$$

$= 74.83\%$

$$\text{BOD of Effluent} = \frac{100 - 74.83}{100} \times 993.6$$

$= 250\ kg/day$ **...Ans.**

(iv) Efficiency of filter 2

$$\eta = \frac{100}{1 + 0.0044\sqrt{\dfrac{Y}{V \times F}}}$$

Y = BOD Remained after filter 1

$= 993.6 - 250 = 743.6\ kg/day$

$V = 0.09\ ha.m$

$$\eta = \frac{100}{1 + 0.0044 \times \sqrt{\dfrac{743.6}{0.09 \times 1.89}}}$$

$\eta = 77.46\ \%$

$$\text{BOD of effluent} = \frac{100 - 77.46}{100} \times 743.66$$

$= 167.6\ kg/day$

Sewage volume $= 3312\ m^3/day$ **...Ans.**

(v) BOD concentration of effluent from filter $-$ 1

$$= \frac{250 \times 10^6}{3312 \times 10^3}$$

$= 75.50\ Mg/l$ **...Ans.**

BOD concentration of effluent from filter $-$ 2

$$= \frac{167.6 \times 10^6}{3312 \times 10^3}$$

$= 50.60\ Mg/l$ **...Ans.**

Example 1.31 : *Design a high rate single stage trickling filter for treating domestic sewage flow of 8 MLD using N.R.C. formula. Use following data.*   **(Aug. 2017, 6M)**

(i)   BOD5 of raw sewage = 240 mg/L

(ii) BOD removed during primary treatment = 30%

(iii) Organic loading rate = 0.8 Kg/m³/d,

(iv) Hydraulic loading rat = 15 m³/m³/d

(v)   Recirculation ratio = 2.

Determine,

(1) Volume of filter media

(2) Dimensions of trickling filter

(3) Efficiency of trickling filter

Solution : Total BOD of raw sewage

$= 240\ mg/l \times 8\ MLD$

$= (10 \times 10^6) \times (240 \times 10^{-6})$

$= 1920\ kg/day$

$\therefore$ Required filter area

$$= \frac{\text{Total BOD of raw sewage}}{\text{Permissible BOD Loading}}$$

$$= \frac{1920}{15000} \times 10^4$$

$= 1280\ m^2$

Volume of tank

$= 1280 \times 1.8$

$= 2304\ m^3$

(Assume depth of tank is 1.8 m)

Let us assume that primary clarifier removes 40% of BOD.

$\therefore$ BOD of influent applied to filter

$$= 0.6 \times 1280$$

$$= 768 \text{ kg/day}$$

The efficiency of filter,

$$E = \frac{100}{1 + 0.44\sqrt{\dfrac{W}{V \times F}}}$$

Here,   W = 768 kg/day

V = 2304 m^3

$$F = \frac{1 + R}{(1 + 0.1\,R)} = \frac{1 + 2}{(1 + 0.1 \times 2)}$$

$$= 2.5 \qquad \text{...Ans.}$$

Substituting,

$$E = \frac{100}{1 + 0.44\sqrt{\dfrac{768}{2304 \times 2.5}}}$$

$$E = 86.15\% \qquad \text{...Ans.}$$

Total BOD of effluent

$$= (1 - 0.8615) \times 768$$

$$= 106.368 \text{ kg/day}$$

BOD concentration of effluent

$$= \frac{\text{Total BOD}}{\text{Sewage Vol}^m}$$

$$= \frac{106.368 \times 10^6}{10 \times 10^6}$$

$$= 10.63 \text{mg/lit} \qquad \text{...Ans.}$$

1.27 OXIDATION PONDS (STABILISATION PONDS)   (Nov. 16)

1.27.1 General

- A **Stabilisation Pond** is simply a shallow body of water contained in an earthen basin, open to sun and air. The **Oxidation Pond**, often used is synonymous. The detention period of these ponds is long. These ponds may be considered to be completely mixed biological reactors without solids return.

- Due to their low construction and operating cost, these ponds are widely used in rural areas. Now-a-days, these ponds are also used for treatment of various industries like dairies, oil refineries, poultry-processing plants etc.

1.27.2 Classification of Ponds   (May 10, 11, Nov. 15)

The ponds are classified according to the nature of the biological activity which takes place within the pond as :

1.   Aerobic
2.   Facultative (aerobic – anaerobic) and
3.   Anaerobic.

1.   Aerobic Stabilisation Ponds (Algae Ponds) :

General :

- These are large shallow earthen basins that are used for the treatment of wastewater by natural processes involving the use of algae and bacteria. Therefore these ponds are also called as **Algae Ponds.**

Applications :

- This type of ponds are used for nutrient removal, treatment of soluble organic wastes, conversion of wastes etc.

Process Description :

- As per the name of the pond, aerobic condition is prevailed throughout the depth of pond. This pond contains algae and bacteria in suspension.

- The following Table 1.15 shows depth of aerobic ponds related to different uses.

Table 1.15 : Depth of Aerobic Ponds

Sr. No.	Uses	Depth
1.	Treatment of irrigation return water or any other industrial waste (Remove the nitrogen by algal growth).	Shallow depth of 0.15 m to 0.45 m.
2.	Treatment of domestic waste water.	1 m to 1.2 m

- The length to width ratio is generally kept as 3 : 1.
- For better results, their contents must be mixed periodically using pumps or surface aerators.

Process Microbiology :

- In aerobic stabilisation pond, the oxygen is supplied by natural surface aeration and by algal photosynthesis. Except for the algale population, the microbiological population present in the ponds are similar to that in activated sludge system.

Bacterial-Algae Symbiosis (May 10, 11, Dec. 10) :

- The sewage containing organic material is a necessary food for aerobic population like bacteria which stabilises the putrescible matter by oxidising it and releases carbon dioxide (CO_2) which is taken up by the algal for their growth, algae produce more algal cells and oxygen which help in maintaining the aerobic condition of the pond. This cycle is known as 'bacterial-algae symbiosis'. This cyclic symbiotic relationship is shown in Fig. 1.43.

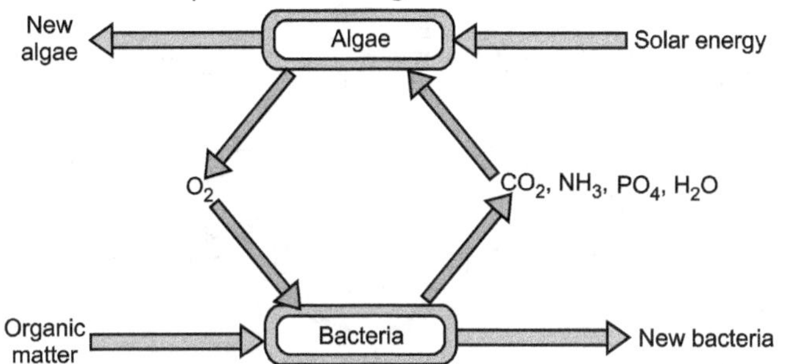

Fig. 1.43 : Symbiosis relationship between algae and bacteria

- As we have seen that micro-organisms like bacteria and algae are predominating in this system. Also protozoa and rotifers are present under certain loading conditions. The main function of protozoa and rotifers is to polish the effluent.

- The presence of these micro-organisms in aerobic pond depends on pH, nutrients, temperature, degree of pond mixing, sunlight etc.

2. Facultative (Aerobic – Anaerobic) Stabilisation Ponds (May 16)

General :

- The stabilisation of wastes in facultative ponds is brought about by a combination of aerobic, anaerobic and facultative bacteria.

Process Description :

- Three zones exist in this type of ponds :

1. **Top Zone :** This is an aerobic zone in which the algae photosynthesis and aerobic biodegradation takes place.

2. **Bottom Zone :** This is an anaerobic zone, the wastewater settle and undergo anaerobic decomposition.

3. **Intermediate Zone :** This is partly aerobic and partly anaerobic zone, the decomposition of organic wastes are done by the facultative bacteria.

   Fig. 1.44 shows elevation diagram of facultative pond.

- The nuisance associated with the anaerobic reactions are eliminated due to the presence of top aerobic zone, the end products of anaerobic decomposition are foul smelling one which are carried to the top zone by mixing currents and are oxidized, hence the maintenance of an aerobic layer in the facultative pond is important.

- The aerobic condition in top layer is maintained by the presence of algae or by surface aerators. When surface aerators are used to maintain aerobic condition in upper layer, algae are not required.

- The depth of pond is generally 1 to 1.5 m.

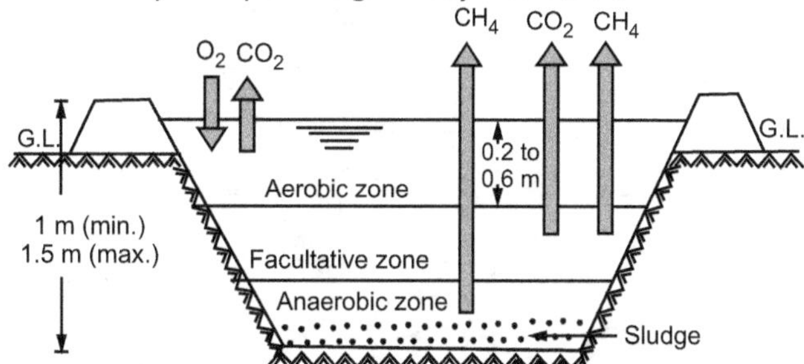

Fig. 1.44 : Elevation diagram of facultative pond

Process Microbiology :

- The micro-organisms present in upper aerobic layer are similar to that of an aerobic pond i.e. giving rise to bacterial algae symbiosis and in the bottom layer of the pond are facultative and anaerobic bacteria. The treatment effected by this type of pond is comparable to that of conventional secondary treatment processes. So facultative ponds are most commonly used for treatment of sewage.

- In this pond, the organic matter is stabilised by bacterial oxidation in top layer and by methane fermentation in the bottom zone. In facultative zone, facultative bacteria oxidize incoming organics as well as the end products of anaerobic decomposition of the bottom anaerobic zone. When the sewage is loading in the pond, the suspended organic matter and bioflocculated organic matter settle down to the bottom of the pond.

- The settled sludge of the bottom undergoes anaerobic fermentation in absence of the dissolved oxygen. Due to the anaerobic fermentation it librates the methane (CH_4), this indicates the removal of BOD (Generally, 0.25 gm of methane is being liberated for every gramm of ultimate BOD utilized.) In the liquid layers (above bottom sludge layer), algae are present under favourable conditions.

- During day light, algae utilizes CO_2 for photosynthesis, and liberating oxygen, which helps to maintain aerobic condition in the top layer of the pond. In presence of this oxygen, aerobic bacteria oxidise organic matter. So it means that there is an interdependence between algae and bacteria. This is shown in Fig. 1.45.

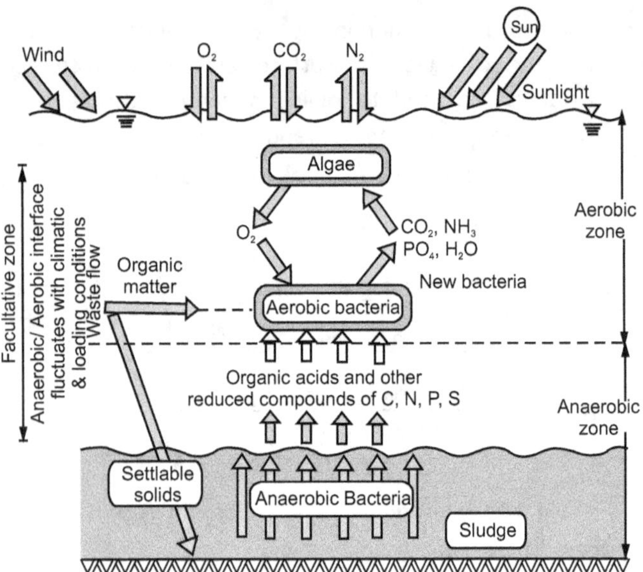

Fig. 1.45 : Diagram of facultative pond reaction

- The boundary as shown in Fig. 1.45 between the aerobic and anaerobic zones is not stationary. Aerobic area may extend downward due to mixing (by wind action) and penetration (by sunlight). Conversely, calm waters and weak lighting result in the anaerobic layer rising towards the surface. Diurnal changes in light conditions may lead in diurnal fluctuations in the aerobic – anaerobic interface.

3. Anaerobic Stabilisation Ponds :

- In this type of ponds, the entire depth is in anaerobic condition except an extremely shallow top layer. Generally, these ponds are used in series after facultative pond for complete treatment of a waste.

Process Description :

- In this system, free dissolved oxygen is not available to the sewage, so anaerobic decomposition, called putrefaction occur. An anaerobic bacteria is survived by extracting and consuming the bounded molecular oxygen present in compounds like sulphates (SO_4) and nitrates (NO_3).

Process Microbiology :

- Firstly, acid producing bacteria produce organic acids like acetic, butyric and propionic by decomposition of dissolved organic waste. Further the organic acids converted into methane gas ($CH_4 \uparrow$), carbon dioxide gas ($CO_2 \uparrow$) etc. by the methane-producing bacteria, which is represented by the following equation and also shown in Fig. 1.46.

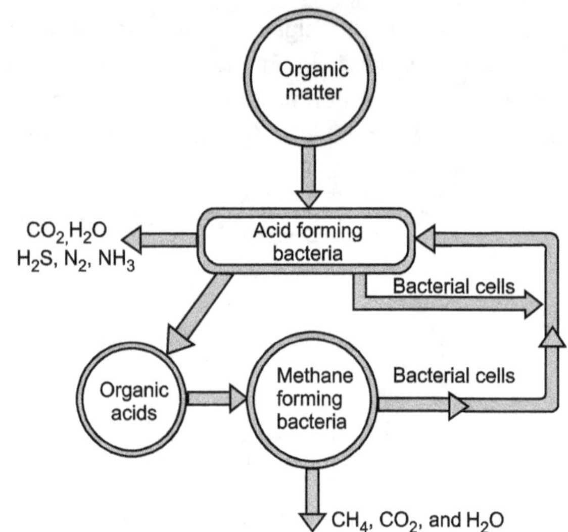

Fig. 1.46 : Anaerobic decomposition

$$\text{Organic acids} \xrightarrow[\text{anaerobic bacteria}]{\text{Methane forming}} CH_4 \uparrow + CO_2 \uparrow + \text{Heat}$$

- Sludge or solids build up is much less in this anaerobic system because some quantity of waste is used by the anaerobic biosystem as source of energy and in the synthesis of new bacterial cells. If the contents of anaerobic pond are in the black colour, it indicates proper functioning of anaerobic pond.

Solids Removal

- Removal of suspended solids, and sometimes dissolved solids, may be necessary in advanced wastewater treatment systems. The solids removal processes employed in advanced wastewater treatment are essentially the same as those used in the treatment of potable water, although application is made more difficult by the overall poorer quality of the wastewater.

Suspended Solids Removal

- As an advanced treatment process, suspended solids removal implies the removal of particles and floes too small or too lightweight to be removed in gravity settling operations. These solids may be carried over from the secondary clarifier or from tertiary systems in which solids were precipitated. Several methods are available for removing residual suspended solids from wastewater.

- Removal by centrifugation, air flotation, mechanical microscreening, and granular media filtration have all been used successfully. In current practice, granular media filtration is the most commonly used process. Basically, the same principles that apply to filtration of particles from potable water apply to the removal of residual solids in wastewater.

- Differences in operational modes for application of these principles to wastewater filtration vs. potable water filtration may range from slight to drastic, however, and the most commonly used wastewater filtration techniques are discussed below. Sand filters have been used to polish effluents from septic tanks, Imhoff tanks, and other anaerobic treatment units for decades. Because they are alternately dosed and allowed to dry, the term intermittent sand filters has been applied to this type of unit.

- The process is essentially the slow sand filter. More recently, this type of filter has been applied to the effluent from oxidation ponds with considerable success. Effluent concentrations of less than 10 mg/L of BOD and suspended solids have been reported at filtering rates of 0.37 to 0.56 m^3/m^2.day. Use of intermittent sand filters in tandem with conventional secondary treatment has not been very successful. The nature of the solids from these processes results in rapid plugging at the sand surface, necessitating frequent cleaning and thus high maintenance costs. The use of intermittent filters for tertiary treatment is usually restricted to plants with small flows.

- Granular media filtration is usually the process of choice in larger secondary systems. Dual or multimedia beds prevent surface plugging problems and allow for longer filter runs. Loading rates depend on both the concentration and nature of solids in the wastewater. Filtering rates ranging from 12 to 30 m^/m2 day have been used with filter runs of up to 1 day. Other recent innovations in filtration practices hold promise for advanced wastewater treatment.

- Moving bed filters have been developed which are continuously cleaned, and the rate of cleaning can be adjusted to match the solids loading rate. Another modification called the pulsed bed filter, uses compressed air to periodically break up the surface mat deposited on a thin bed of fine filter media. Only after a thick suspension of solids has accumulated on the bed, requiring frequent pulsing, is the filter backwashed. Both the moving bed and the pulsed bed filters have the capability of filtering raw wastewater.

- A much higher percentage of solids can be removed by filtration than can be removed in primary settling. The filter effluent, containing lower levels of mostly dissolved organics, responds very well to conventional secondary treatment. The filtered solids can be thickened and treated by anaerobic digestion with a resultant increase in overall methane production, a possible source of energy for use within the plant.

Dissolved Solids Removal

- Both secondary treatment and nutrient removal decrease the dissolved organic solids content of wastewater. Neither process, however, completely removes all dissolved organic constituents, and neither process removes significant amounts of inorganic dissolved solids. Further treatment will be required where substantial reductions in the total dissolved solids of wastewater must be made.

- Ion exchange, microporous membrane filtration, adsorption, and chemical oxidation can be used to decrease the dissolved solids content of water. These processes were developed to prepare potable water from a poor quality raw water. Their use can be adopted to advanced wastewater treatment if a high level pretreatment is provided.

- The removal of suspended solids is necessary prior to any of the processes. Removal of the dissolved organic material (by activated carbon adsorption) is necessary prior to microporous membrane filtration to prevent the larger organic molecules from plugging the micropores. Advanced wastewater treatment for dissolved solids removal is complicated and expensive. Treatment of municipal wastewater by these processes can be justified only when reuse of the wastewater is anticipated.

Nutrient Removal

- Biological Nutrient Removal (BNR) is a process used for nitrogen and phosphorus removal from wastewater before it is discharged into surface or ground water.

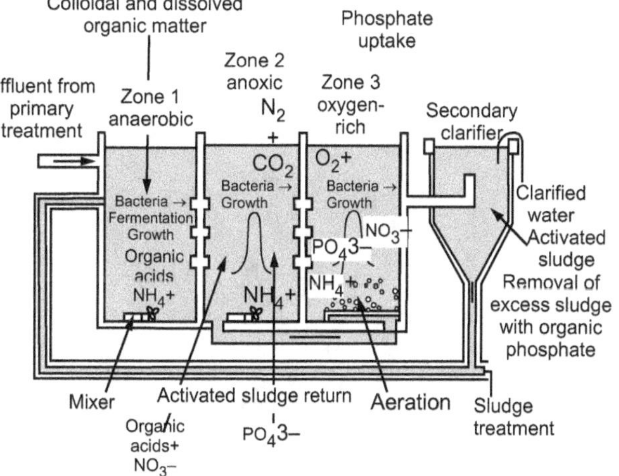

Fig. 1.47

- Biological Nutrient Removal (BNR) is a process used for nitrogen and phosphorus removal from wastewater before it is discharged into surface or ground water.

- The rising concentration of harmful nutrient compounds – specifically nitrogen and phosphorus – in municipal wastewater treatment plant discharge causes cultural eutrophication (nutrient enrichment due to human activities) in surface waters. Summer algal blooms are a familiar example of this eutrophication, and can present problems for ecosystems and people alike: low dissolved oxygen, fish kills, murky water, and depletion of desirable flora and fauna.

- Because conventional biological processes designed to meet secondary treatment effluent standards typically do not remove total nitrogen (TN) and total phosphorus (TP) to the extent needed to protect receiving waters, wastewater treatment facilities are increasingly being required to implement processes that reduce effluent nutrient concentrations to safe levels. This can be a challenge for wastewater treatment plants because it usually involves major process modifications to a plant, such as making a portion of the aeration basin anaerobic and/or anoxic, which reduces the aerobic volume and limits nitrification capacity.

EXERCISE

1. What is necessity of wastewater treatment?
2. Classify wastewater treatment with suitable examples.
3. Distinguish between Low rate and High rate Trickling Filter
4. Derive Equation for Sludge age.
5. Explain working of Imhoff tank with neat sketch.
    (i) Classify Screen chambers.
    (ii) Explain Significance of
    (iii) Screen chamber
    (iv) Grit chamber
6. Explain steps in Digestion process
7. What is bacteria algal symbiosis? What is its role in Aerobic ponds?
8. Explain in detail activated sludge process.
9. Explain SBR system
10. Explain modifications in ASP.

LOW COST WASTEWATER TREATMENT METHODS

2.1 INTRODUCTION

2.1.1 Design Criteria of Oxidation Pond

(Dec. 10, Nov. 15, May 10, 11)

The stabilisation ponds design parameters are not well defined. Due to the simultaneous involvement of the operations like sedimentation, oxidation, digestion, photosynthesis, evaporation, seepage etc., the mathematical modelling of the process of stabilisation in the pond is difficult.

Here, two separate methods are described, the first method applicable for aerobic pond only and the second applicable for aerobic and facultative ponds.

1. **First Method (Applicable for Aerobic Pond) :**

- In this method, the oxygen resources of the pond are equated to the applied organic loading. The principle source of the oxygen is photosynthesis and is also dependent on solar energy. The solar energy is related to geographical, astronomical and meteorological phenomena, and varies with time in the year and the latitude of the place. Based on studies conducted by the National Environmental Engineering Research Institute (NEERI), Nagbur, Table 2.1 shows the values of yield of photosynthetic oxygen in different latitudes IS : 5611 recommends the BOD loading equal to the yield of photosynthetic oxygen, as shown in column 4 of Table 2.1.

- The values shown in Table 2.1 may be modified for elevation above mean sea level (MSL) by dividing by a factor $(1 + 0.003\ H)$, where H is the elevation of the pond site above MSL in hundred meters. Other correction is the pond volume which has to be made when the sky is clear for less than 75% of the days at the rate of 3% for a fall of every 10%.

Table 2.1 : Yield of Photosynthetic Oxygen and Recommended BOD Loading

Sr. No.	Latitude (°N)	Yield of Photosynthetic O_2 (kg/ha/day)	BOD_5 Loading (kg/ha/day)
(1)	(2)	(3)	(4)
1.	8	325	325
2.	12	300	300
3.	16	275	275

...Conti.

Sr. No.	Latitude (°N)	Yield of Photosynthetic O_2 (kg/ha/day)	BOD_5 Loading (kg/ha/day)
4.	20	250	250
5.	24	225	225
6.	28	200	200
7.	32	175	175
8.	36	150	150

(**Ref. :** Manual on sewerage and sewage treatment, 2nd Edition)

If the amount of solar energy is cal/m²/day and the efficiency of the conversion of light energy to fixed energy in the form of algal cells are known, yield of photosynthetic oxygen can be calculated directly.

As we know,

$$\text{Flow} \times \text{Detention time} = \text{Depth} \times \text{Surface area}$$

Detention time can be calculated by using the following equation.

$$t = \frac{1}{K_1} \log_{10}\left(\frac{L_a}{L_a - Y}\right) \qquad \text{... (2.1)}$$

where, L_a = BOD of the effluent entering the pond

Y = BOD removed (say 90% of L_a or 95% of L_a)

The organic surface loading in kg of BOD per hectare per day can be estimated using the following equation :

$$L_O = 10\left(\frac{d}{t}\right) BOD_L \qquad \text{... (2.2)}$$

where, L_O = Organic loading in kg/ha/day

d = Depth of pond in m

t = Detention time in days

BOD_L = Ultimate soluble BOD in mg/lit

The equation (2.1) is also useful to calculate oxygen requirement for the aerobic decomposition of the waste in kg/ha/day in aerobic ponds.

Design Steps (First Method) :

The following are the steps of design of aerobic ponds (with slight modification and ponds can be designed as facultative ponds by first method, see the following note) :

1. Calculate ultimate BOD (BOD_L).

2. Then calculate yield of photosynthetic oxygen related to latitude from Table 2.1.

3. From equation (2.1), calculate (d/t) ratio.

4. Assume suitable depth of pond (d).

5. By knowing value of d, calculate t.

6. Detention time can be calculated by using equation (2.1)

7. Calculate the required surface area, by using the following equation :

$$\therefore \text{ Required surface area} = \frac{\text{Flow} \times \text{Detention time}}{\text{Depth of pond}}$$

[**Note :** The above method may be applied with slight modification in the **Design of Facultative Ponds**, taking into consideration the BOD stabilization of solids in the bottom zone by anaerobic reaction. In this case, the oxygen requirement is corrected to take into account the non-settleable portion of BOD which undergoes the aerobic decomposition in the aerobic zone.] Normally, a correction factor of 0.5 is taken. In other words, in step 3 modify equation (2.1) as follows :

$$L_O = 10 \left(\frac{d}{t}\right) BOD_L \times 0.5 \qquad \text{... (2.3)}$$

Design of facultative pond by first method.

2. Second Method

- In this method, the principle of biological treatment kinetics are applied. For the removal of BOD in the pond, assuming a retarded first order reaction kinetics, the following equation (2.4) gives the relationship between the efficiency of the treatment in BOD removal, firstorder BOD removal rate constant, and the detention time in a complete mix system.

$$\frac{S_1}{S_0} = \frac{1}{1 + K\,(V/Q)} \qquad \text{... (2.4)}$$

where, S_0 = Influent BOD (substrate) concentration

S_1 = Effluent BOD concentration

Q = Waste flow rate

V = Volume of the reactor

K = Firstorder BOD removal rate constant

- But in actual practice the stabilisation ponds are never the complete mix type. The hydraulic regime within the tank assumes an intermediate condition in between plug flow and complete mix system i.e. partially mix system therefore, these may be considered as plug flow system with certain amount of axial dispersion of the materials.

- Wehner and Wilhelm have developed an equation for partially system correlating the efficiency, axial dispersion, detention time, first order BOD removal rate constant etc.

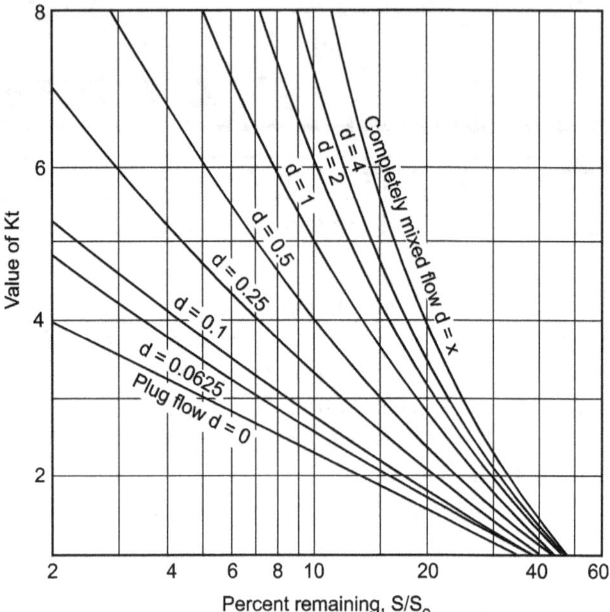

Fig. 2.1

- Thirumurthi developed design curves related to that equation as shown in Fig. 2.1, which gives the product of firstorder BOD removal rate constant (K) and detention time (t) corresponding to percent BOD remaining (S_1/S_0) for different values of dispersion factors.

- The values of dispersion factors and BOD removal rate constant are given in the following Table 2.2 and Table 2.3 respectively.

Table 2.2 : Values of Dispersion Factors

Sr. No.	Description	Value of Dispersion Factor
1.	For idealized complete mix system	∞
2.	For idealized plug flow	zero
3.	For most stabilization ponds	0.1 to 1.0
4.	For aerobic ponds	1.0
5.	For facultative ponds	0.3 to 1.0 (typical value 0.5)

Table 2.3 : Values of BOD Removal Rate Constant

Sr. No.	Description	Value of BOD Removal Rate Constant
1.	In stabilisation ponds (temperature 20to 35°C)	0.05 to 1.0 per day
2.	For primary ponds (temperature 20°C)	0.22 per day
3.	For secondary ponds (temperature 20°C)	0.1 per day
4.	For tertiary ponds (temperature 20°C)	0.06 per day

[**Note :** Generally, most ponds are bound to receive some amount of settleable organics and at the bottom layer anaerobic action will develop due to thick layer of accumulated sludge.]

Design Steps (Second Method) :

The following are steps of design for facultative ponds :

1. Calculate % BOD_5 remaining (if not given) by assuming desired effluent BOD_5 and considering 50% of the BOD is stabilised by anaerobic decomposition alone.

2. Calculate Kt from Fig. 2.1, by knowing value of dispersion factor and BOD_5 removal efficiency.

3. Calculate the value of K for different temperatures (for winter and summer or maximum and minimum as per requirement) from the following equation :

$$(K)_T{}^{\circ}{}_C = (K)_{20} \cdot (\theta_K)^{T-20}$$
$$= K_{20} \cdot (1.06)^{T-20}$$

4. Calculate value of t, by substituting value of K (which is calculated from step 3) in step 2.

5. Calculate the surface area from the following relationship for different temperatures and select higher value of surface area. (Generally, minimum temperature gives higher value of surface area, so the pond can be designed for minimum or winter temperature).

6. If surface aerators are provided, then the power and number of aerators are required to be calculated.

2.1.2 Design Parameters of Various Ponds

The following Table 2.4 shows typical design parameters for ponds.

Table 2.4 : Typical Design Parameters for Ponds

Sr. No.	Parameters	Aerobic pond	Anaerobic pond	Facultative pond
1.	BOD_5 loading, kg/ha/day	150 to 200 (for winter season)	500 to 1000 (winter) 1000 to 2000 (summer)	See table 2.1
2.	Depth, m	See Table 2.1	2.5 to 7	1 to 1.5
3.	Detention time, day	7	2 to 5	At least 6.5 days
4.	BOD removal efficiency	80 to 95% under normal conditions, decrease during winter months	45 to 65% during winter. 65 to 80% during summer	80 to 90%

2.1.3 Construction Details of Pond   (Dec. 15)

The following Table 2.5 shows construction details of pond.

Table 2.5 : Construction Details of Pond

Sr. No.	Item	Description
1.	Shape of Pond	• Round, square or rectangular ponds with length not exceeding three times the width are acceptable. • Avoid narrow or elongated portions. • Adopt maximum basin length of 750 m. • To minimise accumulations of floating matter and to avoid dead pockets, keep rounded corners.

2.	Embankment	• Prepared well compacted embankment of soil.
		 • Top width –1.5 to 3 m for large ponds
		 • Freeboard –minimum 0.5 m
		   – For larger ponds, 1.5 times the wave height.
		 • Outer slopes –2 : 1 to 2.5 : 1
		 • Inner slopes –1 : 1 to 1.5 : 1, if fully pitched
		   – 2 : 1 to 3 : 1, if face isunprotected.

3.	Inlets (a) When sewage is discharged by pump (b) When sewage is discharged under gravity	• Pipeline should extend into the pond atleast 15 to 20 m from the water edge. • When the sewage is to be pumped, the pipeline should be laid at bottom of the pond to discharge through an upward inclined 90° bend. • In case of gravity flows, the outfall sewer terminate at a manhole. • The invert of the manhole should be atleast 0.2 m above the MWL of the pond.
4.	Outlets	• To avoid short circuiting, outlet should be so located with reference to inlets. • Provide one outlet for every 0.5 ha pond area.

...Conti.

5.	Multiple Units of Ponds	• Multiple units should be provided when the required area of pond exceeds 0.5 ha. • These units can be either in series/in parallel/ in series-parallel system. • **Parallel System :** – Provide better distribution of settleable solids. – Also, from operation point of view, one unit at a time can be taken out of operation temporarily for desludging.
		• **Series System :** – Reduced algal concentration in the effluent. – Coliform removal efficiency is less. – Implies a high BOD loading in the primary cell, they should have 65 to 75% of the total surface area requirements to avoid anaerobic conditions in these cells.
		• **Parallel - Series System :** – In this system, getting advantages of both parallel and series operations. (See Fig. 2.2 for arrangement of pond for these three different systems).
6.	Pond Interconnections	• Pond interconnections are needful in case ponds are designed in multiple cells in series. (See Fig. 2.3 for arrangement of pond interconnections).

(a) Parallel system **(b) Series system**

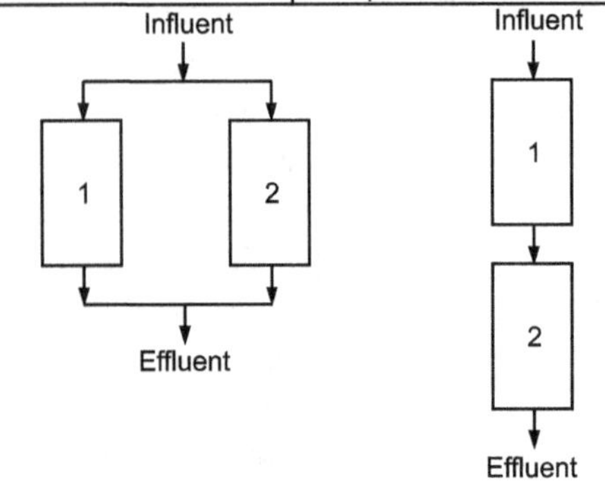

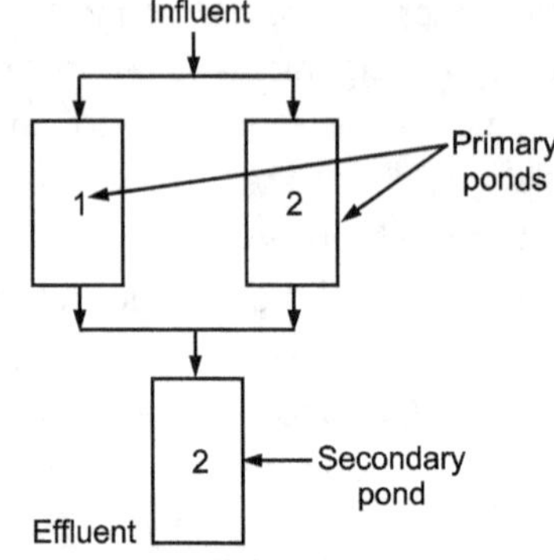

(c) Parallel-series system

Fig. 2.2 : Arrangement of ponds for different systems

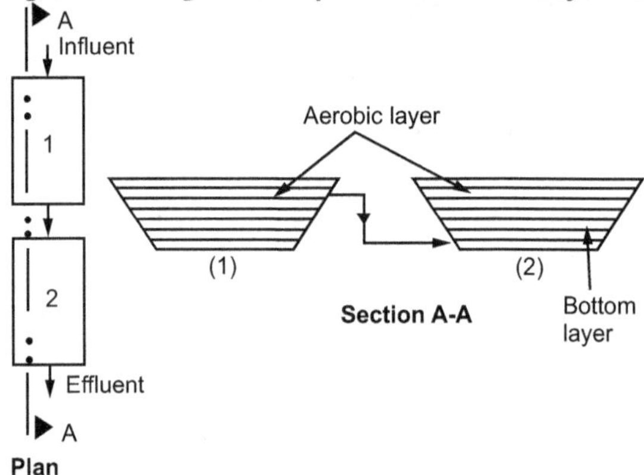

Fig. 2.3 : Arrangement of pond interconnections for series system

In section A-A it is seen that, the effluent from first cell withdrawn from the aerobic layer and introduced at the bottom of next cell. Following Fig. 2.4 shows a neat diagram of stabilisation pond.

(a) Plan

(b) Sectional elevation

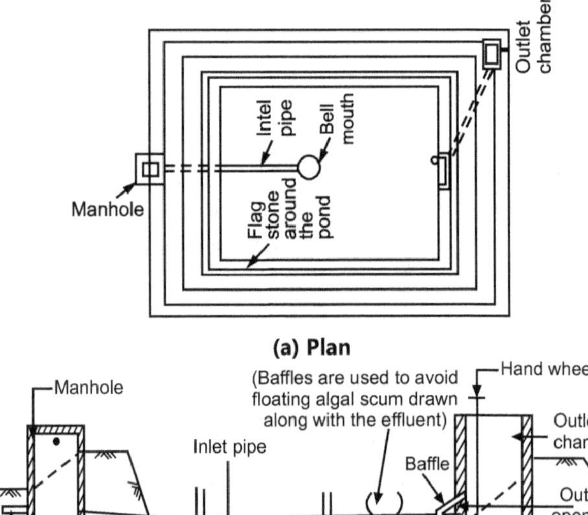

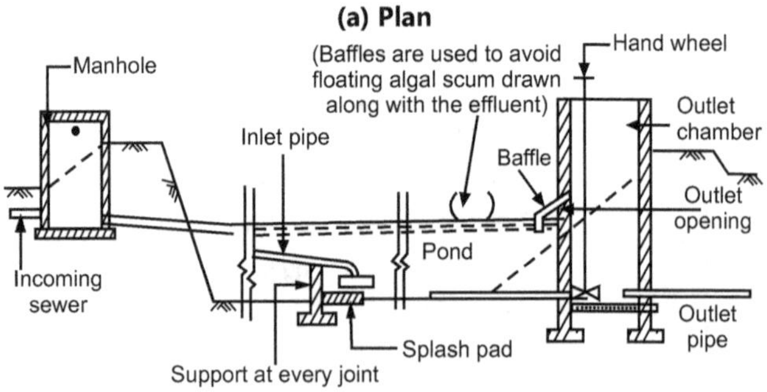

Fig. 2.4 : Stabilisation pond

2.1.4 General Consideration of Ponds

Table 2.6

Specifications	Assumptions
BOD loading to maintain aerobic layer in the top at all times (In India).	– 330 to 560 kg/ha/day
Accumulation of sludge and grit in the pond.	– 2 to 5 cm depth per year/minimum 0.3 m adopted for all the time. OR 0.06 to 0.09 m^3/capita/year (Typical value for design 0.07 m^3/capita/ year)
Cleaning of ponds.	– Once in 6 years for 1.2 m deep pond and once in 12 years for 1.5 m deep pond.
If the surface area of tank is too large.	– Provide two or more number of ponds.
If the soil is too pervious.	– The bottom and dikes should be sealed to prohibit seepage, a commonly used sealing agent is bentonite.

2.1.5 Advantages of Oxidation Ponds

The following are the advantages of oxidation ponds :

- They are suitable in hot countries like India.
- In small cities or towns, they are more advantageous one where large land areas are cheaply available.
- Capital cost is less as compared to conventional system (activated sludge process or trickling filters).
- Operating cost is less as no skilled supervision is required at any stage of operation or construction.

2.1.6 Disadvantages of Oxidation Ponds

The following are the disadvantages of oxidation ponds :

- Effluent standards of 30 mg/litfor suspended solids are not met.
- The main disadvantage is the mosquito breeding and bad odour.
- Only useful in rural areas where land costs are less.

2.2 AERATED LAGOONS　　　(May 16, Nov. 17)

2.2.1 General　　　(Nov. 16)

Aerated lagoons are one sort of deep oxidation ponds in which the oxygen is introduced by means of surface aerators. For aerated lagoons, required less area and detention time as compared to oxidation pond. In this system, wastewater is treated either on a flow through basis or with solids recycle.

2.2.2 Process Description　　　(Nov. 15)

Aerated lagoons are a simple holding earthen basins with a continuous supply of oxygen by surface aerators. This will keep the contents of the basin in suspension and they are initially seeded by the same type of micro-organisms which are used in biodegradation of activated sludge process.

2.2.3 Process Microbiology

Aerated lagoon's process is same as the activated sludge process, the microbiology is also similar. Some differences occur due to the large surface area of the aerated lagoons can cause more significant temperature effects than are normally happening in the conventional activated sludge process.

2.2.4 Design Criteria of Aerated Lagoons(Nov. 15,16)

Aerated lagoon may be designed as a complete mix biological reactor without cell recycle using micro-organism growth kinetics.

BOD Removal :

Generally, in the aerated lagoons, it is assumed that BOD removal can be described by the first order removal function. From a steady state mass balance of the organic materials across the complete mix single lagoon. Based on that analysis, the equation for a single aerated lagoon is

$$\frac{S_1}{S_0} = \frac{1}{1 + K X_1 (V/Q)} \qquad \text{... (2.5)}$$

where,　S_1 = Effluent soluble BOD_5, mg/lit

$\quad\quad S_0$ = Influent soluble BOD_5, mg/lit

$\quad\quad X_1$ = Micro-organism mass concentration in the reactor, mg/lit

$\quad\quad K$ = Specific soluble BOD_5 removal rate coefficient, l /mg/day

$\quad\quad V$ = Volume of the reactor, litre

$\quad\quad Q$ = Rate of flow of waste, l/day

The following Fig. 2.5 shows general flow diagram of aerated lagoon.

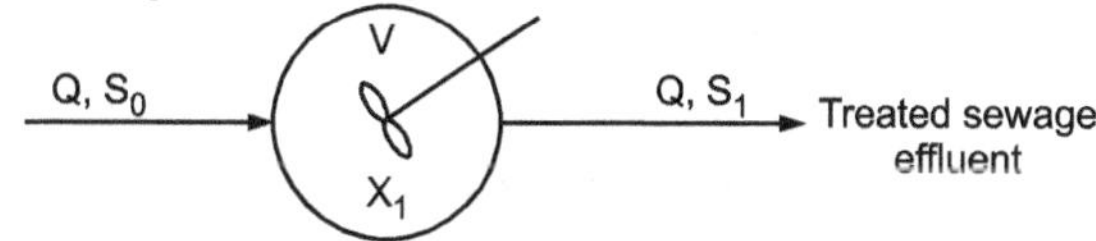

Fig. 2.5 : Flow diagram for aerated lagoon

The value of X_1 in equation (2.5) can be calculated by the following relationship which is developed from the mass balance for the mass of micro-organisms across the system in a steady state :

$$X_1 = \frac{Y(S_0 - S_1)}{1 + K_d(V/Q)} \qquad ...(2.6)$$

where,　Y = Yield coefficient

K_d = Specific decay coefficient, day^{-1}

For design purpose, the following values of K, Y and K_d can be used, where temperature rarely falls below 15°C.

$K = 0.05$ l/mg/day at 20°C, $Y = 0.5$ and $K_d = 0.05$ per day.

The values of K at other temperatures can be calculated as :

$$K_T = K_{20}\,\theta^{T-20} \qquad ...(2.7)$$

where,　θ = Temperature coefficient

= 1.060

= 1.097 (for cold climates)

But at first approximation, X_1 can be calculated as follows :

$$X_1 = Y(S_0 - S_1) \qquad ...(2.8)$$

Temperature : The temperature of the aerated lagoon can be calculated by the following relationship (for complete mix system) :

$$T_w = \frac{Q\,T_i + Af\,T_a}{Q + Af} \qquad ...(2.9)$$

where,　T_w = Lagoon water temperature, °C

T_i = Temperature of influent, °C

T_a = Ambient air temperature, °C

f = Proportionality factor, 0.5

A = Surface area of lagoon, m^2

Q = Flow rate through the lagoon, m^3/day

Oxygen Requirement : Oxygen requirement may be calculated by using the following equation :

Weight of oxygen required/day

$$= Q(S_0 - S_1)(1.47 - 1.42\,Y) + 1.42\,K_d\,X_1\,V \quad ...(2.10)$$

Field transfer rate N can be calculated by the following relationship :

$$N = N_o\left[\frac{C_s - C_o}{C_w}(1.024)^{T-20}\cdot \alpha\right] \qquad ...(2.11)$$

where,　N = kg of O_2/W/hr transfer under field conditions

N_o = kg of O_2/W/hr transferred in water at 20°C and zero DO

C_s = Saturation oxygen concentration for waste at operating temperature and altitude (see Fig. 2.6).

C_o = Operating oxygen concentration

C_w = Saturation oxygen concentration for water at 20°C

= 9.17 mg/lit

T = Temperature of lagoon in °C

and　α = Salinity - surface tension correction factor, generally 0.8 to 0.85.

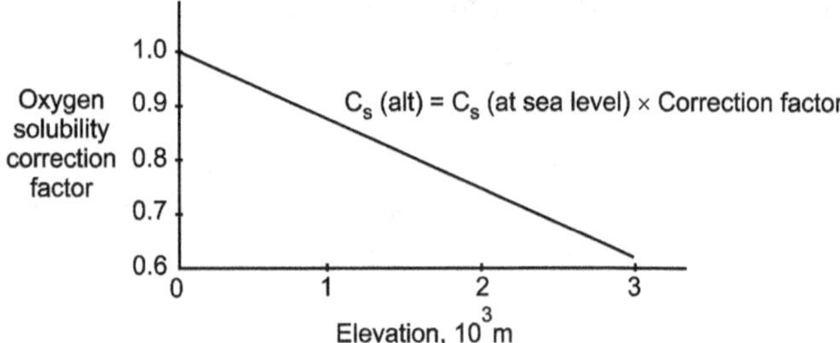

Fig. 2.6 : Oxygen-solubility correction factor versus elevation

2.2.5 Design Steps for Design of Aerated Lagoons

(May 11)

1. Calculate approximate X_1 by using equation (2.8).

2. Calculate the approximate hydraulic retention time $\left(\dfrac{V}{Q}\right)$, from equation (2.4).

$$\frac{V}{Q} = \frac{S_0 - S_1}{S_1\,K\,X_1} \qquad ...(2.12)$$

3. Calculate required volume.

4. Calculate surface area.

5. Calculate lagoon water temperature from equation (2.9).

6. Calculate X_1 from equation (2.6).

Again use equation (2.12) and calculate $\left(\dfrac{V}{Q}\right)$.

7. Calculate required volume related to step 7.

8. Now keeping the surface area same as calculated in step 4; calculate depth of aerated lagoon.

9. Calculate oxygen requirement.

10. Calculate surface aerators power requirement.

11. Calculate the field transfer rate, N.

2.2.6 Design Parameters of Aerated Lagoons

The following Table 2.7 shows design parameters of aerated lagoons.

Table 2.7 : Design Parameters of Aerated Lagoons

Sr. No.	Parameters	Values
1.	Flow regime	Completely mixed.
2.	Detention time, days	2 – 3
3.	Depth, m	2.5 – 4
4.	BOD removal	50 – 60%

(Ref. Manual on sewerage and sewage treatment 2nd edition).

2.2.7 Advantages of Aerated Lagoons (May 10,11,16)

The following are the advantages of lagoons :

- Sludge production is low.
- Low sensitivity to toxicity, pH disturbance and load variations.
- Odour problem is less.

2.2.8 Disadvantages of Aerated Lagoons

(May 10, 11, 16)

The following are the disadvantages of aerated lagoons :

- Energy is required to run aerators.
- It requires large area.
- It is sensitive to cold climate.

SOLVED EXAMPLES ON STABILISATION POND (OXIDATION POND)

Example 2.1 : *Design a facultative stabilisation pond to treat a domestic sewage flow of 3 MLD, at a place, the latitude of which is 24°N. The 5 day 20°C BOD of the sewage is 250 mg/lit. Assume necessary data.*

Solution :

Given : Domestic sewage flow

$$= 3 \text{ mld} = 3 \times 10^6 \text{ litres/day}$$
$$= 3 \times 10^3 \text{ m}^3/\text{day}$$

Latitude $= 24°N$

$\therefore$ Yield of photosynthetic oxygen

$$= 225 \text{ kg/ha/day}$$

BOD_5 of the sewage

$$= 250 \text{ mg/lit}$$

Assume, 1st stage BOD removal constant,

$$K = 0.23 \text{ per day}$$

The facultative pond can be designed by both methods (first and second).

(A) First Method :

1. Determine the Ultimate BOD (L_a) :

As we know,

$$BOD_5 = L_a [1 - e^{-K \cdot t}]$$

$\therefore$ Ultimate BOD,

$$L_a = \frac{250}{[1 - e^{-0.23 \times 5}]} = 365 \text{ mg/lit}$$

2. Determine the Detention Period :

Now, from equation (2.2),

$$L_O = 10 \left(\frac{d}{t}\right) BOD_L$$
$$= 10 \times \frac{d}{t} \times 365 \text{ kg/ha/day}$$

Assuming that (i) 50% of this load is non-settleable and, (ii) it undergoes aerobic decomposition in the top layer,

Oxygen requirement

$$= 10 \left(\frac{d}{t}\right) \times 365 \times 0.5 \text{ kg/ha/day}$$

$\therefore \quad 225 = 10 \times \left(\frac{d}{t}\right) \times 365 \times 0.5$

$\therefore \quad \dfrac{d}{t} = 0.123$

Now providing the depth of pond, $d = 1.5$ m

$\therefore \quad \dfrac{1.5}{t} = 0.123$

$\therefore \quad t = \dfrac{1.5}{0.123}$

$$t = 12.20 \text{ days}$$

3. Determine the Surface Area :

Now,

$$\text{Surface area} = \frac{\text{Flow} \times \text{Detention time (t)}}{\text{Depth}}$$
$$= \frac{3 \times 10^3 \times 12.20}{1.5}$$
$$= 24.40 \times 10^3 \text{ m}^2$$

Providing two ponds in parallel, area of each

$$= 12.20 \times 10^3 \text{ m}^2.$$

Assuming length (L)/width (B) ratio $= 2$,

$\therefore \quad 2B^2 = 12.20 \times 10^3$

$\therefore \quad B = 78 \text{ m}$

$\therefore$ The size of each pond $= 78 \text{ m} \times 156 \text{ m}$

and overall depth $= (1.5 + 1) = 2.5$ m

$\therefore$ Size of pond $= 7.8 \text{ m} \times 156 \text{ m} \times 2.5$ **...Ans.**

(B) Second Method :

1. Determine the Detention Time :

Assuming desired BOD_5 effluent = 30 mg/lit and 50% of the BOD is stabilized by anaerobic decomposition alone,

$$\% \ BOD_5 \text{ remaining} = \frac{S}{S_O} = \frac{30}{250/2} = 24\%$$

Assuming a dispersion factor in the facultative pond $= 0.5$

$\therefore$ From Fig. 2.1,

$$Kt = 2.1$$

Now, assuming winter temperature in the pond = 15°C, and the value of K in the pond $= 0.22$ per day at 20°C,

$\therefore \quad K_{15°C} = 0.22 \ (1.06)^{15-20}$

$$= 0.1645 \text{ per day}$$

$$\therefore \quad t = \frac{2.1}{0.1645}$$

$$= 12.77 \text{ days}$$

2. Determine the Surface Area :

$$\text{Now, Surface area} = \frac{\text{Flow} \times \text{Detention time}}{\text{Depth}}$$

$$= \frac{3 \times 10^3 \times 12.77}{1.5}$$

$$= 25.53 \times 10^3 \, m^2$$

Providing two ponds, area of each

$$= 12.76 \times 10^3 \, m^2.$$

$\therefore$ Size of each pond = 80 m $\times$ 160 m

and the overall depth = (1.5 + 1) = 2.5 m.

$\therefore$ Size of pond = 80 m $\times$ 160 m $\times$ 2.5 m   **...Ans.**

Example 2.2 : *Design a facultative pond to treat a wastewater flow of 4000 m³/day. As the ponds are to be installed near a residential area, surface aerators will be used to maintain oxygen in the upper layers of the pond. Use the following data :*

Influent suspended solids = 220 mg/lit

Influent BOD$_5$ = 200 mg/lit

Summer liquid temperature = 37ºC

Winter liquid temperature = 15ºC

Overall first order BOD$_5$ removal rate constant = 0.25 d^{-1} at 20ºC

Temperature coefficient, θ_K = 1.06

Depth of pond, d = 1.8 m

Pond dispersion factor = 0.5

Overall BOD$_5$ removal efficiency = 80%

Solution :

1. From Fig. 2.1, Determine the Value of K$_t$ for a dispersion factor of 0.5 and a BOD$_5$ removal efficiency of 80 percent i.e. % BOD$_5$ remaining is 20%.

$$\therefore \qquad Kt = 2.4$$

2. Determine the Temperature Coefficient for Summer and Winter Conditions :

(a) Winter :

$$K_{15} = (0.25) \, [(1.06)^{15-20}] = 0.187 \, d^{-1}$$

(b) Summer :

$$K_{37} = (0.25) \, [(1.06)^{37-20}] = 0.673 \, d^{-1}$$

3. Determine the Detention Time for Winter and Summer Conditions :

(a) Winter :

As, $Kt = 2.4$

$$\therefore \quad 0.187 \times t = 2.4$$

$$\therefore \qquad t = 12.8 \text{ days}$$

(b) Summer :

$$0.673 \times t = 2.4$$

$$t = 3.57 \text{ days}$$

4. Determine the Surface Area Requirement :

(a) Winter :

$$\text{As,} \quad \text{Surface area} = \frac{\text{Flow} \times \text{Detention time}}{\text{Depth}}$$

$$\therefore \quad \text{Surface area} = \frac{4000 \times 12.8}{1.8}$$

$$= 28.444 \times 10^3 \, m^2 \qquad \text{... (i)}$$

(b) Summer :

$$\text{Surface area} = \frac{4000 \times 3.57}{1.8}$$

$$= 7.93 \times 10^3 \, m^2 \qquad \text{... (ii)}$$

Selecting the larger of (i) and (ii),

$\therefore$ The surface area = 28.444 $\times$ 10³ m² (2.8 ha).

(It means winter conditions control the design).

Providing two ponds,

    Area of each pond = 14.22 $\times$ 10³ m²

$\therefore$ Size of each pond = 84 m $\times$ 168 m and overall depth

$$= (1.8 + 1) = 2.8 \text{ m.}$$

$\therefore$ Size of pond = 84 m $\times$ 164 m $\times$ 2.8 m

5. Determine the Surface Loading :

kg BOD$_5$/ha·d

$$= \frac{\text{Flow} \times \text{Influent BOD}_5}{\text{Surface area}}$$

$$= \frac{4000 \times 200}{2.8} \times \frac{1}{10^3}$$

$$= 285.7 \text{ kg/ha/day}$$

6. Determine the Power Requirements and Number of Surface Aerators :

Assume that the capacity of the aerators in terms of oxygen transferred is equal to double the value of the BOD$_5$ applied per day and that a typical aerator will transfer about 24 kg O$_2$/kW · day.

$\therefore$ kg O$_2$/d required = 2 $\times$ Flow $\times$ Influent BOD$_5$

$$= \frac{2 \times 4000 \times 200}{10^3} = 1600 \text{ kg/d}$$

$$kW = \frac{1600 \text{ kg/d}}{24 \text{ kg/kW·d}}$$

$$= 66.67 \text{ kW} \cong 70 \text{ kW}$$

Use seven 10 kW units.

7. Check the Power Input (if Mixing will Occur) :

$$kW/10^3\,m^3 \;=\; \frac{70\;kW}{28.44 \times 10^3\,m^3}$$

$$= 2.46\;kW/10^3\,m^3 \qquad \textbf{...Ans.}$$

This is insufficient power for surface aerator to mix the pond contents, as minimum requirement is 3 kW/10^3 m^3.

Example 2.3 : *Design an oxidation pond for treating sewage from a hot climatic residential colony with 6000 persons. The sewage generation is about 130 litres per capita per day. The BOD$_5$ of sewage is 350 mg/lit.*

Solution :

Given :Total number of persons = 6000

The sewage generation rate = 130 l/capita/d

BOD$_5$ of sewage = 350 mg/lit

BOD loading or organic loading or yield of photosynthetic oxygen in hot climates (assume) = 300 kg/ha/d

1. Determine the Quantity of sewage :

The quantity of sewage to be treated

$$= 6000 \times 130$$

$$= 0.78 \times 10^6 \text{ litres/day}$$

$$= 780\ m^3/day$$

2. Determine the Ultimate BOD (L_a) :

Assuming 1st stage BOD removal rate constant,

$$K = 0.23 \text{ per day.}$$

As we know,

$$BOD_5 \;=\; L_a\,[1 - e^{-Kt}]$$

∴ Ultimate BOD,

$$L_a \;=\; \frac{350}{[1 - e^{-\,0.23 \times 5}]}$$

$$= 512.17 \text{ mg/lit}$$

3. Determine the Detention Period :

Now, from equation (2.2),

$$L_O \;=\; 10\left(\frac{d}{t}\right) \times BOD_L$$

$$= 10 \times \frac{d}{t} \times 512.17 \text{ kg/ha/day}$$

Assuming that (i) 50% of this load is non-settleable and, (ii) it undergoes aerobic decomposition in the top layer.

Oxygen requirement

$$= 10\left(\frac{d}{t}\right) \times 512.17 \times 0.5$$

∴ $$300 \;=\; 10 \times \frac{d}{t} \times 512.17 \times 0.5$$

∴ $$\frac{d}{t} = 0.117$$

Now, providing the depth of pond, d = 1.5 m

∴ $$\frac{1.5}{t} = 0.117$$

∴ $$t = 12.82 \text{ days}$$

4. Determine the Surface Area :

Now,

$$\text{Surface area} \;=\; \frac{\text{Flow} \times \text{Detention time (t)}}{\text{Depth}}$$

$$= \frac{780 \times 12.82}{1.5}$$

$$= 6.6 \times 10^3\ m^2$$

Assuming length (L)/width (B) ratio = 2

∴ $$2B^2 = 6.6 \times 10^3$$

∴ $$B = 57.73\ m \text{ say } 58\ m$$

∴ $$L = 116\ m$$

∴ The size of pond is 58 m × 116 m and overall depth = (1.5 + 1) = 2.5 m.

∴ Size of pond = 58 m × 116 m × 2.5 m   **...Ans.**

5. Design of Inlet Pipe :

Assuming an average velocity of sewage as 0.9 m/sec. and daily flow for 8 hours only,

$$\text{Discharge (cumecs)} \;=\; \frac{\text{Sewage flow (m}^3\text{/day)}}{8 \times 60 \times 60}$$

$$= \frac{780}{8 \times 60 \times 60}$$

$$= 0.027 \text{ cumecs}$$

∴ Area of inlet pipe required (m^2)

$$= \frac{\text{Discharge (cumecs)}}{\text{Velocity (m/sec)}}$$

$$= \frac{0.027}{0.9}$$

$$= 0.03\ m^2$$

∴ Diameter of inlet pipe

$$= \sqrt{\frac{4 \times 0.03}{\pi}}$$

$$= 0.195\ m$$

$$= 195\ mm \text{ say } 200\ mm \qquad \textbf{...Ans.}$$

6. Design of Outlet Pipe :

Assume the diameter of outlet pipe is 1.5 times the diameter of inlet pipe,

∴ Diameter of outlet pipe

$$= 1.5 \times 200$$

$$= 300\ mm$$

Example 2.4 : *Design an oxidation pond to treat 10 m^3/d of sewage discharge.*

Assume : BOD of raw sewage = 210 mg/lit

Design temperature = 10°C

BOD loading = 140 kg/ha/d

Sketch the details.

Solution :

1. Determine Ultimate BOD (L_a) :

Assume, 1st stage BOD removal constant,

$$K = 0.23 \text{ per day}$$

As we know,

$$BOD_5 = L_a\, [1 - e^{-K \cdot t}]$$

$$\therefore \quad L_a = \frac{210}{[1 - e^{-0.23 \times 5}]}$$

$$= 307.3 \text{ mg/lit}$$

2. Determine the Detention Period :

Now, from equation (2.2),

$$L_O = 10\left(\frac{d}{t}\right) \times BOD_L = 10 \times \frac{d}{t} \times (307.3)$$

Assuming that (i) 50% of this load is non-settleable and, (ii) it undergoes aerobic decomposition in the top layer.

Oxygen requirement

$$= 10\left(\frac{d}{t}\right) \times 307.3 \times 0.5 \text{ kg/ha/d}$$

$$\therefore \quad 140 = 10\left(\frac{d}{t}\right) \times 307.3 \times 0.5$$

$$\frac{d}{t} = 0.09$$

Now, providing the depth of pond, d = 1.5 m

$$\therefore \quad \frac{1.5}{t} = 0.09$$

$$\therefore \quad t = 16.67 \text{ days}$$

3. Determine the Surface Area :

Now,

$$\text{Surface area} = \frac{\text{Flow} \times \text{Detention time (t)}}{\text{Depth}}$$

$$= \frac{10 \times 16.67}{1.5}$$

$$= 111.11 \text{ m}^2$$

Assuming length/width ratio = 2

$$\therefore \quad 2B^2 = 111.11$$

$$\therefore \quad B = 7.45 \text{ m} \quad \text{say } 7.5 \text{ m}$$

$$\therefore \quad L = 15 \text{ m}$$

$\therefore$ The size of the pond is 7.5 m × 15 m and the overall depth = (1.5 + 1) = 2.5 m.

$\therefore$ Size of pond = 7.5 m × 15 m × 2.5 m   **...Ans.**

4. Sketch of Pond :

Refer Fig. 2.4 and show dimensions on the same.

Example 2.5 : *Design an oxidation pond for the following data :*

Sewage flow=10 m^3/d

BOD of raw sewage = 300 mg/lit

Mean monthly temperature =30°C maximum and 10°C minimum

Desired effluent BOD = 30 mg/lit

Location =20°C latitude

Yield of photosynthetic O_2= 250 kg/ha/d.

Solution :

1. Determine the Detention Time :

Design the pond for minimum temperature i.e. 10°C.

$$t = \frac{1}{K_1} \log_{10}\left(\frac{L_a}{L_a - Y}\right)$$

$$(K_1)_{10°C} = 0.1\,(1.047)^{10-20} = 0.06317 \text{ per day}$$

$$\therefore \quad t = \frac{1}{0.06317} \log_{10}\left(\frac{300}{30}\right)$$

$$(L_a - Y = 30 \text{ mg/lit i.e. BO.D. remaining})$$

$$\therefore \quad t = 15.83 \text{ days} \approx 16 \text{ days}$$

2. Determine the Surface Area :

(Consider depth of pond is 1.5 m)

Now,

$$\text{Surface area} = \frac{\text{Flow} \times \text{Detention time (t)}}{\text{Depth}}$$

$$= \frac{10 \times 16}{1.5 \text{ m}}$$

$$= 106.67 \text{ m}^2$$

Assuming length/width ratio = 2

$$\therefore \quad 2B^2 = 106.67$$

$$\therefore \quad B = 7.3 \text{ m} \quad \text{say } 7.5 \text{ m}$$

$$\therefore \quad L = 15 \text{ m}$$

$\therefore$ The size of pond is 7.5 m × 15 m and

overall depth = (1.5 + 1) = 2.5 m.

$\therefore$ Size of pond = 7.5 m × 15 m × 2.5 m

[**Note :** Depth of pond can be determined by proper calculations (see example 2.6).]

Example 2.6 : *Design an oxidation pond for 8000 population and a sewage flow of 150 lit/head/day with a BOD of 300 mg/lit. The BOD of effluent should not be more than 30 mg/lit. Assume temperature as 24°C. Assume sludge accumulation rate as 0.05 m³/capita/year and desludging interval as 5 years. Assume*
$K_{20} = 0.20/day$, *given* $K_T = K_{20}(1.047)^{T-20}$.

Solution :

Given : Population = 8000

Sewage flow   = 150 lit/head/day

$\quad\quad\quad\quad$ = 150 × 8000 lit/day

$\quad\quad\quad\quad$ = 1.2×10^3 m³/day

BOD of sewage = 300 mg/lit

Temperature   = 24°C

BOD of effluent $\not> $ 30 mg/lit

Sludge accumulation rate = 0.05 m³/capita/year

Desludging interval = 5 years

Assume BOD loading = 300 kg/ha/day

1. Determine Total BOD Load :

BOD per capita/day = (150 × 300) × 10^{-6}

$\quad\quad\quad$ = 0.045 kg/day

∴   Total BOD load   = 8000 × 0.045

$\quad\quad\quad$ = 360 kg/day

2. Determine Pond Area :

$$\text{Pond area} = \frac{\text{Total applied BOD}_5}{\text{BOD loading}}$$

$$= \frac{360 \text{ kg/d}}{300 \text{ kg/ha/d}}$$

$\quad\quad\quad$ = 1.2 ha

$\quad\quad\quad$ = 1.2×10^4 m²

3. Determine Detention Period :

By using equation (2.1),

$$t = \frac{1}{K_1} \log_{10}\left(\frac{L_a}{L_a - Y}\right)$$

Now, $(K_1)_{24°C} = K_{20}(1.047)^{T-20}$

$\quad\quad\quad$ = $0.2 (1.047)^{24-20}$

$\quad\quad\quad$ = 0.24 per day

∴ $\quad\quad t = \dfrac{1}{0.24} \log_{10}\left(\dfrac{300}{30}\right)$

$(L_a - Y$ = 30 mg/lit i.e. BOD remaining)

$\quad\quad\quad$ t = 4.2 days

4. Determine Pond Liquid Volume and Pond Liquid Depth :

Pond liquid volume = Flow × Detention time

$\quad\quad\quad$ = $1.2 \times 10^3 \times 4.2$ = 5040 m³

Liquid depth = $\dfrac{5040 \text{ m}^3}{1.2 \times 10^4 \text{ m}^2}$

$\quad\quad\quad$ = 0.42 m

Also, provide free board of 0.6 m.

Total pond area is 1.2 ha. Adopt a **Parallel-Series** system of 6 ponds with 4 primary ponds and 2 secondary ponds of equal area, with 2 primary ponds feeding a secondary pond in each set as shown in the following Fig. 2.7. This would give the primary pond area as 66.67% which is within the required range of **65 to 75%** of the total pond area.

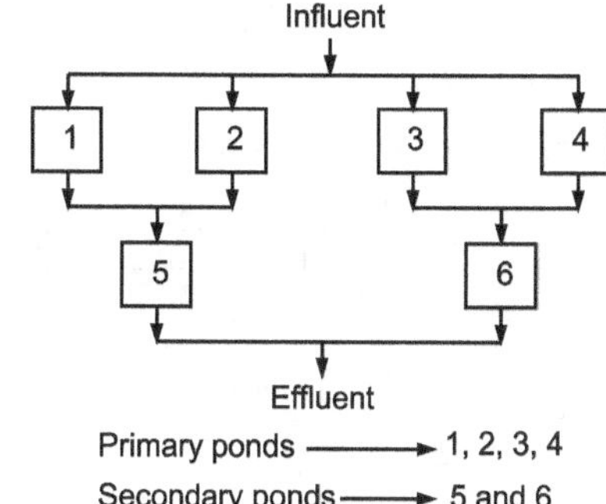

Fig. 2.7

$$\text{Area of each pond} = \frac{1.2 \times 10^4}{6} = 2000 \text{ m}^2$$

Provide rectangular ponds with length to breadth ratio as 2.0.

∴ $\quad\quad 2B^2$ = 2000

∴ $\quad\quad B$ = 31.62 m ≈ 32 m

∴ $\quad\quad L$ = 32 × 2 = 64 m

∴ Actual area of each pond

$\quad\quad\quad$ = 32 × 64 = 2048 m²

5. Determine Volume of Sludge and Depth of Sludge

Volume of sludge = 0.05 m³/capita/year

Desludging interval = 5 years

∴   Adopt minimum 0.3 m depth for sludge accumulation.

Example 2.7 : *Design an oxidation pond for the following data :*

Location = 20°N

Elevation = 1000 m above MSL

Mean monthly temperature = 35°C maximum & 10°C minimum

Population to be served = 9000

Sewage flow = 160 l pcd

Desired effluent BOD₅ = 30 mg/lit

Pond removal constant at 20°C = 0.1/d

Sky is clear for 15% of the days.

BOD of sewage = 300 mg/lit.

Solution :

1. **Determine Total BOD Load :**

   BOD per capita/day = $(160 \times 300) \times 10^{-6}$

   = 0.048 kg/d

   $\therefore$ Total BOD load = 9000×0.048 = 432 kg/d

2. **Determine Permissible Area/BOD Loading :**

   Area/BOD loading at 20°N = 250 kg/ha/day.

   (Refer Table 2.2)

   Correction factor for elevation

   $$= 1 + 0.003 \times \frac{1000}{100}$$

   $$= 1.03$$

   Correction factor for sky clearance

   $$= \frac{100}{100 + 3 \times \dfrac{15}{10}} = \frac{100}{104.5} = 0.96$$

   $\therefore$ Corrected area/BOD loading

   $$= \frac{250}{1.03} \times 0.96$$

   $$= 232.27 \text{ kg/ha/d}$$

   $$\approx 232 \text{ kg/ha/d}$$

3. **Determine Pond Area :**

   $$\text{Area of pond} = \frac{\text{Total applied BOD}_5}{\text{Area/BOD loading}}$$

   $$= \frac{432 \text{ kg/day}}{232 \text{ kg/ha/d}}$$

   $$= 1.86 \text{ ha}$$

   $$= 1.86 \times 10^4 \text{ m}^2$$

4. **Determine the Detention Period :**

   Design for minimum temperature 10°C.

   By using equation (2.1),

   $$t = \frac{1}{K_1} \log_{10}\left(\frac{L_a}{L_a - Y}\right)$$

   $$(K_1)_{10°C} = 0.1 \, (1.047)^{10-20}$$

   $$= 0.06317 \text{ per day}$$

   $\therefore$
   $$t = \frac{1}{0.06317} \log_{10}\left(\frac{300}{30}\right)$$

   ($L_a - Y$ = 30 mg/lit i.e. BOD remaining)

   $$t = 15.83 \text{ days} \approx 16 \text{ days}$$

5. **Determine the Volume and Depth of Pond :**

   Pond volume = Flow $\times$ Detention time

   $$= (160 \times 9000) \times 16$$

   $$= 23040 \times 10^3 \text{ litres/day}$$

   $$= 23040 \text{ m}^3/\text{day}$$

$$\text{Depth of pond} = \frac{23040}{1.86 \times 10^4}$$

$$= 1.23 \text{ m} \approx 1.25 \text{ m}$$

Provide a depth of 1.25 m, also provide a F.B. of 0.6 m.

Provide a parallel-series system of 6 ponds, as same is provided in example 2.6.

$$\text{Area of each pond} = \frac{1.86 \times 10^4}{6} = 3100 \text{ m}^2$$

Assume rectangular ponds with length to breadth ratio as 2.

$\therefore$ $\quad 2B^2 = 3100$

$\therefore$ $\quad B = 39.34$ m say 39.5 m

$\quad$ Length = 2×39.5

$\quad\quad\quad\quad = 79$ m

Actual area of each pond

$$= 39.5 \times 79$$

$$= 3120.5 \text{ m}^2 \text{ say } 3120 \text{ m}^2. \quad \textbf{... Ans.}$$

Example 2.8 : *Design an oxidation pond for treating sewage from a residentialcolony having population of 10,000 with sewage flow rate of 120 lpcd with the following data.* **(Nov. 15, 8 M)**

> *BOD_5 of raw sewage = 300 mg/l*
>
> *Desired effluent BOD_5 = 30 mg/l*
>
> *Location – 28° N*
>
> *Elevation – 200 m above sea level*
>
> *Temperature – 25°C*
>
> *Sky clearance factor – 60%*
>
> *BOD removal rate constant for the pond at 20° C as 0.1/d*
>
> *Assume permissible organic loading at 28°N as 200 kg/ha.d.*

Solution :

1. **Determine Total BOD Load :**

   BOD per capita/day = $(120 \times 300) \times 10^{-6}$

   = 0.036 kg/day

   Total BOD load = 10000×0.036 = 360 kg/day

2. **Determine Pond Area :**

   $$\text{Pond area} = \frac{\text{Total applied BODs}}{\text{BOD loading}}$$

   $$= \frac{360}{200} = 1.8 \text{ Ha} = 1.8 \times 10^4 \text{ m}^2$$

3. **Determine Detention Period :**

   $$t = \frac{1}{k_1} \log_{10}\left(\frac{L_a}{L_1 - Y}\right)$$

Now,

$$(k_1)\ 24°C = k_{20}\ (1.047)^{T-20}$$

$$= 0.1\ (1.047)^{24-20} = 0.120 \text{ per day}$$

$$\therefore \quad t = \frac{1}{0.120}\ \log_{10}\left(\frac{300}{30}\right)$$

$$= 8.34 \text{ days} \approx 8 \text{ days}$$

4. Determine the Volume and Depth of Pond

Pond volume = Flow × Detention time

$$= (120 \times 10000) \times 8$$

$$= 9600 \times 10^3 \text{ lit/day}$$

$$= 9600 \text{ m}^3/\text{day}$$

$$\text{Depth of pond} = \frac{9600}{1.8 \times 10^4} = 0.54 \text{ m}$$

Provide a depth of 0.54 m, also provide freeboard of 0.6 m, provide parallel series system of 6 ponds, same is provide in example (2.6)

$$\text{Area of each pond} = \frac{1.8 \times 10^4}{6} = 3000 \text{ m}^2$$

Assume rectangular ponds with length to breadth ratio as 2.

$$2B^2 = 3000$$

$$B = 38.72 \text{ m say 39 m}$$

$$\text{Length} = 2 \times 39 = 78 \text{ m}$$

Actual area of each pond = 39 × 78 = 3042 m^2 **...Ans.**

Example 2.9 : *Design an oxidation pond for a colony of 3000 population. The sewage flow is 100 l/c/d. BOD$_5$ is 300 mg/l. Assume necessary required data.* **(Nov. 16, 8 M)**

Solution :

Assume BOD loading or yield of photosynthetic O_2 in hot climates = 300 kg/ha/day.

1. Determine the Quantity of Sewage :

$$= 3000 \times 100 = 0.3 \times 10^6 \text{ lit/day}$$

$$= 300 \text{ m}^3/\text{day}$$

2. Determine the Ultimate BOD (L_a) :

Assuming 1st stage BOD removal rate constant, k = 0.23 per day,

We know, $\text{BOD}_5 = L_a\ [1 - e^{-kt}]$

$$300 = L_a\ [1 - e^{-0.23 \times 5}]$$

$$L_a = 439.01 \text{ mg/lit}$$

3. Determine the Detention Period :

$$L_o = 10\left(\frac{d}{t}\right) \times \text{BOD}_L$$

$$= 10\left(\frac{d}{t}\right) \times 439.01 \text{ kg/ha/day}$$

Assuming that (i) 50% of this load is non-settleable and (ii) It undergoes aerobic decomposition in the top layer.

$$O_2 \text{ requirement} = 10\left(\frac{d}{t}\right) \times 439.01 \times 0.5$$

$$300 = 10 \times \frac{d}{t} \times 439.01 \times 0.5$$

$$\frac{d}{t} = 0.137$$

Now, providing the depth of pond, d = 1.5 m

$$\frac{1.5}{t} = 0.137$$

$$t = 10.97 \text{ days}$$

4. Determine the Surface Area :

$$\text{Surface area} = \frac{\text{Flow} \times \text{detention time (t)}}{\text{Depth}}$$

$$= \frac{300 \times 10.97}{1.5} = 2194 \text{ m}^2$$

Assume, $\frac{L}{B}$ ratio = 2 $\therefore$ L = 2B

$$2B^2 = 2194 \therefore B = 33.12 \text{ m , Say 33 m}$$

$$L = 2 \times 33 = 66 \text{m}$$

$\therefore$ The size of pond is 33 m × 66 m and

Overall depth = 1.5 + 1 = 2.5 m

$\therefore$ Size of pond = 33 m × 66 m × 2.5 m **...Ans.**

5. Design of Inlet Pipe :

Assuming velocity of sewage as 0.9 m/s and daily flow for 8 hrs only.

$$\text{Discharge (m}^3\text{/s)} = \frac{300}{8 \times 60 \times 60} = 0.0104 \text{ m}^3/\text{sec}$$

$\therefore$ Area of inlet pipe required (m^2)

$$= \frac{0.0104}{0.9} = 0.012 \text{ m}^2$$

$\therefore$ Diameter of inlet pipe $= \sqrt{\dfrac{4 \times 0.012}{\pi}}$

$$= 0.123 \text{ m} = 123.60 \text{ m, say 125 mm}$$

6. Design of Outlet Pipe :

Assume the diameter of outlet pipe is 1.5 times the diameter of inlet pipe.

Diameter of outlet pipe = 1.5 × 125 = 187.5 mm $\approx$ 200 mm

Example 2.10 : *Design an oxidation pond for following data.*

(i) Location : 24° Latitude

(ii) BOD loading at 24° Latitude : 225 kg/ha/d

(iii) Elevation : 900 m above sea level

(iv) Mean monthly temperature :30° maximum and 15° minimum

(v) Population to be served : 10000

(vi) Sewage flow : 100 pcd

(vii) Desired effluent BOD5 : 20 mg/L

(viii)Pond removal constant at 20 °C : 0.1/d

(May 17, 8M)

Solution :

$$K_{30} = K_{20} (1.047)^{30-20}$$
$$= 0.1(1.047)^{10} = 0.158$$
$$Y_t = L (1 - 10^{-k.t})$$
$$L - 20 = L (1 - 10^{-0.158 \times 5})$$
$$L - 20 = 0.84 L$$
$$\therefore \quad L = 125 \text{ mg/l}$$

1. Surface Area :

$$\text{Sewage to be treated/day} = 1000 \times 100$$
$$= 1 \times 10^6 \text{ lit}$$
$$= 1000 \text{ m}^3$$
$$\text{BOD content /day} = 1 \times 10^6 \times 125$$
$$= 125 \text{ kg/day}$$
$$\therefore \quad \text{Surface area} = \frac{\text{BOD content/day}}{\text{organic loading}}$$
$$= \frac{125 \text{ kg/day}}{225 \text{ kg/ ha/day}} = 0.56 \text{ ha}$$
$$= 5600 \text{ m}^2$$

2. Dimensions of Tank :

Assume length of tank 'L'

$$= \text{Twice of breadth 'b'}$$
$$L = 2 B$$
$$L \times B = \text{Surface area}$$
$$2 B \times B = 5600$$
$$B^2 = 2800$$
$$\text{Say } B = 52.91$$
$$\therefore \quad B = 53 \text{m}$$
$$\therefore \quad L = 2 \times 53 = 106 \text{ m}$$

Assume the effective depth, 'h' = 1.2 m

$$\therefore \quad \text{capacity of tank} = L \times B \times h$$
$$= 106 \times 53 \times 1.2$$
$$= 6739.8 \text{ m}^3$$

3. Detention Time (D.T.) :

$$= \frac{\text{Capacity of tank}}{\text{Sewage flow per day}}$$
$$= \frac{96739.8}{1000} = 6.74$$
$$\approx 7 \text{ days} \qquad \text{...Ans.}$$

4. Diameter of the Pipe

Let the average velocity of sewage = 0.09 m/sec

Daliy flow = 8 hrs

$$\text{Area of pipe} = \frac{\text{Discharge}}{\text{Velocity}}$$
$$Q = \frac{1000}{60 \times 60 \times 8} = 0.035 \text{ m}^3/\text{sec}$$
$$= \frac{0.035}{09} = 0.0386 \text{ m}^2$$
$$\therefore \quad \frac{\pi}{4} d^2 = 386$$
$$D = 22.16/\text{cm}$$
$$D = 22.17 \text{ cm} \approx 24 \text{ cm}$$

Dia. of outtlet pipe = 1.5 × d
$$= 1.5 \times 24$$
$$= 36 \text{ cm}$$

Therefore, provide tank of length = 106 m

$$\text{Width} = 53 \text{ m}$$
$$\text{Depth} = 1.2 \text{ m}$$

Detention time = 7 days

Inlet pipe diameter = 24 cm   **...Ans.**

Outlet pipe diameter = 35 cm   **...Ans.**

SOLVED EXAMPLES ON AREATED LAGOONS

Example 2.11 : *Design a complete mix aerated lagoon system to treat a domestic sewage flow of 2.5 MLD. Use the following data :*

Influent suspended solids = 250 mg/lit

Effluent suspended solids after settling ≯ 20 mg/lit

Influent BOD$_5$ = 220 mg/lit

Effluent BOD ≯ 20 mg/lit

Volatile suspended solids = 80% of total solids produced.

Summer temperature = 38°C

Winter temperature = 15°C

Wastewater temperature = 22°C

Oxygen concentration to be maintained in the lagoon = 1.5 mg/lit

Lagoon depth = 3.0 m

Specific substrate removal coefficient, K = 0.05 l/mg/day at 20°C; Growth yield coefficient, Y = 0.5; Decay coefficient, K$_d$ = 0.05 per day,; f = 0.5.

Oxygen transfer coefficient of the aerator, α = 0.8.

Temperature coefficient, θ = 1.065

Elevation = 1000 m.

Solution :

  Given : Flow $= 2.5\ mld\ = 2.5 \times 10^3\ m^3/day$,

$$S_O = 220\ mg/lit$$

Now the total effluent $BOD_5 = 20$ mg/lit and effluent SS after settling $= 20$ mg/lit.

$$X = 0.80 \times 20 = 16\ mg/lit.$$

                  (As VSS $= 80\%$ of total SS)

1. Determine X_1 :

  $\therefore$ $S_1 = $ Soluble BOD_5 in the effluent

                $= (20 - 0.54 \times 16)$

                  (Assuming ESS BOD_5 @ 0.54 kg/kg)

                $= 11.36\ mg/lit$

As first approximation X_1 can be calculated from Eqn. 2.8,

  $\therefore$ $X_1 = Y\ (S_0 - S_1) = 0.5\ (220 - 11.36)$

                $= 104.32\ mg/lit$

2. Determine Retention Time :

Now,

$$\boxed{\dfrac{V}{Q} = \dfrac{S_0 - S_1}{S_1\ K\ X_1}}$$

$$= \frac{220 - 11.36}{11.36 \times 0.05 \times 104.32} = 3.5\ days$$

3. Determine Surface Area :

     Volume $=$ Flow $\times$ Retention time

            $= 2.5 \times 10^3 \times 3.5 = 8.75 \times 10^3\ m^3$

  Surface area $= \dfrac{8.75 \times 10^3}{2.5}$

               ... (Assuming a depth $= 2.5$ m)

            $= 3.5 \times 10^3\ m^2$

4. Determine Waste Water Temperature of the Water in the Aerated Lagoon :

Summer and winter temperature of the water in the aerated lagoon can be calculated by equation (2.9).

  (a) Summer temperature,

$$T_W = \frac{Q\ T_i + AFT_a}{Af + Q}$$

$$= \frac{2.5 \times 10^3 \times 22 + 3.5 \times 10^3 \times 0.5 \times 38°C}{3.5 \times 10^3 \times 0.5 + 2.5 \times 10^3}$$

  $\therefore$ $T_W = 28.58°C$

  (b) Winter temperature,

$$T_W = \frac{Q\ T_i + AfT_a}{Af + Q}$$

$$= \frac{2.5 \times 10^3 \times 22 + 3.5 \times 10^3 \times 0.5 \times 15°C}{3.5 \times 10^3 \times 0.5 + 2.5 \times 10^3}$$

            $= 19.11°C$

[**Note :** As related to winter temperature, the value of surface area becomes larger than related to summer temperature, therefore, always design the lagoon for winter temperature.]

     Value of K at temperature 19.11°C

            $= 0.05 \times (1.065)^{19.11 - 20}$

            $= 0.047\ l/mg/day$

5. Determine Value of X_1 :

$$X_1 = \frac{Y\ (S_0 - S)}{1 + K_d\ (V/Q)} = \frac{0.5\ (220 - 11.36)}{1 + 0.05 \times 3.5}$$

            $= 88.78\ mg/lit$

6. Determine (V/Q) :

$$\frac{V}{Q} = \frac{S_0 - S_1}{S_1\ K\ X_1}$$

$$= \frac{220 - 11.36}{11.36 \times 0.047 \times 88.78} = \frac{208.64}{474}$$

            $= 4.4\ days \approx 4.5\ days$

Now, modifying our design, after using

$$\frac{V}{Q} = 4.5\ days$$

     Volume required

            $= 4.5 \times 2.5 \times 10^3$

            $= 11.25 \times 10^3\ m^3$

7. Determine the Depth of Lagoon :

Now keeping the surface area same as in the 1st trial i.e. $3.5 \times 10^3\ m^2$

  $\therefore$ The depth required

$$= \frac{11.25 \times 10^3}{3.5 \times 10^3}$$

            $= 3.2\ m$

  $\therefore$ Provide the depth of aerated lagoon $= 3.2$ m.

8. Determine the Oxygen Requirement :

By using equation (2.10),

     $O_2\ kg/day = Q\ (S_0 - S_1)\ (1.47 - 1.42\ Y)$

                     $+ 1.42\ K_d\ X_1 V$

            $= [2.5 \times 10^3\ (220 - 11.36)$

               $(1.42 - 1.42 \times 0.5)$

               $+ 1.42 \times 0.05 \times\ 88.78 \times 11.25 \times 10^3]$

               $\times 10^{-3}$

            $= 441.25\ kg/day$

9. Determine the Ratio of Oxygen Required to BOD_5 Removed :

$$\frac{O_2\ required}{BOD_5\ removed} = \frac{441.25\ kg/day}{[(220 - 11.36)] \times 2.5 \times 10^3 \times 10^{-3}}$$

            $= 0.85$

10. Determine the Surface Aerator Power Requirements :

Assume that the aerators to be used are conservatively rated at 2.0 kgO$_2$/kW·h.

Determine the field transfer rate for surface aerators for summer condition by using Oxygen saturation concentration at 28.58°C

$$= 7.84 \text{ mg/lit} \qquad \textbf{(Refer Appendix A)}$$

Oxygen saturation concentration at 28.58°C corrected for altitude

$$= 7.84 \times 0.9 = 7.056 \text{ mg/lit}$$

Now,
$$N = N_O \left[\frac{C_S - C_O}{C_W} (1.024)^{T-20} \cdot \alpha \right]$$

$$= 2.0 \text{ kg O}_2/\text{kW} \cdot$$

$$h \left[\frac{7.056 - 1.5}{9.17} \times (1.024)^{28.58-20} \times 0.8 \right]$$

$$= 1.19 \text{ kg O}_2/\text{kW·h}$$

The amount of O$_2$ transferred per day per unit is equal to 28.56 kg O$_2$/kW·d.

The total power required,

$$kW = \frac{441.25 \text{ kg O}_2/\text{d}}{28.56 \text{ kg O}_2/\text{kW·d}}$$

$$= 15.45 \text{ kW}$$

11. Check the Energy Requirements for Mixing :

Assume the power requirement for a completely mixed flow regime is 15 kW/10^3 m^3

$$\text{Volume of lagoon} = 11250 \text{ m}^3$$

Power required

$$= (15 \text{ kW}/10^3 \text{ m}^3) \times (11250 \text{ m}^3)$$

$$= 168.75 \text{ kW}$$

$$\cong 169 \text{ kW} \qquad \textbf{...Ans.}$$

Use six 30 kW surface aerators.

Example 2.12 : *Design an aerated lagoon to treat a wastewater (W/W) discharge of 1200 m³/d, with the following data :*

BOD of influent W/W = 200 mg/lit.

BOD of treated W/W effluent to be limited to 20 mg/lit.

Growth constants, Y = 0.6 and K_d = 0.06/day.

BOD removal rate constant = K_{20} = 2.5/day.

Mean temprature of W/W = 25°C.

Temperature coefficient = 1.06.

Aeration constants, α = 0.85 and β = 1.0.

Minimum D.O. to be maintained = 2 mg/lit.

SRT = 4 days.

Solution :

Please refer example 2.11 for the procedure.

Note : Here SRT = 4 days is given.

So 1st trial is not required, the example can be solved from step 3 onwards.

Example 2.13 : *A wastewater flow is 10000 m³/d, BOD$_5$ is 200 mg/L, design an aerobic flow through type lagoon to serve a town of 50000 persons, using a ideal complete mixing model. Since the lagoon is proposed to be followed by another treatment unit, its size can be restricted to give a detention time of only 3 days.*

Given data : K = 0.015 per day at 20°C, Y = 0.5, k_d = 0.07 per day.   ***(May 2017, 8M)***

Solution : 1. Lagoon Size :

$$\text{Volume of lagoon} = 10000 \times 3 = 30000 \text{ m}^3$$

Assume depth of lagoon = 4 m

$$\therefore \quad \text{Area of lagoon} = \frac{30000}{4} = 7500 \text{ m}^2$$

Provide square shaped lagoon.   **...Ans.**

$\therefore$ Dimensions are 86.6 m 86.6 m

2. Lagoon Temperature in Water

$$\frac{t}{h} = \frac{T_1 - T_2}{f(T_2 - T_a)}$$

where   f = 0.49 m/day for aerated/lagon

$$\frac{3}{4} = \frac{20 - T_w}{0.49 (T_w - 10)}$$

$$\therefore \qquad T_w = 17.3°$$

3. BOD Removal Rate in Winter

$$k (17.3°C) = k (20°C) \, \theta^{17.3 - 20}$$

$$= 0.015 \times 1.035^{(17.3 - 20)}$$

$$= 0.0137 \text{ day}$$

4. Efficient BOD in Winter

$$s = \frac{1 + kdt}{Yk't} = \frac{1 + (0.07 \times 3)}{0.5 \times 0.019 \times x_3}$$

$$= 58.88 \text{ mg/L} \qquad \textbf{...Ans.}$$

2.3 SEPTIC TANKS

- Septic tanks are horizontal continuous flow, small sedimentation tanks. The sewage is allowed to flow slowly so that the suspended sewage solids settle to the bottom of the tank. The settled solids are digested anaerobically and settled effluent is disposed off safely. The thick layer of scum is formed at the surface which helps to maintain anaerobic conditions into the tank. The anaerobic bacteria convert the organic sewage solids into gaseous, liquid and sludge. The digested sludge is removed from the tank at regular intervals usually once every 1–5 years.

- To maintain anaerobic conditions, the tank must be air-tight and water-tight. The septic tank is constructed into masonry or RCC. The shape of the tank is usually rectangular. The bottom of the tank should be 1.2 m above the ground water table. The septic tank should be provided in such a area where site condition is favourable for effluent disposal preferably at the lowermost contour. The effluent from septic tank is highly offensive with bad odour and dark in colour.
- In rural areas and the fringe areas of suburban towns and also in cases of isolated buildings and institutions such as hotels, hospitals, schools, small residential colonies, septic tanks are provided. Septic tanks are recommended for individual homes and small communities and institutions whose contributory population does not exceed 300.

2.4 DESIGN AND CONSTRUCTION FEATURES OF SEPTIC TANK   (Dec. 10, May 10, 11)

Since a septic tank is a settling cum digestion tank, its rational design is based on the following three functions:

1. Sedimentation to remove maximum possible amounts of suspended solids from sewage.
2. Digestion of settled sludge resulting in a much reduced volume of dense and digested sludge's.
3. Storage of sludge and scum accumulating in between successive cleanings, thereby preventing their escape.

Hence, the dimensions of tank should be such that the above mentioned requirements are fulfilled.

2.4.1 Sewage Flow

- The maximum sewage flow does not depend on number of users but it depends on the number of plumbing fixtures discharging simultaneously.
- Tables 2.8, 2.9, 2.10, 2.11 give the idea about the determination of sewage flow.

Table 2.8 : Equivalent Fixture Units

Sr. No.	Facility	Equivalent Fixture Unit
1.	Water closet	1
2.	Bath kitchen	1/2
3.	Wash basin/kitchen sink	1/2
4.	Urinal (autoflush)	1
5.	Urinal (ordinary)	1/2
6.	Slop tank	1
7.	Lab sink	2
8.	Combination fixture	1
9.	Shower bath	1
10.	Bath tub	2
11.	Drinking fountain	1/2
12.	Ablution tap	1/2
13.	Dish washer	1/2

Table 2.9 : Estimated Peak Discharges for Small Tanks upto 50 Users

No. of Users	No. of Fixtures	Probable No. of Fixture Units Discharging Simultaneously	Probable peak Discharge (*l*pm)
5	1	1	10
10	2	2	20
15	3	2	20
20	4	3	30
25	5	4	40
30	6	4	40
35	7	5	50
40	8	6	60
45	9	6	60
50	10	7	70

Table 2.10 : Estimated Peak Discharge for Residential Housing Colonies

No. of Users	No. of Households	No. of Fixture Units	ProbablePeak Discharge (Based on 60% Fixture Units Discharging Simultaneously) (*l*pm)
100	20	40	240
150	30	60	360
200	40	80	480
300	60	120	720

Table 2.11 : Estimated Peak Discharge for Eating Establishments, Boarding Schools etc.

No. of Users	W.C.	Baths	Wash Basin / kitchen Sinks	No. of Fixture Units	ProbablePeak Discharge (Based on 70% Fixture Units Discharging Simultaneously) (*l*pm)
50	6	6	6	12	84
100	12	12	12	24	168
150	19	19	19	38	266
200	25	25	25	50	350
300	37	37	37	74	518

2.4.2 Dimensions of Tank

The capacity of tank depends on the functions to be performed i.e. sedimentation, sludge digestion and sludge storage.

1. Sedimentation :

Design criteria to determine surface area is 0.92 m² for every 10 *l*pm peak flow rate. A minimum depth of 25–30 cm is necessary. The length is maintained 2–4 times the breadth.

2. Sludge Digestion :

The volume of fresh sludge can be considered as 0.00083 m³/head/day.

The volume of digested sludge = 0.0002 m³/head/day. In septic tank, the digestion zone contains both fresh sludge and space for digestion. Hence, the total volume required will be equal to 0.000515 m³/head/day. Assuming the digestion period of 63 days for average sludge, the capacity of sludge digestion is 63 × 0.000515 = 0.032 m³/head.

3. Space for Sludge and Scum :

Volume of digested sludge = 0.0002 m³/head/day. The capacity for 100 persons with the cleaning period of one year works out to be 0.0002 m³/head/day × 365 days × 100 persons = 7.3 m³.

The total capacity will be sum of the above three requirements.

2.4.3 Detention Period   (Dec. 10, May 10)

A detention period of 24 to 48 hrs is provided based on the design criteria discussed above.

2.4.4 Sludge Withdrawal

The period of sludge withdrawal may vary from 6 months to 3 years.

2.4.5 Construction Details

- Fig. 2.8 shows the details of a septic tank. The material of construction may be brick, stone or concrete. The watertight and airtight cover is provided to maintain anaerobic conditions. Ventilating pipe, extending at least 2 m above the top of the highest building with a radius of 20 mm is provided. Only one compartment for smaller capacities and two compartments for larger capacities are provided. The partition wall at a distance of about 2/3 the length from the inlet is provided.

- The inlet and outlet are located at the two opposite ends. The baffles are generally provided at both inlet and outlet and should dip 25 to 30 cm into and project 15 cm above the liquid level. The invert of outlet pipe should be placed at a level of 5 to 7 cm below the invert level of inlet pipe. The bottom of the tank will have slope towards the sludge outlet.

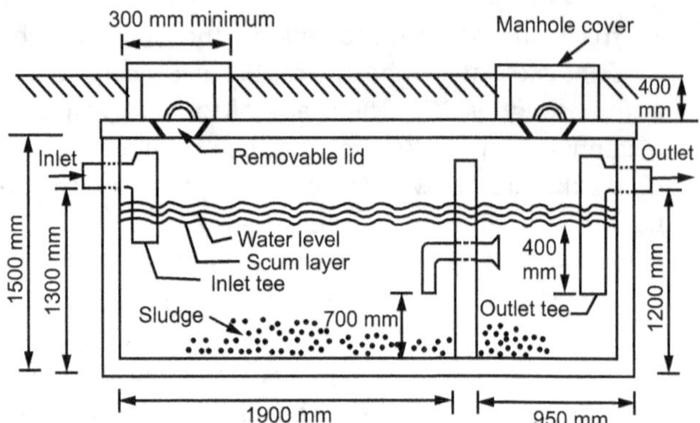

Fig. 2.8 : Septic tank

2.5 TREATMENT AND DISPOSAL OF SEPTIC TANK EFFLUENT (SOAKAGE SYSTEM)

The effluent from septic tank generally is septic and malodorous. It is necessary to dispose off this effluent safely because of presence of pathogens. The effluent can be disposed off by one of the following methods :

1. By subsurface irrigation.
2. By surface irrigation.
3. By discharging into nearby water courses.
4. By soil absorption system.

Soil Absorption System :

It has two types :

1. Seepage pit or soak pit.
2. Dispersion trenches.

2.5.1 Seepage Pit or Soak Pit

- These are circular pits more than 1 meter in diameter and 1 meter depth below the invert of the inlet pipe. These pits are lined with dry bricks or stone and are filled with brick bats or coarse aggregate more than 7.5 cm in size. In the case of large pits, the top portion is reduced in size for the reduction in the size of R.C.C. cover. The lining above the inlet level should be finished with mortar. The details are shown in Fig. 2.9.

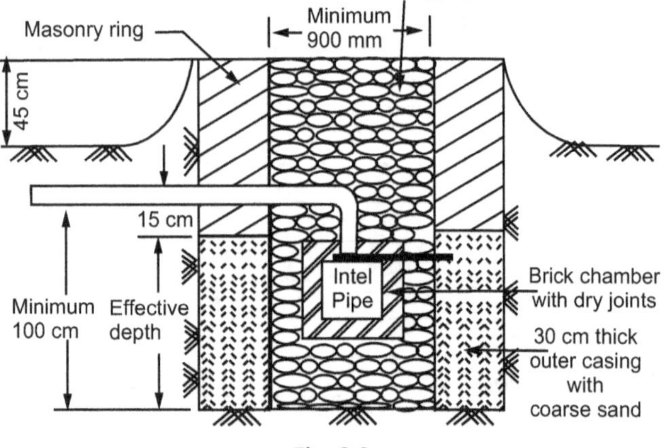

Fig. 2.9

- Soak pit or seepage pit can be used in all porous soils where percolation rate is below 25 min/cm and depth of water table is 180 cm or more from ground level.

- The total surface soil area required for the soak pits or dispersion trenches is given by the following empirical relation :

$$Q = \frac{130}{\sqrt{t}}$$

where, Q = Maximum rate of effluent application in lpd/m^2 of leaching surface

t = Standard percolation rate for the soil, in minutes per cm.

- While calculating the effective leaching area required, only area of trench bottom in case of dispersion trenches, and effective side wall area below the inlet level for soak pits should be taken into account.

2.5.2 Dispersion Trenches

- In this system, the septic tank effluent is uniformly distributed into a large area of subsoil through open jointed or perforated tile drains, each housed in a dispersion trench as shown in Fig. 2.10. Dispersion trenches consist of relatively narrow and shallow trenches about 0.5 to 1 m deep and 0.3 to 1 m wide excavated to a slight gradient of about 0.25%.

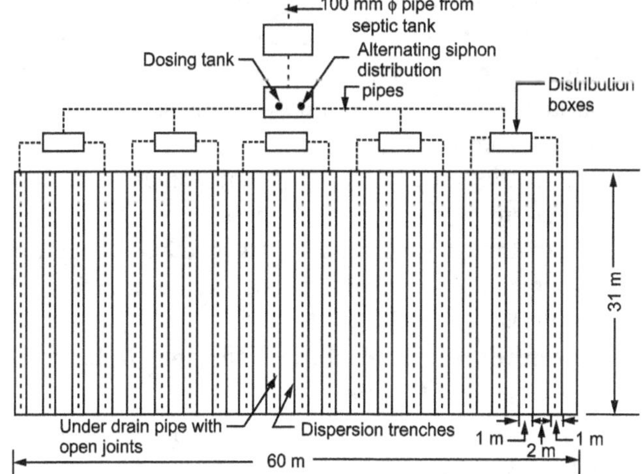

(a) Typical soil absorption system with dispersion trenches

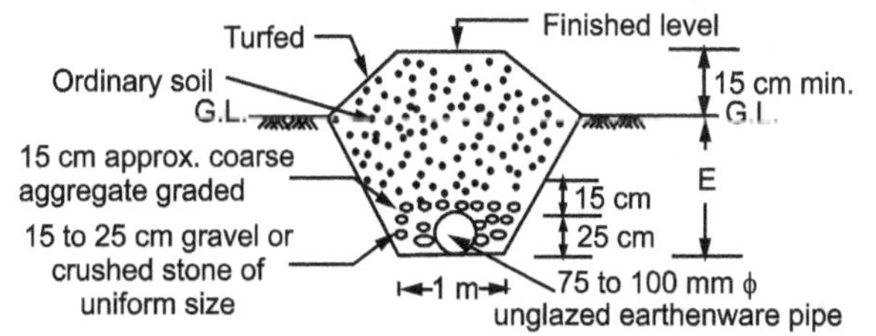

(b) Enlarged section through filled dispersion system

Fig. 2.10 : Dispersion trench system

- Open jointed earthen ware or concrete pipes of 70 mm to 100 mm dia. Are laid in the trenches over a bed of 15 to 25 cm of washed gravel or crushed stone. The maximum length of each trench is kept as 30 m and these are spaced not closer than 2 m apart. One distribution box is provided for a group of about 3 to 4 trenches.

2.6 ADVANTAGES AND DISADVANTAGES OF SEPTIC TANKS

Advantages :

- Easy to construct.
- No skilled supervision is necessary.
- The quantity of sludge produced is relatively low and with no odour.
- The cost of construction, operation and maintenance is affordable to private house holders.
- No moving parts are required for operation.
- The effluent can be disposed off easily.
- When once installed, it gives long carefree service.

Disadvantages :

- The size required is large and uneconomical.
- The functioning of tank is eratic.
- The effluent is dark and foul smelling with high BOD.
- Leakage of gases from top cause bad smell.
- Periodic cleaning, removal and disposal of sludge is often tedious.

SOLVED EXAMPLES ON DESIGN OF SEPTIC TANK

Example 2.14 : *Design a septic tank for a hostel housing 150 persons. Also design the soil absorption system for the disposal of the septic tank effluent, assuming the percolation rate as 15 minutes per cm.*

Solution :

(i) From table the estimated peak discharge for 150 persons is equal to 360 *lpm*.

Let us assume sludge withdrawal once in a year.

Surface area of tank at 0.92 m² for every 10 lpm

$$= \frac{0.92 \text{ m}^2}{10 \text{ lpm}} \times 360 \ lpm = 33.12 \text{ m}^2$$

Let us assume depth for sedimentation = 0.3 m.

Also provide a free board of 0.3 m. The total volume of tank will be as follows :

1. For sedimentation ,
   $33.12 \times 0.3 = 9.936$ m³

2. For digestion,
   $0.032 \times 150 = 4.8$ m³

3. For sludge storage

$$\frac{7.3}{100} \times 150 = 10.95 \text{ m}^3$$

4. Free board

$$33.12 \times 0.3 = 9.936 \text{ m}^3$$
$$\text{Total} = 35.622 \text{ m}^3$$

∴ Total depth of tank,

$$= \frac{35.622}{33.12} = 1.075 \text{ m}$$

Provide depth = 1.1 m

Let us keep L/B ratio as 2.5.

∴ [2.5 B] [B] = 33.12

∴ B = 3.64 m

and L = 9.1 m

Hence, tank dimensions are 9.1 m × 3.64 m × 1.1 m

(ii) For absorption system,

$$Q = \frac{130}{\sqrt{t}}$$

Here, t = 15 min/cm

∴ $$Q = \frac{130}{\sqrt{15}} = 33.56 \text{ } lpd/\text{m}^2$$

Let us assume sewage flow at 135 $l$pcd.

∴ Total flow per day = 150 × 135

$$= 20250 \text{ lpd}$$

∴ Total trench area required

$$= \frac{20250}{33.56} = 603.39 \text{ m}^2$$

Let us assume trench width as 1 m and separation between trenches as 2.0 m, total land area required

$$= 3.9 \text{ m} \times 603.39 = 1810.17 \text{ m}^2.$$

Let us provide length of each trench = 30 m.

Hence, width of land $= \dfrac{1810.17}{30} = 60.34$ m

Hence, provide $\dfrac{60.34}{3.0} = 20$ trenches   **...Ans.**

∴ Actual percolation area provided

$$= 20 \times 30 \times 1$$
$$= 600 \text{ m}^2 \quad \textbf{...Ans.}$$

2.7 EFFLUENT DISPOSAL AND STREAM SANITATION

- The sanitary engineer can design a treatment plant to accomplish as much removal of pollutants as may be required. Ultimate disposal of wastewater effluents will be by :
- Dilution i.e. disposal in larger bodies like lakes, rivers, estuaries or ocean; and

- Disposal on land.
- Disposal by dilution is most common method.

2.8 DISPOSAL BY DILUTION

- Disposal by dilution is the process whereby the treated wastewater or effluent from treatment plants is discharged either in large static water bodies like lake or ocean or in moving water bodies like streams or rivers.
- The discharged sewage, in due course of time, is purified by what is known as self purification process of natural waters. The limit of effluent discharge and the degree of treatment of wastewater depend not only upon the quality of raw sewage but also upon the self purification capacity of the river-stream as well as the intended use of the water body.

2.8.1 Conditions Favouring Disposal by Dilution (Without Treatment)

The dilution method without treatment for disposing of the sewage can favourably be adopted under the following conditions :

- Where the wastewater is quite fresh, i.e. it is discharged within 2 to 3 hours of its collection.
- Where the diluting water has a high dissolved oxygen (DO) content.
- Where receiving water body has large volume in comparison to the volume of untreated wastewater.
- Where receiving water is not used for the purpose of navigation or water supply immediately to the downstream side.
- Where the wastewater does not contain industrial wastewater having toxic substances.
- Where the flow currents of the diluting waters are favourable, causing no deposition, destruction of aquatic life.

2.8.2 Conditions Essential for Treatment before Disposal by Dilution

The treatment is essential before disposal of wastewater in the following cases :

- Where the wastewater contain industrial wastes having toxic substances.
- Where the industrial wastewater is in warm condition.
- Where receiving water body has less volume in comparison to the volume of the untreated wastewater. i.e. the volume of diluting water is insufficient.
- Where receiving water is used for drinking purpose.
- Where the receiving water is used for inland navigation.

2.8.3 Standards of Dilution for Discharge of Wastewaters into Rivers

$$\therefore \text{Dilution factor} = \frac{\text{The quantity of the diluting water or receiving water}}{\text{The quantity of the wastewater or effluent discharge}}$$

Related to dilution factor, the Royal Commission Report on sewage disposal has laid down certain standards which are indicated in Table 2.12.

Table 2.12 : Standards of Dilution

Sr. No.	Dilution Factor	Standards of Purification Required
1.	Above 500	No treatment is required. The raw sewage or wastewater can be discharged directly in the receiving water.
2.	Between 300 to 500	Primary treatment such as plain sedimentation is essential, and the effluents should not contain suspended solids more than 150 ppm.
3.	Between 150 to 300	Treatment such as sedimentation, screening and chemical precipitation are required. The effluents should not contain suspended solids more than 60 ppm.
4.	Less than 150	The sewage should be treated thoroughly. The effluent should not contain suspended solids more than 30 ppm and its 5 days BOD at 18.3°C should not exceed 20 ppm.

2.9 STREAM AND EFFLUENT STANDARDS

- Once the criteria necessary for the protection of the various beneficial uses have been established, it is possible to set standards for surface waters with stipulation that no discharge shall create conditions that violate them. These standards are known as receiving - water or stream standards.

- The standards shown in Table 2.12 have been operated in England since 1912, and has also been followed in India without much variance.

- As the increasing pollution of surface streams by discharging domestic and industrial wastewaters without bothering to look into the available dilution ratios, it has become essential to limit the concentrations of various pollutants being discharged into the surface water sources along with the sewage and industrial effluents.

- Therefore, the tolerance limits for such constituent pollutants have been prescribed by various countries, including India. These limits are dependent upon the treatment to domestic and industrial wastewater upto minimum level of "Secondary Treatment".

- The Bureau of Indian Standards (BIS), previously known as Indian Standards of Institution (ISI), has therefore laid down its guiding standards for sewage effluents, vide **IS 4764 – 1973** and for industrial effluents vide **IS 2490, 1974** (Refer Table 2.12).

- Column No. 4 of Table 2.12 shows tolerance limits for industrial effluents discharged into public sewers prescribed by **IS 3306 – 1974** and Column No. 5 of Table 2.12 shows tolerance limits for inland surface water, when used as raw water for public water supplies and bathing ghats prescribed by **IS 2296 – 1974.**

DISPOSAL METHODS

2.10 DISPOSAL OF WASTEWATER INTO SEA WATER

- Sea disposal is typically accomplished by submarine outfalls that consists of a long section of pipe to transport the wastewater some distance from shore and in the best examples, a diffuser section to dilute the waste with wastewater. Diffusers are one of the most efficient methods of providing initial dilution of a waste in any waterway.

- The saturation concentration of DO in water decreases with increasing salt content. Because of this, sea water normally contains 20% less oxygen than that contained in fresh water of a river stream.

Sleek :

- The specific gravity of sea water is larger than that of sewage and temperature of sea water is lesser than the sewage temperature. Due to this reason, when the sewage is discharged into the sea water, the lighter and warmer sewage will rise upto the surface, this will result in spreading of the sewage at the top, surface of sea in a thin film or 'sleek'.

Sludge Banks :

- Sea water contains a large amount of dissolved matter which chemically react with the sewage solids, when thrown into sea water, resulting in precipitating some of the sewage solids, giving a milky appearance to sea water and forming **Sludge Banks.**

- These sludge banks and the thin milky layer produce offensive hydrogen sulphide (H_2S) gas by reacting with the sulphate rich water of the sea. The reaeration in sea water is slower, also we know that, the oxygen content of sea water is less than that of fresh water and the sea contains too larger volumes of water, most of these deficiencies can be overcome if the sewage is discharged deep into the sea much away from the coast line. To avoid the backing up and spreading of sewage on the sea shore, the sewage should not be disposed during high tides.

- In such a case, to hold the sewage during high tides, the large sized tanks may therefore be constructed near the outfall.

The Following Points should be considered while Discharging Sewage into the Sea:

- The sewage should be discharged in deep sea (Generally 1 to 1.5 km away from the shore). As we have seen earlier, this is accomplished by outfall that consists of a long section of pipe to transport the wastewater some distance from the shore. Such outfalls are placed on a firm rocky foundation so it protects from wave action, floating debris, etc.

- Fig. 2.11 shows the cast iron pipe used for this purpose which is encased in thick special type of concrete. (For making such a special type of concrete, generally pozzolona cement is used).

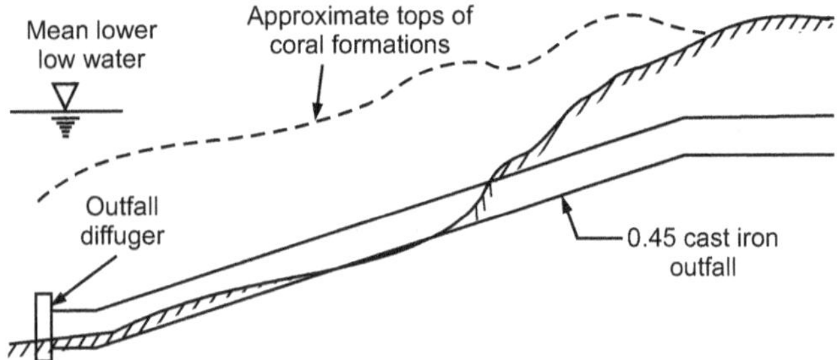

Fig. 2.11 : General layout of outfall

- The diffuser should be provided at the end of section for the proper dilution of waste with sea water.

- The sewage should be released at a minimum depth of 3 to 5 metres below the water level.

- At time of deciding the position of outfall point, the direction of wind, velocity, sea currents etc. should be carefully taken into consideration.

The following Table 2.13 shows standards for wastewater effluents to be discharged into marine coast. So while discharging industrial effluent into sea water the standards shown in the following table should be considered to control the effluent in respect of the quality of the effluents.

Table 2.13 : BIS (ISI) Standards for Wastewater Effluents to be Discharged into Marine Coasts (IS : 1968 – 1976)

Sr. No.	Constituent Pollutant Contained in the Wastewater Effluent	Tolerance Limit
1.	pH	5.5 to 9.0
2.	BOD_5	100 mg/lit
3.	COD	250 mg/lit
4.	Total suspended solids	100 mg/lit
5.	Oil and grease	20 mg/lit

Table 2.13 shows the comparison of sewage dilution by sea and stream.

Table 2.14 : Comparison between Sea and Stream Water Dilution

Sr. No.	Item	Sea Water	Stream Water
1.	DO	20% less than stream water	More DO
2.	Specific gravity	High	Low
3.	Quantity of solids in suspension	Large	Small
4.	Maximum sewage load	No limit	Depends on stream discharge.

2.11 DISPOSAL OF WASTEWATER INTO LAKES

- In many inland locations where nearby streams are not available, it may be necessary to discharge treated wastewater into lakes.

- Actually, disposal of wastewaters in lakes is much more harmful than its disposal in flowing streams and rivers. So a study of the lake system is essential. The study of lake pollution is called limnology.

- The major lake pollutant is phosphorus which is largely contained in industrial as well as domestic wastewaters. As phosphorus is a major pollutant, there needs a special study of phosphorous in water quality management of lakes.

Stratification in Large Lakes :

- For large lakes the complete mixing assumption cannot be applied. In such cases, a different model should be used based on the physical phenomena found to exist in the water system under study.

- Particularly significant is the vertical stratification common during certain seasons of the year. A complete mix model would not be a good representation of a stratified lake because waste would not normally distribute itself over the entire lake volume.

The water of a lake gets stratified during summers and winters in the following way :

- During summer season, due to sunlight and warm air, the surface water of lake gets heated. Such heated water being lighter, therefore remains in upper layers near the surface, until mixed downward by turbulence from waves, winds, boats etc.

- Since, such turbulence is only to a limited depth from below the water surface, so the top layers of water in the lake becomes well mixed and aerobic. This top layer is called epilimnion zone. In other words, epilimnion zone is fairly uniform in temperature because of mixing by wind action.

- The lower depth, which remains cooler, poorly mixed and anaerobic, is called hypolimnion zone. The intermediate zone or a dividing line is called thermocline. The thermocline is a zone of significant temperature change and is extremely resistant to mixing. Fig. 2.12 shows summer stratification in a lake.

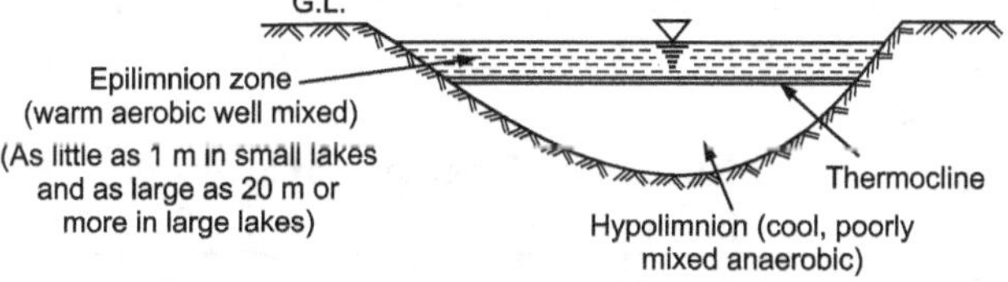

Fig. 2.12 : Summer stratification in a lake

- In colder season, the epilimnion cools, until it is more dense than the hypolimnion. The surface water, then sinks, causing 'over turning or turn over'. In other words, during the fall, temperatures drops, decreasing the amount of stratification until wind action may again completely mix the lake waters.

- This phenomenon is known as the fall turnover. In regions of freezing temperatures, when the temperature drops below 4°C, the above process of turnover stops, because water is most dense at this temperature. Fig. 2.13 shows winter stratification in a lake.

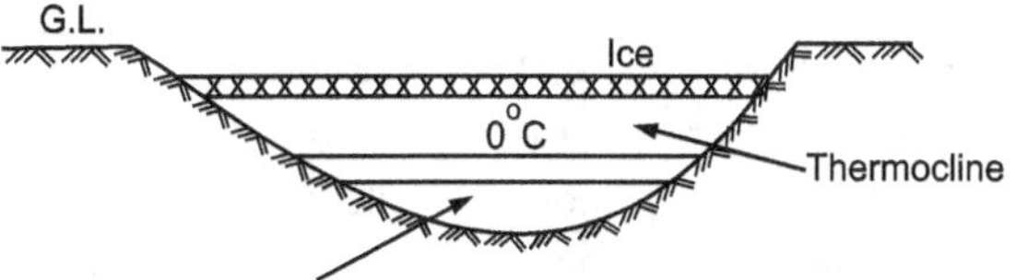

Fig. 2.13 : Winter stratification in a lake

2.12 DISPOSAL BY LAND TREATMENT

- In this method, the sewage effluent, either raw or partly treated is generally disposed on land, this method is called disposal by land treatment. The percolating water may either join the water-table or is collected below by a system of under drains. This method can then be used for irrigating crops. The sewage adds to the fertilising value of the land and crops can be profitably raised on such land. Due to this, the disposal by land treatment is also sometimes called as sewage farming.

- The three principal processes of land treatment are as follows :

  1. Broad irrigation/sewage farming;

  2. Rapid infiltration and

  3. Overland runoff.

- These three principal processes of land treatment are shown in Fig. 2.14.

1. **Effluent Irrigation (or Broad Irrigation) / Sewage Farming :**

(a) **Effluent Irrigation :** The chief consideration in 'effluent irrigation' is the successful disposal of sewage. The raw or settled sewage is discharged on a vacant land, which is provided underneath, with a system of properly laid under-drains. These under-drains usually consist of 15 to 20 cm diameter porous tile pipes laid open jointed at a spacing of 12 to 30 m. The effluent filtered through the soil pores which is collected in these drains generally of small quantity and well stabilised, and can be easily disposed into some natural water courses, without any further treatment.

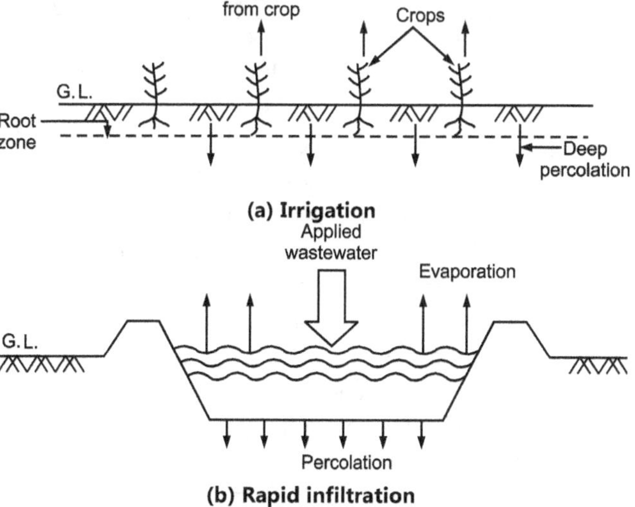

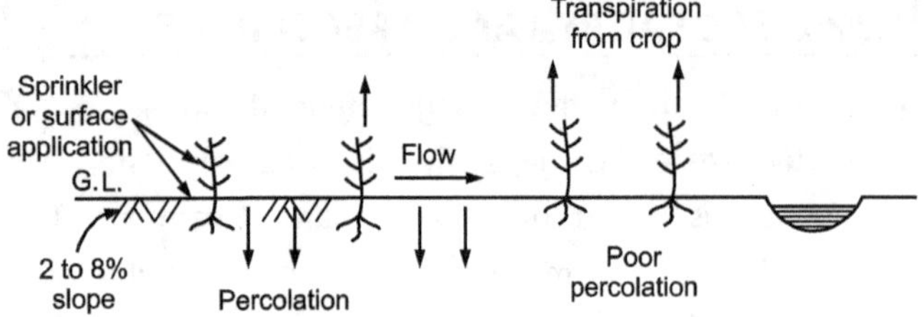

(c) Overland runoff

Fig. 2.14 : Three principal processes of land treatment

(b) Sewage Farming : The chief consideration in 'sewage farming' is the successful growing of the crops. In this case, the stress is laid upon the use of sewage effluents for irrigating crops and increasing the fertility of the soil. The pre-treatment in this case is necessary to remove ingradients which may prove harmful and toxic to the plants.

- So for all practical purposes, both these terms (effluent irrigation and sewage farming) are used as synonyms.

Recommended Doses for Sewage Farming : When raw or partly treated sewage is applied on to the land, a part of it evaporates, and the remaining portion percolates through the ground soil. While percolating through the soil, the suspended particles present in the sewage are caught in the soil voids. If proper aeration of these voids is maintained, the organic sewage solids get oxidised by aerobic process. Such aerobic condition will more likely prevail if the soil is sufficiently porous and if the soil is of sticky and fine grained materials, then the void spaces will soon get choked up and due to this developing anaerobic decomposition of organic matter and evolution of foul gases. So dosing of sewage depends on type of soil. Table 2.15 shows recommended doses for sewage farming.

Table 2.15 : Recommended Doses for Sewage Farming

Sr. No.	Type of Soil	Doses of Sewage in Cubic Metres per Hectare Per Day	
		Raw Sewage	Settled Sewage
1.	Sandy soil	120 – 150	220 – 250
2.	Sandy loam	90 – 100	150 – 200
3.	Loam	60 – 80	100 – 150
4.	Clayey loam	40 – 50	50 – 100
5.	Clayey	30 – 45	30 – 50

Irrigation process is discussed in detail under article 2.12.2.

2. **Rapid Infiltration :** The rapid infiltration process involves spreading wastewater in shallow, unlined earthen basins and allowing the liquid to pass through the porous bottom and percolate towards the ground water. Rapid infiltration may be used for waste disposal, ground water recharge or both. Many of the rapid infiltration systems in current use were designed primarily to dispose of unwanted wastewater. But now-

a-days, the process has been used as a means of aquifer recharge or as an advanced wastewater treatment, with the percolated wastewater being collected for reuse. Fig. 2.15 shows different techniques of rapid infiltration of wastewater. See rapid infiltration technique in Fig. 2.15 (a).

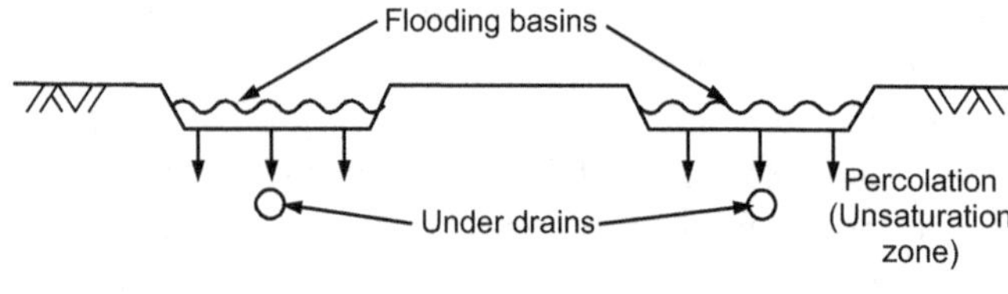

(a) Recovery by under-drain tile

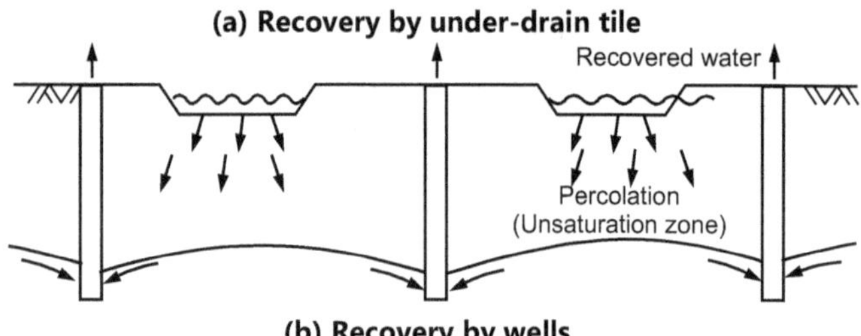

(b) Recovery by wells

Fig. 2.15 : Rapid infiltration of wastewater

3. **Overland Runoff :** This technique is applied when soils have poor permeability.

2.12.1 Quality Standards

IS : 3307 – 1965 laid down the standards of sewage effluents for their discharge on land for irrigation. Table 2.16 shows Indian Standards of sewage for sewage farming.

Table 2.16 : Indian Standards of Sewage Effluents for Sewage Farming

Sr. No.	Characteristics of Wastewater	Prescribed Limit
1.	pH	5.5 to 9.0
2.	BOD$_5$	500 mg/lit
3.	Total dissolved solids	2100 mg/lit
4.	Oil and grease	30 mg/lit
5.	Chlorides (as Cl)	600 mg/lit
6.	Sulphates	1000 mg/lit
7.	Percentage of sodium with respect to total content of sodium, calcium, magnesium and potassium	60%
8.	Radioactive materials (i) α-emitters (ii) β-emitters	$10^{-9}\ \mu C/ml$ $10^{-8}\ \mu C/ml$

Favourable Conditions for Land Treatment :

The effluent irrigation method for disposal of sewage can be adopted favourably under the following conditions :

- Land treatment is the only alternative, when natural rivers or streams are not located in the vicinity.

- Cash crops can be easily grown on sewage farms.
- The use of sewage for irrigating crops is good in case of irrigation where water is scarcely available.
- Land treatment is much favoured, when the land available for disposal is porous, such as sandy, loamy or alluvial soil, or soft moorum. Such soils are easily aerated and it is easy to maintain aerobic conditions in them. It should not be made of heavy retentive soils like clay, etc. which prevent easy aeration of the soil, voids, and thus creating anaerobic conditions.
- This method will prove useful in areas where rainfall is low. Due to this, it is easy to maintain good absorption capacity of the soil.
- Where large areas of open land are available, broad irrigation with the help of sewage effluent may be practised.
- This treatment is favoured when subsoil water table is low, where rate of percolation may be quite high.

2.12.2 Methods of Application of Sewage Effluent to Farms

Sewage effluent can be applied to land by the following methods :

1. Surface Irrigation (Broad Irrigation)
2. Sub-surface Irrigation.
3. Sprinkler or Spray Irrigation.

The sewage effluents can be used for irrigating farms similarly as irrigation water is used for farming. Fig. 2.16 shows different irrigation techniques.

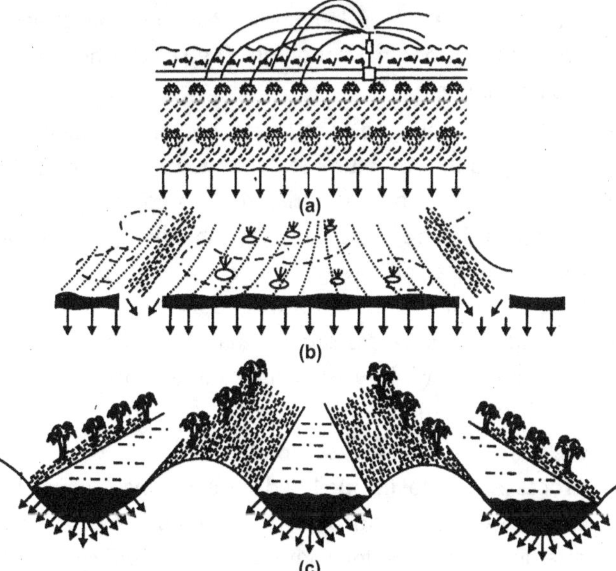

Fig. 2.16 : Irrigation techniques (a) Sprinkler, (b) Flooding, (c) Ridge and furrow

1. **Surface Irrigation or Broad Irrigation :** In this method, sewage is applied in different ways on the surface of the land. The following are the different types related to mode of application.

- Free flooding
- Border flooding
- Check flooding
- Furrow irrigation method and Basin flooding.

2. **Sub-Surface Irrigation :** In this method, sewage effluent is applied through open-jointed pipes to root zone of crops.

3. **Sprinkler or Spray Irrigation :** In this method, sewage is spread over the land through nozzles, which are fixed at the tips of the pipe carrying sewage under pressure. This method generally not used in India, as this process is costly one.

2.12.3 Sewage Sickness

Due to continuous application of sewage on land, the total pores of the soil get clogged with sewage matter retained in them, thus preventing oxidation and causing noxious smells. The time taken for such a clogging will depend on the type of soil (Porous sandy soils will clog less as compared to heavy clayey soils) and on the load of sewage. So due to continuous loading land is unable to take any further load of sewage. This phenomenon of soil is called as 'sewage sickness' of land.

To Avoid 'Sewage Sickness' of a Land, the Following Preventive Measures may be Adopted :

- **Choice of Land :** The land that may be chosen should be having higher permeability such as sandy or loamy, clayey land should be avoided.

- **Primary Treatment of Sewage :** To avoid clogging of soil pores, sewage should be disposed of, only after primary treatment (screening, grit removal and sedimentation). Due to this settlable solids are removed and also BOD load will be reduced by about 30%.

- **Under Drainage of Soil :** The sewage is being disposed of, can be better drained, if a system of under-drains is laid below to collect the effluent. Due to this, minimise the possibility of sewage sickness.

- **Provision of ExtraLand :** Extra land is useful as reserve or stand-by. So the land being used for disposal should be given rest, periodically and using reserve land for diverting the sewage during the period of the first land is at rest. During the rest period, the sick land should be properly ploughed so that it is broken up and aerated.

- **Rotations of Crops :** By planting different crops in rotation instead of growing single type of crop, minimise the chances of sewage sickness.

- **Applying Shallow Depths :** Sewage should be applied in thin layers. Greater depth of sewage on a land does not allow the soil to receive the sewage satisfactorily and therefore, chances of sewage sickness increases.

2.13 COMPARISON OF DILUTION AND LAND DISPOSALMETHOD FOR DISPOSAL OF SEWAGE

Table 2.17 : Comparison between Dilution Method and Land Disposal Method

Sr. No.	Item	Dilution Method	Land Disposal Method
1.	Cost	In cities and urban areas which are generally situated near rivers or ocean, this method is economical, as cost of land is very high.	In rural areas, this method is economical as land value is less.
2.	Management	This is a simple method and does not require too much of management.	Good management is required in case if cost of land is high. Due to good management, some return may be available due to sewage farming.
3.	Pumping	Requires nil or small head pumping.	Requires high head pumping.
4.	Hot climatic areas	Not suitable, as DO content due to hot climate is reduced.	This method is generally suitable in hot climatic areas.
5.	Pollution	Rivers are polluted due to sewage.	This method saves rivers from pollution.

STREAM SANITATION

2.14 SELF PURIFICATION OF NATURALSTREAMS

(May 10, 11, 18, Aug. 15, Nov. 16)

- When pollutants are discharged into a stream, a succession of changes in water quality take place, in the downstream side of the point of pollution. The resulting pattern of change along the stream establishes a well defined profile of pollution and self purification, which again changes with seasons and hydrography.

- Whenever a single, heavy charge of putrescible organic matter is added into a clean stream, depending on the hydrography of the stream, the suspended matter is either settled at the bed near the point of discharge, or is carried along with the water to the downstream side.

- If the wetted surface of the river bed is sufficiently large, a major portion of the organic load is also removed from the main stream by adsorption. At the same time, the aerobic micro organisms, which utilize the organic pollutants as the source of their food and energy, grow till the food supply is adequate for them, and thus the organic matter is stabilized under aerobic condition. The removal of organics are accomplished by (i) settling and adsorption, and (ii) micro-biological activities.

- The intensity of the life activities of the micro-organisms is reflected by the biochemical oxygen demand (BOD). Due to the microbial activities, the oxygen resources of the water are heavily drawn upon; in an overloaded stream, the dissolved oxygen (DO) may be completely exhausted due to these activities.

- In course of time and flow, the food supply gets exhausted. The life activities of the microbial population come to an end; and as such the BOD is decreased. The rate of reaeration or the absorption of oxygen from the atmosphere, which at first has lagged behind the rate of oxygen consumption by the micro-organisms, assumes a momentum and very soon takes the lead. The water becomes clear and the stream returns to its original condition. The self purification is thus complete.

- In short, when the wastewater or the effluent is discharged into a natural stream, the organic matter is broken down by bacteria to ammonia, nitrates, sulphates, carbon dioxide, etc. In this process of oxidation, the dissolved oxygen content of natural water is utilised. Due to this, deficiency of DO is created. As the excess organic matter is stabilized, the normal cycle will be reestablished in a process known as self-purification.

- As stated earlier, the self-purification is a very slow process. A heavily polluted stream may have to traverse quite a long distance for many days for the attainment of a significant degree of purification. It may also be noted that, besides the factor noted earlier, many other natural forces play an important role in the natural purification either in favour or against the process.

The various natural forces of purification which affect the process of self purification of stream are summarised below :

(A) Physical forces :

    1. Dilution and dispersion (due to currents),

    2. Sedimentation,

    3. Sunlight.

(B) Bio-chemical forces (chemical forces aided by biological forces)

    1. Oxidation,

    2. Reduction.

(A) Physical Forces :

1. **Dilution and Dispersion :** When the wastewater is discharged into the large volume of water, it gets rapidly dispersed and diluted. Of course, dispersion depends upon currents. Due to dispersion and dilution, preventing locally high concentration of pollutants and the potential nuisance of sewage is also reduced.

High velocity of currents improves reaeration which reduces the concentration of pollutants.

The concentration of the mixture (concentration of sewage and river) is given by

$$C_{Se} \cdot Q_{Se} + C_R \cdot Q_R = C (Q_{Se} + Q_R)$$

or $\quad C = \dfrac{C_{Se} Q_{Se} + C_R Q_R}{Q_{Se} + Q_R}$ ... (2.13)

where, C_{Se} = Concentration of sewage

         C_R = Concentration of river

         Q_{Se} = Rate of flow of sewage of concentration C_{Se}

         Q_R = Rate of flow of river of concentration C_R

The equation (2.13) is applicable separately to concentrations of different impurities such as oxygen content, suspended solid, BOD and other characteristic content of sewage.

When the dilution ratio is high, large quantities of DO are always available which will reduce the chances of putrefaction and pollutional effects.

2. **Sedimentation :** The settlable solids, if present in sewage effluents, will settle down into the bed of the river (when the stream velocity is lesser than the scour velocity of particles). It will help in self-purification process.

3. **Sunlight :** In presence of sunlight, certain micro-organisms absorbing carbon dioxide and releasing oxygen by a process known as photosynthesis.

Sunlight acts as a disinfectant and stimulates the growth of algae which produces oxygen during daylight but utilise the same at night. Due to evolution of oxygen in river water will help in the self purification process.

(B) Bio-Chemical Forces :

1. **Oxidation :** An aerobic bacteria oxidises the organic matter which is present in sewage effluent by utilising dissolved oxygen (DO) of the river. The process of oxidation will continue till the organic matter has been completely oxidised. This is the very important action to help in self-purification of rivers.

2. **Reduction :** Reduction occurs in the river or streams due to hydrolysis of the organic matter settled at the bottom either biologically or chemically. Anaerobic bacteria will split the organic matter into liquids and gases and thus paving the way for their ultimate stabilization by oxidation.

The following are the various factors on which these natural forces of purification depend :

- **Available Dissolved Oxygen :** If the larger amount of dissolved oxygen is available in water, then the self purification will occur in better manner.

- **Amount and Type of Organic Matter Biological Growth Present :** Due to this also rate of self purification will affect. e.g. Algae.

- **Temperature :** At low temperature, the activity of bacteria is less and at higher temperature it is more. At higher temperature the DO concentration is low, so that the self purification takes lesser time and this will lead to anaerobic conditions.

- **Turbulence :** It will help in maintaining aerobic conditions in the river stream. But too much turbulence is not desirable, because it scours the bottom sediment, leads to increasing the turbidity and retards algae growth.

2.14.1 Zones of Pollution in the Stream

(May 10; Dec. 10, Aug. 15, 17)

The self-purification process of polluted stream can be divided into the following four zones :

    1. Zone of degradation or zone of pollution

    2. Zone of active decomposition

    3. Zone of recovery and

    4. Clear water zone.

Table 2.18 is itself explanatory about zones of pollution along a river stream.

Table 2.18 : Zones of Pollution along a Stream

Parti-culars	Zones of Pollution				
Zones	Clear water	Zone of degradation	Zone of active decomposition	Zone of recovery	Clear water zone
1	2	3	4	5	6
Dissolved oxygen sag curve	100% Saturation level 40% 0%				100% DO
Physical indices	Clear water, No colour, No bottom sludge	Floating solids, Bottom sludge present, Colour getting turbid	Darker and greyish colour, evolution of gases like CH_4, CO_2, H_2S etc. Lot of sludge coming to the surface forming an ugly scum layer at top.	Turbid with bottom sludge	Clear water with no bottom sludge.
Fish presence	Ordinary fish like game, pan, food and for age etc. present.	Tolerant fishes like carps, buffalo, gary, etc. present.	No fish present.	Tolerant fish like carp, buffalo etc. present	Ordinary fish like game, pan, food and forage etc. present.

2.14.2 Oxygen Sag Curve or Oxygen Deficit of a Polluted River Stream

(Dec. 10, Nov. 15, May 10, Aug. 17)

The oxygen sag or oxygen deficit (D) at any time in a polluted river stream is defined as the difference between the actual DO content at that time and the saturation DO content at the water temperature.

Oxygen deficit (D) = Saturation DO – Actual DO

$$... (2.14)$$

The normal saturation DO value for fresh water depends upon the temperature and its value varies from 14.62 mg/lit to 7.6 mg/litfor temperature varying between 0°C to 30°C.

Oxygen deficit must be nil to maintain a river-stream in clean conditions.

Oxygen deficit can be found out by knowing rates of deoxygenation and reoxygenation.

Fig. 2.17 shows oxygen deficit or oxygen sag curve.

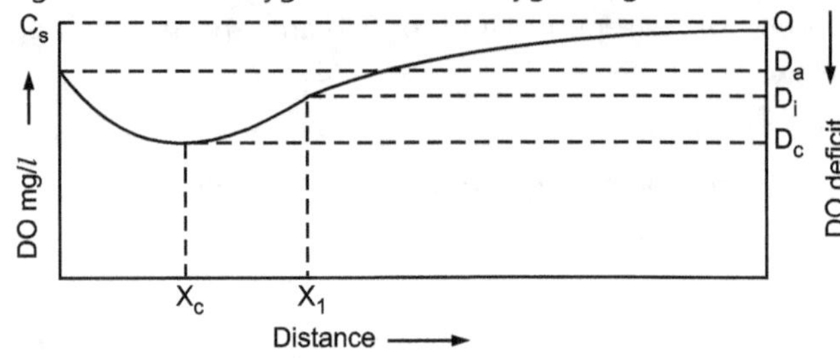

Fig. 2.17 : Oxygen sag curve

Deoxygenation Curve :

- Organic waste normally undergoes aerobic decomposition in the stream only when the rate of supply of oxygen cannot keep pace with the rate of oxygen demand, the condition within the stream becomes anaerobic. The anaerobic condition however not desirable, as while the aerobic receiving water look reasonably clean and is free from odour, the anaerobic condition makes it black, unsightly and malodourous.

- The biochemical reactions within the stream exert BOD, resulting in the deoxygenation of the stream. The rate of deoxygenation depends upon the amount of the organic matter remaining to be oxidised at the given time as well as on the temperature of reaction. Deoxygenation curve (refer curve I of Fig. 2.18) is similar to the first stage BOD curve.

- Apart from the above, a certain portion of the biodegradable organics get deposited at the bed of the stream.

- They undergo anaerobic and benthic decomposition. The products of such decomposition are organic acids and reduced gases. These are further stabilized by the aerobic microorganisms in the upper layer, thereby increasing the BOD of the stream. A small amount of oxygen is also utilized by the higher animals for their respirations.

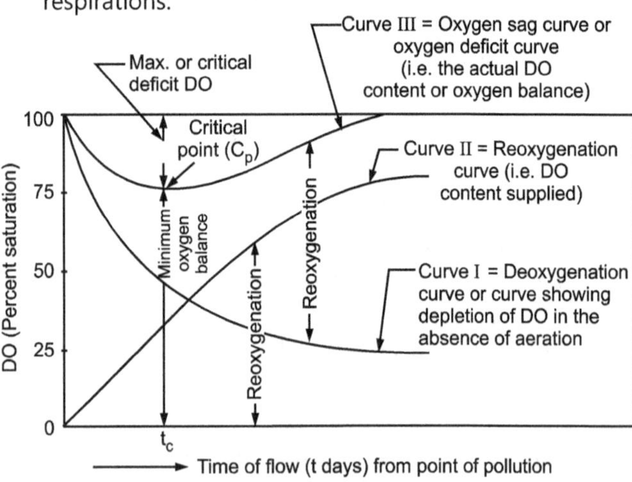

Fig. 2.18

Reoxygenation Curve : (Dec. 10, May 11)

- The simultaneous replenishment of oxygen in the stream occurs due to the absorption of oxygen from the atmosphere and also due to the release of oxygen by the green plants during photosynthesis. In other words, though the DO content of the stream is gradually consumed due to BOD load, atmosphere supplies oxygen continuously to the water and the process is known as reaeration or reoxygenation of the stream.

- The rate of reoxygenation depends upon :
  - ➢ Depth of the receiving water (more for shallow depth).
  - ➢ The temperature of water.
  - ➢ The oxygen deficit below saturation DO.

- The velocity of flow in the stream (rate is more in running stream).

- Fig. 2.18 shows curve II called reoxygenation curve. Depending on above factors, the rate of reoxygenation can also be expressed mathematically.

Oxygen Deficit Curve :

- The interplay between deoxygenation and reaeration produces a well defined profile (oxygen deficit curve) of the dissolved oxygen in the stream as shown in Fig. 2.18. (Refer curve III). If deoxygenation is more rapid than the reoxygenation, an oxygen deficit results. While algebraically adding the deoxygenation and reoxygenation curves, the resultant curve so obtained is known as the oxygen sag curve or the oxygen deficit curve.

- From Fig. 2.18, we have concluded that when the reoxygenation rate is less than the deoxygenation rate, oxygen deficit will increase and at the point CP (critical point) these two rates are equal and then finally the rate of reoxygenation increases, the oxygen deficit goes on decreasing till it becomes zero.

2.14.3 Streeter–Phelp's Equation (May 16, 17, Nov. 17)

- The entire analysis of super-imposing the rates of deoxygenation and reoxygenation have been carried out mathematically, as suggested by Streeter-Phelp's analysis. This is the classic Streeter-Phelp's oxygen sag equation, which is most commonly used in river analysis. This equation is applied to channels of uniform cross section only, where effects of algae and sludge deposits are negligible.

- The following is the form of well-known Streeter-Phelp's equation :

$$D_t = \frac{K_1 L_a}{K_2 - K_1} [(10)^{-K_1 \cdot t} - (10)^{-K_2 \cdot t}] + [D_a \times (10)] \quad \ldots(2.15)$$

(By using different values of t in above equation (2.15), the oxygen sag curve can be plotted easily.)

where, D_t = DO deficit in mg/lit.at any time t

L_a = Initial BOD_L or ultimate first stage BOD of the mix at the point of discharge in mg/lit.

D_a = Initial DO deficit, mg/lit, at the point of waste discharge, at time t = 0.

K_1 = Deoxygenation constant per day; which can be considered as equal to the BOD rate constant.

Also K_1 varies with temperature as per the following relationship :

$$\therefore \quad (K_1)_T = (K_1)_{20} \cdot (\theta_{K_1})^{T-20} \quad \ldots (2.16)$$

Where,

$(K_1)_{20}$ = Deoxygenation constant at a temperature of T°C, and θ_{K_1}

= A temperature coefficient = usually 1.047 for K_1 in the temperature range of 15°C to 30°C.

The value of K_1 may vary from 0.12 per day for treated wastewater to 0.39 per day for a strong waste; while values of $(K_1)_{20}$ vary between 0.1 to 0.2, generally taken as 0.1

K_2 = Reaeration or Reoxygenation coefficient per day. It can be calculated by the field tests by using equation :

$$(K_2)_{20} = 4.75 \, V/(H)^{3/2} \quad \ldots (2.17)$$

where, $(K_2)_{20}$ = Reaeration coefficient at 20ºC per day

V = Mean velocity in m/sec.

and H = Average depth of the stream in m.

K_2 varies with temperature as per the equation :

$$\therefore \quad (K_2)_T = (K_2)_{20} \cdot \theta_{K_2}^{(T-20)} \quad \ldots (2.18)$$

where, θ_{K_2} = Reaeration temperature coefficient, usually 1.0241.

Typical values of reaeration constants are given in Table 2.19.

Table 2.19 : Reaeration Constants at 20°C

Sr. No.	Type of Water Body	$(K_2)_{20}$ Per Day
1.	Small ponds and back waters	0.05 – 0.10
2.	Large lakes and sluggish streams	0.10 – 0.15
3.	Large streams of low velocity	0.15 – 0.20
4.	Large streams of normal velocity	0.20 – 0.30
5.	Swift streams	0.30 – 0.50
6.	Rapids and waterfalls	> 0.50

Critical Time (t_c) :

The critical time is a time after which the minimum DO occurs and the same can be found by differentiating equation (2.15) and equating it to zero. Thus, we obtain

$$\therefore \quad t_c = \left[\frac{1}{K_2 - K_1}\right] \log\left[\left\{\frac{K_1 \cdot L_a - K_2\, D_a + K_1\, D_a}{K_1 \cdot L_a}\right\} \frac{K_2}{K_1}\right] \quad \text{... (2.19)}$$

Critical or Maximum Oxygen Deficit :

The critical or maximum oxygen deficit is given by

$$D_c = \frac{K_1\, L_a}{K_2} [10]^{-K_1 \cdot t_c} \quad \text{... (2.20)}$$

Self Purification Constant or Self Purification Ratio (f) :

It is the ratio of $\dfrac{K_2}{K_1}$ and represented as (f) and is called the self purification constant.

The values of self purification constant are given in Table 2.20.

Table 2.20 : Self Purification Constants

Sr. No.	Type of Water Body	$f = \dfrac{K_2}{K_1}$
1.	Small ponds and back waters	0.5 – 1.0
2.	Large lakes, sluggish streams and impounding reservoirs	1.0 – 1.5
3.	Large streams of low velocity	1.5 – 2.0
4.	Large streams of normal velocity	2.0 – 3.0
5.	Swift streams	3.0 – 5.0
6.	Rapids and waterfalls	above 5.0

By substituting $\dfrac{K_2}{K_1} = f$ in the equation (6.7), we obtain,

$$t_c = \frac{1}{K_1 (f - 1)} \log\left[\left\{1 - (f - 1)\frac{D_a}{L_a}\right\} f\right] \quad \text{... (2.21)}$$

Also from equation (2.20), we obtain,

$$D_c = \frac{L_a}{f} [10]^{-K_1 \cdot t_c} \quad \text{... (2.22)}$$

Taking log of both sides, we obtain,

$$\log D_c = \log \frac{L_a}{f} - K_1 \cdot t_c \quad \text{... (2.23)}$$

Substituting the value of t_c from equation (2.21) in equation (2.23), we obtain,

$$\log D_c = \log \frac{L_a}{f} - \frac{K_1 \cdot 1}{K_1 (f - 1)} \log\left[f\left\{1 - (f - 1)\frac{D_a}{L_a}\right\}\right]$$

$$\text{or} \quad \log D_c = \log \frac{L_a}{f} - \frac{1}{(f - 1)} \log\left[f\left\{1 - (f - 1)\frac{D_a}{L_a}\right\}\right]$$

$$\text{or} \quad (f - 1)\left[\log \frac{L_a}{f} - \log D_c\right] = \log\left[f\left\{1 - (f - 1)\frac{D_a}{L_a}\right\}\right]$$

$$\text{or} \quad \log\left[\frac{\frac{L_a}{f}}{D_C}\right]^{f-1} = \log\left[f\left\{1 - (f - 1)\frac{D_a}{L_a}\right\}\right]$$

$$\text{or} \quad \left(\frac{L_a}{D_c \cdot f_a}\right)^{f-1} = f\left[1 - (f - 1)\frac{D_a}{L_a}\right] \quad \text{... (2.24)}$$

This is important equation, in which L_a is the first stage BOD of mixture of wastewater and stream. The self purification constant (f) corresponds to the temperature of mixture of wastewater and stream at the outfall. The solubility of oxygen in water which affects the reoxygenation, is a function of temperature. AppendixAgives the values of DO in fresh water at various temperatures.

Distance at which Oxygen Deficit Occurs :

The distance, at which the critical DO deficit will occur in the downstream, can be calculated by following equation :

Distance at which oxygen deficit occurs = Velocity of river × Travel time

$$\text{... (2.25)}$$

Limit of Dissolved Oxygen (DO) in the Stream :

To avoid anaerobic decomposition of the organic wastes and also to provide adequate support to the aquatic life, the DO is not allowed to fall below about 3 mg/lit. This information fixes the allowable maximum DO saturation deficit.

2.15 RIVER CLASSIFICATION AS PER MOEF, GOVERNMENT OF INDIA   (Aug. 16)

Table 2.21

Classes	No Development Zone for Any Type of Industries	Only Green and Orange Category of Industries with Pollution Control Devices.	Any Type of Industries (Red, Orange, Green) with Pollution Control Devices
A-I (Origin to Dam)	3 km on the either side of river	From 3 km to 8 km from river (H.F.L.) on either side	Beyond 8 km from river (H.F.L.) on either side.
A-II	1 km on the either side of river.	From 1 km to 2 Km from (H.F.L.) on either side	Beyond 2 km from river (H.F.L.) on either side.

...Conti.

A-III (Fisheries and wildlife)	1/2 km on the either side of river	From 1/2 km to 1 km from river (H.F.L.) on either side	Beyond 1 km from river (H.F.L.) on either side.
A-IV (Agricultural and Industrial usages.)	1/2 km on the either side of river	From 1/2 to 1 km from river (H.F.L.) on either side	Beyond 1 km from river (HFL.) High Flood Line on either side

2.16 NATIONAL RIVER CLEANING PLAN

- The Ganga Action Plan (GAP) was launched by the Government in 1985, for pollution abatement activities in identified polluted stretches of river Ganga and later expanded to include other major rivers under National River Conservation Plan (NRCP). NRCP presently covers 40 rivers in 190 towns spread over 20 States. Pollution abatement schemes implemented under the plan include interception, diversion and treatment of sewage; low cost sanitation works on river banks; electric/ improved wood based crematoria, etc. Sewage treatment capacity of 4664 mld (million litres per day) has been created under the Plan. An expenditure of Rs. 5343.06crore has been incurred so far under NRCP, including the National Ganga River Basin Authority (NGRBA) programme.

- Based on independent monitoring undertaken by reputed institutions, the water quality in terms of BOD (Bio-chemical Oxygen Demand) values for major rivers is reported to have improved as compared to the water quality before taking up pollution abatement works. However, the levels of bacterial contamination in terms of fecal coliform are reported to be exceeding the maximum permissible limit at various monitoring locations.

- In light of experience gained in implementation of the river action plans since 1985, the river conservation strategy was reviewed by the Government. Accordingly, in February 2009, the NGRBA has been constituted as an empowered, planning, financing, monitoring and coordinating authority with the objective to ensure effective abatement of pollution and conservation of the river Ganga by adopting a holistic river basin approach.

- External assistance from bilateral/ multilateral agencies is availed of from time to time by the Government for conservation of rivers. Presently, the Japan International Cooperation Agency (JICA) has extended loan assistance of Yen 13.33 billion for Phase-II and Yen 32.571 billion for Phase-III of the ongoing Yamuna Action Plan (YAP) for pollution abatement of the river.

- JICA is providing loan assistance of 11.184 billion Yen for pollution abatement of the river Ganga at Varanasi. World Bank is also providing loan assistance of US $ 1 billion for abatement of pollution of river Ganga under the NGRBA programme.

- Conservation of rivers is an ongoing and collective effort of the Central and State Governments and this Ministry is supplementing the efforts of the State Governments in abatement of pollution in rivers under NRCP. Further, the Central Pollution Control Board and respective State Pollution Control Boards monitor industries for compliance with respect to effluents discharge standards and take action for non-compliance under the Water (Prevention and Control of Pollution) Act, 1974 and the Environment (Protection) Act, 1986.

SOLVED EXAMPLES ON SEWAGE DISPOSAL IN STREAM OR RIVER

Example 2.15 : *The domestic sewage of a town is to be discharged into a river stream after treatment. Find out the BOD of the diluted water, if the quantity of sewage produced per day is 10 million litres, having BOD as 300 mg/lit, the discharge in the river is 210 litres/sec and it's BOD is 9 mg/lit.*

Solution :

Given : Rate of flow of sewage, $Q_{Se} = 10 \times 10^6$ litres.

$$= \frac{10 \times 10^6}{24 \times 60 \times 60}$$

$$= 115.74 \text{ litres/sec.}$$

Rate of flow of the river,

$$Q_R = 210 \text{ litres/sec}$$

BOD of sewage,

$$C_{Se} = 300 \text{ mg/lit}$$

BOD of river,

$$C_R = 9 \text{ mg/lit}$$

BOD of diluted mixture can be calculated by using equation (6.1).

$\therefore$　BOD of diluted mixture,

$$C = \frac{C_{Se} \cdot Q_{Se} + C_R \cdot Q_R}{Q_{Se} + Q_R}$$

$$= \frac{300 \times 115.74 + 9 \times 210}{115.74 + 210}$$

$$C = 112.4 \text{ mg/lit} \qquad \text{... Ans.}$$

Example 2.16 : *The sewage of a town is to be discharged into a river stream. The population of town is 65000. Determine the maximum permissible effluent BOD and the percentage purification required in the treatment plant, given the following data :*

　D.W.F. of sewage : 160 lit/capita/day

　BOD concentration per capita : 0.075 kg per day.

　Minimum flow of stream : 0.23 m³/sec.

　BOD of stream : 3 mg/lit.

　Maximum BOD of stream on downstream : 5 mg/lit.

Solution :

Given :Rate of flow of sewage,

$$Q_{Se} = 160 \times 65000 \text{ litres/day}$$

$$= \frac{160 \times 65000}{24 \times 60 \times 60 \times 1000}$$

$$= 0.12 \text{ m}^3/\text{sec.}$$

Rate of flow of stream, $Q_R = 0.23$ m³/s.

BOD of sewage, $C_{Se} = ?$

BOD of stream, $C_R = 3$ mg/lit.

BOD of mixture, $C = 5$ mg/lit.

BOD of sewage can be calculated by using equation (2.13).

∴　BOD of the mixture on the downstream is

$$C = \frac{C_{Se} Q_{Se} + C_R Q_R}{Q_{Se} + Q_R}$$

$$5 = \frac{C_{Se} \times 0.12 + 3 \times 0.23}{0.12 + 0.23}$$

$$1.75 = 0.12 \, C_{Se} + 0.69$$

∴　　　$C_{Se} = 8.83$ mg/lit

Now,　BOD per capita per day

$$= 0.075 \times 1000 \times 1000 = 75000 \text{ mg/day}$$

Sewage D.W.F.

$$= 160 \text{ litres/day}$$

∴　Actual BOD of effluent

$$= \frac{75000}{160} = 468.75 \text{ mg/lit}$$

∴　Percentage purification required

$$= \frac{468.75 - 8.83}{468.75} \times 100$$

$$= 98.12\% \qquad\qquad \text{... Ans.}$$

Example 2.17 : *A stream saturated with DO has a flow of 1.4 m³/sec, BOD of 5 mg/lit and rate constant of 0.3 per day. It receives an effluent discharge of 0.29 m³/s having BOD 25 mg/lit, DO 6 mg/lit and rate 0.12 per day. The average velocity of flow of the stream is 0.20 m/sec. Calculate the DO deficit at point 20 km and 50 km downstream. Assume that the temperature is 20°C throughout and BOD is measured at 5 days. Take saturation DO at 20°C as 9.17 mg/lit.*

Solution :

Given :Rate of flow of stream, $Q_R = 1.4$ m³/sec.

BOD of stream, $C_R = 5$ mg/lit.

Rate of flow of effluent, $Q_{Se} = 0.29$ m³/sec.

BOD of effluent, $C_{Se} = 25$ mg/lit.

BOD of mixture can be calculated by using equation (2.13).

∴　BOD of the mixture,

$$C = \frac{C_{Se} Q_{Se} + C_R Q_R}{Q_{Se} + Q_R}$$

$$= \frac{25 \times 0.29 + 5 \times 1.4}{0.29 + 1.4}$$

$$= 8.432 \text{ mg/lit}$$

Now, $Y_5 = C = L_a [1 - 10^{-K_1 \cdot t}]$

$$8.432 = L_a [1 - 10^{-0.13 \times 5}]$$

$$L_a = 10.86 \text{ mg/lit}$$

Again saturation DO of stream at 20°C, $(DO)_R = 9.17$ mg/lit.

DO effluent, $(DO)_{Se} = 6$ mg/lit

∴　$(DO)_{mix} = \dfrac{(DO)_{Se} \cdot Q_{Se} + (DO)_R \cdot Q_R}{Q_{Se} + Q_R}$

$$= \frac{6 \times 0.29 + 9.17 \times 1.4}{0.29 + 1.4}$$

$$= 8.63 \text{ mg/lit}$$

∴　Initial DO deficit (D_a)

$$= 9.17 - 8.63 = 0.54 \text{ mg/lit}$$

1.　DO Deficit at a Point 20 km Downstream :

$$t = \frac{\text{Distance}}{\text{Velocity}} = \frac{20 \times 1000}{0.20 \times 60 \times 60 \times 24}$$

$$= 1.16 \text{ days}$$

Using Streeter-Phelp's equation (6.3),

$$D_t = \frac{K_1 \cdot L_a}{K_2 - K_1} \left[(10)^{-K_1 \cdot t} - (10)^{-K_2 \cdot t} \right]$$

$$+ \left[D_a \times (10)^{-K_2 \cdot t} \right]$$

$$= \frac{0.13 \times 10.86}{0.3 - 0.13} \left[(10)^{-0.13 \times 1.16} - \right.$$

$$\left. (10)^{-0.3 \times 1.16} \right]$$

$$+ \left[0.54 \times (10)^{-0.3 \times 1.16} \right]$$

$$= 8.3\,[0.71 - 0.448] \; + \; [0.54 \times (-0.448)]$$

$$= 2.168 - 0.24$$

$$= 1.93 \text{ mg/lit} \qquad \text{... Ans.}$$

2. DO Deficit at a Point 50 km Downstream :

$$t = \frac{50 \times 1000}{0.20 \times 60 \times 60 \times 24} = 2.894 \text{ days}$$

$$D_t = \frac{0.13 \times 10.86}{0.3 - 0.13}\,[(10)^{-0.13 \times 2.894}$$

$$- (10)^{-0.3 \times 2.894}] \; + \; [0.54 \times (10)^{-0.3 \times 2.894}]$$

$$= 8.3\,[0.42 - 0.135] + [0.54 \times (-0.135)]$$

$$= 2.37 - 0.0729$$

$$= 2.29 \text{ mg/lit} \qquad \text{... Ans.}$$

Example 2.18 : *A city discharges 1700 litres per second of sewage into a stream. The minimum rate of flow of stream is 7000 litres per second. The temperature of sewage as well as water is 20°C. The 5 day BOD at 20°C for sewage is 250 mg/lit and that of the stream is 90% of the saturation DO If the minimum DO to be maintained in the stream is 4.5 mg/lit, find out the degree of sewage treatment required. Assume the deoxygenation coefficient as 0.1 and reoxygenation coefficient as 0.3. Take saturation DO at 20°C as 9.17 mg/lit .*

Solution :

Given :

Rate of flow of sewage, $Q_{Se} = 1700$ litres/sec.

Rate of flow of stream, $Q_R = 7000$ litres/sec

BOD of river, $C_R = 1$ mg/lit.

Deoxygenation constant $(K_1) = 0.1$

Reoxygenation constant $(K_2) = 0.3$

Minimum DO maintained in the stream $= 4.5$ mg/lit

Saturation DO at 20°C $= 9.17$ mg/lit.

DO content of the stream

$$= 90\% \text{ of the saturation DO}$$

$$= \frac{90}{100} \times 9.17$$

$$= 8.25 \text{ mg/lit}$$

DO of the mix at the start point

$$= \frac{(DO)_{Se} \times Q_{Se} + (DO)_R \times Q_R}{Q_{Se} + Q_R}$$

$$= \frac{0 \times 1700 + 8.25 \times 7000}{1700 + 7000}$$

(Assuming DO of sewage as zero)

$$= 6.83 \text{ mg/lit}$$

$\therefore$ Initial DO deficit,

$$D_a = \text{Saturation DO} - \text{DO of mix.}$$

$\therefore \qquad D_a = 9.17 - 6.83$

$\therefore \qquad D_a = 2.34$ mg/lit

Minimum DO to be maintained in the stream

$$= 4.5 \text{ mg/lit}$$

$\therefore$ Maximum permissible saturation deficit (critical DO deficit),

$$D_c = 9.17 - 4.5 = 4.67 \text{ mg/lit}$$

The first stage BOD of mixture of sewage and stream (L_a) can be calculated by using equation (2.24).

$$\left[\frac{L_a}{D_c \cdot f}\right]^{f-1} = f\left[1 - (f-1)\frac{D_a}{L_a}\right]$$

$$f = \frac{K_2}{K_1} = \frac{0.3}{0.1} = 3$$

Now, $\left[\dfrac{L_a}{4.67 \times 3}\right]^{3-1} = 3\left[1 - (3-1)\dfrac{2.34}{L_a}\right]$

$$\left[\frac{L_a}{14.01}\right]^{2} = 3\left[1 - \frac{4.68}{L_a}\right]$$

Solving by hit and trial, we get

$$L_a = 21 \text{ mg/lit}$$

Now using $Y_t = L_a\,[1 - 10^{-K_1 \cdot t}]$

$\therefore$ Maximum permissible 5 day BOD of the mix at 0°C,

$$Y_5 = 21\,[1 - 10^{-0.1 \times 5}] = 14.36 \text{ mg/lit}$$

The permissible BOD_5 of discharged wastewater can be calculated by using equation (6.1).

$$\therefore \qquad C = \frac{C_{Se}\,Q_{Se} + C_R\,Q_R}{Q_{Se} + Q_R}$$

$$14.36 = \frac{C_{Se} \times 1700 + 1 \times 7000}{1700 + 7000}$$

$\therefore \qquad C_{Se} = 69.37$ mg/lit

$\therefore$ Degree of treatment required

$$= \frac{\text{Original BOD of sewage} - \text{Permissible BOD}}{\text{Original BOD}} \times 100$$

$$= \frac{250 - 69.37}{250} \times 100 = 72.25\% \qquad \text{... Ans.}$$

Example 2.19 : *A town discharges 90 cumecs of sewage into a stream having a rate of flow of 1400 cumecs during its lean days with a velocity of 0.12 m/sec. The BOD_5 of sewage at the given temperature is 290 mg/lit. Find when and where the critical DO deficit will occur in the downstream portion of the river and what is its amount ? Assume coefficient of self purification (f) as 3.5 and coefficient of deoxygenation (K_1) as 0.1. Assume saturation DO at given temperature as 9.2 mg/lit.*

Solution :

Given : Rate of flow of sewage, Q_{Se} = 90 cumecs

Rate of flow of stream, Q_R = 1400 cumecs

Velocity of stream = 0.12 m/sec

BOD_5 of sewage = 290 mg/lit

Coefficient of self purification (f) = 3.5

Deoxygenation coefficient (K_1) = 0.1

Saturation DO = 9.2 mg/lit

$$(DO)_{mix} = \frac{9.2 \times 1400 + 0 \times 90}{1400 + 90}$$

(Assuming DO of sewage as zero)

$$= 8.64 \text{ mg/lit}$$

∴ Initial DO deficit, D_a = 9.2 − 8.64 = 0.56 mg/lit

BOD_5 of the mixture can be calculated by using equation (2.13).

$$C = \frac{C_{Se}\, Q_{Se} + C_R\, Q_R}{Q_{Se} + Q_R}$$

$$= \frac{290 \times 90 + 0 \times 1400}{90 + 1400}$$

$$= 17.52 \text{ mg/lit}$$

The ultimate BOD of mix can be calculated by

$$Y_5 = L_a\,[1 - 10^{-K_1 \cdot t}]$$

∴ $\qquad 17.52 = L_a\,[1 - 10^{-0.1 \times 5}]$

∴ $\qquad L_a = 25.62$ mg/lit

Now, critical deficit, D_c can be calculated by using equation (2.24).

$$\left[\frac{L_a}{D_c \cdot f}\right]^{f-1} = f\left[1 - (f-1)\frac{D_a}{L_a}\right]$$

$$\left[\frac{25.62}{D_c \times 3.5}\right]^{3.5-1} = 3.5\left[1 - (3.5-1)\frac{0.56}{25.62}\right]$$

$$\frac{7.32}{D_c} = (3.31)^{1/2.5} = 1.61$$

∴ $\qquad D_c = 4.53$ mg/lit

The critical time can be calculated by using equation (2.21).

$$t_c = \frac{1}{K_1(f-1)}\,\log_{10}\left[\left\{1-(f-1)\frac{D_a}{L_a}\right\}f\right]$$

∴ $\qquad t_c = \dfrac{1}{0.1\,(3.5-1)}\,\log_{10}$

$$\left[\left\{1-(3.5-1)\frac{0.56}{25.62}\right\}3.5\right]$$

$$= \frac{1}{0.25}\,\log_{10}[3.31]$$

$$= 2.08 \text{ days} \qquad\qquad \text{... Ans.}$$

Now,

Distance = Velocity of stream × Travel time

$$= 0.12 \times (2.08 \times 24 \times 60 \times 60)$$

$$= 21551.5 \text{ m}$$

$$= 21.55 \text{ km} \qquad\qquad \text{... Ans.}$$

Example 2.20 : *The population of town is 40,000 and the domestic sewage is 150 litres/capita/day having per capita BOD of 75 gm/day.*

Dairy wastes of 4 million litres per day with BOD of 1200 mg/lit and sugar mill waste of 2.6 million litres per day with BOD of 1600 mg/lit are produced. An overall expansion factor of 12% to be provided. The sewage effluents are to be discharged to a river stream with a minimum dry weather flow of 4900 litre per second and a saturation dissolved oxygen content of 9.0 mg/lit. It is necessary to maintain a dissolved oxygen content of 4 mg/lit in the stream. For design of the treatment plant, determine the degree of treatment required to be given to the sewage. (K_1 = 0.1, K_2 = 0.3).

Solution :

Given : Population = 40,000

Domestic sewage = 150 lit/capita/day

Quantity of dairy waste = 4 million litres

BOD of dairy waste = 1200 mg/lit

Quantity of sugar mill waste = 2.6 million litres

BOD of sugar mill waste = 1600 mg/lit

Rate of flow of stream = 4900 mg/lit

Per capita sewage = 150 litres/day

Amount of domestic sewage

$$= 40000 \times 150$$

$$= 6 \text{ million litres/day}$$

Per capita BOD of domestic sewage

$$= 75 \text{ gm/day}$$

$$= 75 \times 1000 \text{ mg/day}$$

BOD per litre of the domestic sewage

$$= \frac{75 \times 1000}{150} = 500 \text{ mg/lit.}$$

Net BOD of all wastewaters

$$= \left[\frac{6 \times 500 + 4 \times 1200 + 2.6 \times 1600}{6 + 4 + 2.6}\right]$$

$$= 949.21 \text{ mg/lit}$$

Total wastewater discharge in litres/sec

= Volume of wastewaters entering per day

$$= \frac{6 \times 10^6 + 4 \times 10^6 + 2.6 \times 10^6}{24 \times 60 \times 60}$$

$$= 145.83 \text{ litres/sec}$$

Using an expansion factor of 12%,

Total wastewater

$$= 1.12 \times 145.83$$

$$= 163.33 \text{ litres/sec}$$

Initial DO of stream $= 9.0$ mg/lit

$\therefore$ DO of the mixture

$$= \frac{(DO)_{Se} \times Q_{Se} + (DO)_R \times Q_R}{Q_{Se} + Q_R}$$

$$= \frac{0 \times 150 + 9 \times 4900}{150 + 4900}$$

(Assuming DO of sewage as zero)

$$= 8.73 \text{ mg/lit}$$

$\therefore$ Initial DO deficit,

$$D_a = 9 - 8.73$$

$$= 0.27 \text{ mg/lit}$$

Critical DO deficit,

$$D_C = 9 - 4$$

$$= 5.0 \text{ mg/lit}$$

Now, L_a can be calculated by using equation (6.24).

$$\left[\frac{L_a}{D_C\, f}\right]^{f-1} = f\left[1 - (f-1)\frac{D_a}{L_a}\right]$$

Here, $\quad f = \dfrac{K_2}{K_1} = \dfrac{0.3}{0.1} = 3$

$$\therefore \quad \left[\frac{L_a}{5 \times 3}\right]^2 = 3\left[1 - (3-1)\frac{0.27}{L_a}\right]$$

Solving by hit and trial,

$$L_a = 25.5 \text{ mg/lit}$$

Maximum permissible BOD_5 of mix can be calculated by

$$Y_5 = L_a\,[1 - 10^{-K_1 \cdot t}]$$

$$= 25.5\,[1 - 10^{-0.1 \times 5}]$$

$$= 17.44 \text{ mg/lit}$$

Now, maximum permissible BOD_5 of wastewaters can be calculated by using equation (2.13).

$$C = \frac{C_{Se}\, Q_{Se} + C_R\, Q_R}{Q_{Se} + Q_R}$$

$$17.44 = \frac{C_{Se} \times 150 + 0 \times 4900}{150 + 4900} = 587.02 \text{ mg/lit}$$

$\therefore$ Degree of treatment required

$$= \frac{\text{Original BOD of city wastewaters} - \text{Permissible BOD}}{\text{Original BOD of city wastewaters}} \times 100$$

$$= \frac{949.21 - 587.02}{949.21} \times 100 = 38.16\% \qquad \textbf{... Ans.}$$

Example 2.21 : *In the previous example, No treatment is provided. Determine the dilution ratio and river discharge.*

Solution :

When no treatment is provided, the value of maximum permissible BOD_5 should be 949.21 mg/lit. Q_R can be calculated as :

$$17.44 = \frac{949.21 \times 150 + 0 \times Q_R}{150 + Q_R}$$

$$\therefore \qquad Q_R = 8014.08 \text{ litres/sec} \qquad \textbf{... Ans.}$$

$$\text{Dilution ratio} = \frac{Q_R}{Q_{Se}} = \frac{8014.08}{150} = 53.43 \qquad \textbf{... Ans.}$$

Example 2.22 : *A waste water effluent of 560 litres/sec with a BOD = 50 mg/lit, DO = 3.0 mg/lit and temperature of 23°C enters a river where the flow is 28 m³/sec, and BOD = 4.0 mg/lit, DO = 8.2 mg/lit, and temperature of 17°C. K_1 of the waste is 0.10 per day at 20°C. The velocity of water in the river downstream is 0.18 m/sec and depth of 1.2 m. Determine the following after mixing of waste water with the river water :*

(i) Combined discharge; (ii) BOD; (iii) DO; and

(iv) Temperature.   **(Civil Services)**

Solution :

Particulars of Sewage Thrown	Particulars of River
$Q_{Se} = 560$ litres/sec $= 0.56$ m³/sec	$Q_R = 28$ m³/sec
Concentrations (C_{Se}) :	**Concentrations (C_R) :**
BOD $= 50$ mg/lit	BOD $= 4.0$ mg/lit
DO $= 3.0$ mg/lit	DO $= 8.2$ mg/lit
Temperature $= 23$°C	Temperature $= 17°$

K_1 at 20°C $= 0.1$ per day

(i) Combined discharge

$$= Q_{Se} + Q_R = 0.56 + 28$$

$$= 28.56 \text{ m}^3/\text{sec.} \qquad \textbf{... Ans.}$$

Now, using equation (2.13), for concentration of mix as

$$C = \frac{C_{Se} \cdot Q_{Se} + C_R \cdot Q_R}{Q_{Se} + Q_R}$$

(ii) BOD of mix $= \dfrac{50 \times 0.56 + 4.0 \times 28}{0.56 + 28}$

$$= \frac{140}{28.56} = 4.9 \text{ mg/lit} \qquad \textbf{... Ans.}$$

(iii) DO of mix $= \dfrac{3.0 \times 0.56 + 8.2 \times 28}{0.56 + 28}$

$$= 8.098 \text{ mg/lit} \qquad \textbf{... Ans.}$$

(iv) Temperature of mix $= \dfrac{23 \times 0.56 + 17 \times 28}{0.56 + 28}$

$$= 17.12°\text{C} \qquad \textbf{... Ans.}$$

Example 2.23 : *125 cumecs of sewage of a city is discharged in a perennial river which is fully saturated with oxygen and flows at a minimum rate of 1600 cumecs with a minimum velocity of 0.12 m/sec. If the 5 day BOD of the sewage is 300 mg/lit, find out where the critical DO will occur in the river. Assume*

   (i)   The coefficient of purification of the river as 4.0,

   (ii) The coefficient of DO as 0.11, and

   (iii) The ultimate BOD as 125% of the 5 day BOD of the mixture of sewage and river water.

Solution :

Assume saturation DO concentration of the given river = 9.2 mg/lit

The DO of the river at the mixing point after disposal of sewage (D)

$$= \frac{125 \times 0 + 1600 \times 9.2}{125 + 1600}$$

$$= 8.53 \text{ mg/lit}$$

Initial DO deficit $(D_a) = D_s - D = 9.2 - 8.53$

$$= 0.67 \text{ mg/lit}$$

BOD_5 of the river at the mixing point after disposal of ewage (Y_5)

$$= \frac{125 \times 300 + 1600 \times 0}{125 + 1600}$$

$$= 21.74 \text{ mg/lit}$$

The ultimate BOD of river (mix) at mixing point (L_a)

$$= 125\% \ BOD_5$$

[as per given in assumption (iii)]

$$= 1.25 \times 21.74 = 27.17 \text{ mg/lit}$$

Now, $BOD_5 = L_a [1 - (10)^{-K_1 \times 5}]$

or $21.74 = 27.17 [1 - (10)^{-K_1 \times 5}]$

or $0.8 = [1 - (10)^{-5K_1}]$

or $(10)^{-5K_1} = 0.20$

or $- 5K_1 \log 10 = \log 0.20$

or $K_1 = 0.14$

The coefficient of DO or BOD (K_1) is given in assumption No. (ii) to be 0.11, as against its value of 0.14 computed above on the basis of assumption (iii). Eventually, there is some inconsistency in the given data, and the Examiner should have given only one of the two assumptions, i.e. either (ii) and (iii), which would have suffice the purpose.

Under such a difficult situation, we may solve the example by using both the values of K_1, i.e. 0.11 as well as 0.14. The K_1 value of 0.14 will, however, give more DO deficit and will displace the critical point upstream; and will thus provide more conservative design values :

Case 1 : When K_1 = 0.11 : Using equation (2.21) as

$$t_c = \frac{1}{K_1 (f-1)} \log \left[\left\{ 1 - (f-1) \frac{D_a}{L_a} \right\} f \right]$$

We get $t_c = \frac{1}{0.11 \ (4-1)} \log \left[\left\{ 1 - (4-1) \frac{0.67}{27.17} \right\} 4 \right]$

$$= 1.723 \text{ days}$$

The distance along the river, where the critical DO deficit will occur,

$$S = \text{Velocity} \times \text{Time}$$

$$= 0.12 \times (1.723 \times 24 \times 3600)$$

$$= 17.86 \text{ km; Say 18 km}$$

Hence, critical DO deficit will occur at 18 km downstream of the sewage disposal point.   **... Ans.**

Case 2 : When K_1 = 0.14 :

$$t_c = \frac{0.11}{0.14} \times 1.723 = 1.354 \text{ days}$$

$$S = 17.86 \times \frac{1.354}{1.723} = 14.04 \text{ km}$$

Hence, critical DO deficit will occur at 14 km downstream of sewage disposal point.   **... Ans.**

Example 2.24 : *A wastewater treatment plant disposes its effluents into a stream at a point A. Characteristics of the stream at a location fairly upstream of A and of the effluent are as below :*

Item	Units	Effluent	Stream
Flow	m³/s	0.20	0.50
Dissolved oxygen	mg/lit	2.00	8.00
Temperature	°C	26	22
BOD_5 at 20°C	mg/lit	40	3

Assume that the deoxygenation constant K_1 at 20°C (base e) = 0.20 d^{-1} and the reaeration constant K_2 at 20°C (base e) = 0.40 d^{-1} for the mixture. Equilibrium concentration of dissolved oxygen (C_s) for the fresh water is as follows :

Temperature, °C	18	20	22	23	24	25	26
C_s(mg/lit)	9.54	9.17	8.99	8.83	8.53	8.38	8.22

The velocity of the stream downstream of the point A is 0.2 m/sec. Determine the critical oxygen deficit and its location. [Use temperature coefficients of 1.04 for K_1 and 1.02 for K_2]

                                        (Civil Services)

Solution :

K_1 at 20°C (base e) = 0.2 d^{-1}

$\qquad$ = 0.2 per day

$\therefore$ K_1 at 20°C (base 10)

$$= \frac{K_1}{2.3} = 0.434 \, K_1$$

$$= 0.434 \times 0.2 \text{ per day}$$

$$= 0.087 \text{ per day}$$

Similarly, K_2 at 20°C

$$= 0.434 \times 0.4 \text{ d}^{-1}$$

$$= 0.174 \text{ per day}$$

The formulae to be used in this example for converting K_1 and K_2 at any other temperature (T°C) will be

$$K_{1(T°)} = K_{1(20°)} \, [1.04]^{T° - 20°}$$

and $\qquad K_{2(T°)} = K_{2(20°)} \, [1.02]^{T° - 20°}$

(i) We will now determine DO, BOD and temperature of mixture as below :

$$DO \text{ of mixture} = \frac{DO \text{ of sewage} \times Q_{Se} + DO \text{ of river} \times Q_R}{Q_{Se} + Q_R}$$

$$= \frac{2 \times 0.20 + 8 \times 0.50}{0.20 + 0.50}$$

$$= 6.29 \text{ mg/lit}$$

BOD_5 of mixture (i.e. 5 day BOD at 20°C)

$$= \frac{40 \times 0.20 + 3 \times 0.50}{0.20 + 0.50}$$

$$= 13.57 \text{ mg/lit}$$

Temperature of mixture

$$= \frac{26 \times 0.20 + 22 \times 0.50}{0.20 + 0.50}$$

$$= 23.14°C$$

(ii) Ultimate BOD of mixture (L_a)

$$L_a = \frac{Y_5 \text{ (i.e. 5 day BOD of mixture at 20°C)}}{1 - (10)^{-K_1 \times 5}}$$

where, K_1 is at 20°C = 0.087 per day

$$= \frac{13.57}{1 - (10)^{-0.087 \times 5}}$$

$$= \frac{13.57}{0.633}$$

$$= 21.45 \text{ mg/lit}$$

(iii) Initial DO deficit of mixture,

DO of mixture = 6.29 mg/lit

Saturation DO at mixture temperature of 23.14°C

= 8.79 (interpolated from given values)

$\therefore \qquad D_a$ = DO deficit = 8.79 − 6.29

$$= 2.50 \text{ mg/lit}$$

(iv) Corrected values of K_1 and K_2 are :

$$K_{1(23.14°)} = K_{1(20°)} \, [1.04]^{T - 20}$$

$$= 0.087 \, [1.04]^{3.14}$$

$$= 0.098$$

$$K_{2(23.14°)} = K_{2\,(20°)} \, [1.02]^{T - 20}$$

$$= 0.174 \, [1.02]^{3.14}$$

$$= 0.185$$

(v) The time (t_c) after which critical DO deficit (D_c) occurs is given by equation (2.21) as,

$$t_c = \frac{1}{K_1 (t - f)} \log_{10} \left[\left\{ 1 - (f - 1) \frac{D_a}{L} \right\} f \right]$$

where, $\qquad K_2$ = 0.185

$\qquad\qquad K_1$ = 0.098

$\therefore \qquad f = \dfrac{K_2}{K_1} = \dfrac{0.185}{0.098} = 1.888$

$\qquad\qquad L_a$ = 21.45 mg/lit

$\qquad\qquad D_a$ = 2.5 mg/lit

$\therefore \qquad t_c = \dfrac{1}{0.098\,(1.888 - 1)} \log_{10}$

$$\left[\left(1 - \frac{0.888 \times 2.5}{21.45} \right) 1.888 \right]$$

$$= \frac{1}{0.098\,(0.888)} \times 0.228$$

$$= 2.625 \text{ days}$$

(vi) Now, Distance = Velocity × Travel time

$$= 0.2 \times (2.625 \times 24 \times 60 \times 60)$$

$$= 45.36 \text{ km} \qquad \text{... Ans.}$$

(vii) D_c is now given by equation (2.24) as

$$\left(\frac{L_a}{D_c \cdot f} \right)^{f - 1} = f \left(1 - (f - 1) \frac{D_a}{L_a} \right)$$

or $\quad \left(\dfrac{21.45}{D_c \times 1.888} \right)^{0.888} = 1.888 \left(1 - \dfrac{0.888 \times 2.5}{21.4} \right)$

or $\quad \dfrac{21.45}{1.888 \, D_c} = (1.692)^{\frac{1}{0.888}} = (1.692)^{1.126} = 1.808$

or $\qquad D_c = \dfrac{21.45}{1.888 \times 1.808} = 6.28 \text{ mg/lit}$

Hence, the critical DO deficit equal to 6.28 mg/lit occurs at 45.36 km downstream of A, after 2.625 days.

$$\text{... Ans.}$$

SOLVED EXAMPLES ON SEWAGE DISPOSAL ON LAND

Example 2.25 : *A town having population of 70000 dispose sewage by land treatment. The water supply from the waterworks is 150 lit/capita/day. The land used for sewage disposal can absorb 70 m³ of sewage per hectare per day. Determine the land area required.*

Solution :

Given :   Population = 70000

Rate of water supply = 150 lit/capita/day

Total water supplied per day

$$= 70000 \times 150$$
$$= 10.5 \times 10^6 \text{ litres}$$
$$= 10500 \text{ cu.m.}$$

Assuming that sewage generation is 80% of this water,

∴   The quantity of sewage generated per day

$$= \frac{80}{100} \times 10500$$
$$= 8400 \text{ cu.m.}$$

∴   Area of land required for disposing of sewage

$$= \frac{8400}{70}$$
$$= 120 \text{ hectares}$$

Providing 50% extra land for rest and rotation, we have

Total land area required

$$= 1.5 \times 120 = 180 \text{ hectares} \qquad \textbf{... Ans.}$$

Example 2.26 : *A town having a population of 45000 and the rate of water supply as 140 lit/capita/day, disposes sewage by land treatment. The area of sewage farm is 160 hectares. The area included an extra provision of 50% for rest and rotation. If 80% of the water is converted into sewage, determine the consuming capacity of soil.*

Solution :

Quantity of water produced per day

$$= 45000 \times 140$$
$$= 6.3 \times 10^6 \text{ litres}$$
$$= 6300 \text{ cu.m.}$$

Quantity of sewage produced per day

$$= \frac{80}{100} \times 6300$$
$$= 5040 \text{ cu.m./day}$$

Area of farm land provided

$$= 160 \text{ ha(including 50\% for rest and rotation)}$$

∴   Actual available land for application

$$= \frac{160}{1.5} = 106.67 \text{ ha}$$

∴   Consuming capacity of soil

$$= \frac{5040}{106.67}$$
$$= 47.25 \text{ cu.m/ha/day} \qquad \textbf{... Ans.}$$

2.17   RECYCLING OF SEWAGE

- With 80 countries and 40% of the world's population facing chronic water problems and with the demand for water doubling every two decades, these extracts mentioned above merit action. The largest source of reuse resides in agriculture and the equally largest misplaced resource is sewage in the habitations. In the "Handbook on Service Level Benchmarking" by MoUD, reuse and recycling of sewage is defined as the percentage of sewage recycled or reused after appropriate treatment in gardens and parks, irrigation, etc. and, is to be at least 20% to begin with. The objective of this chapter is to bring out guiding principles for practice in India.

2.17.1 Current Practices in India

In India treated sewage is being used for a variety of applications such as

1. Farm Forestry,
2. Horticulture,
3. Toilet flushing,
4. Industrial use as in non-human contact cooling towers,
5. Fish culture and
6. Indirect and incidental uses.

They are briefly mentioned hereunder.

- The CMWSSB has been promoting the growth of farm forestry in Chennai from the 1980s and this helps to promote a micro climate in a city environment.
- The Indian Agricultural Research Institute, Karnal has carried out research work on sewage farming and has recommended an irrigation method for sewage fed tree plantations.
- The University of Agricultural Sciences, Dharwad, Karnataka has found that sewage could be used in producing vermicompost to be used for tree plantations provided its details with respect to composition of toxic substances are known.
- Chandigarh is using treated sewage for horticulture needs of its green areas.
- Delhi has put in place planned reuse of treated sewage for designated institutional centres.

- The Government of Karnataka has issued an official directive to take all necessary steps to ensure that only tertiary treated sewage is used for non-potable purposes, like all gardening including parks, resorts and golf course. The Bangalore Water Supply and Sewerage Board will make all arrangements including construction of filling points, installation of vending machines at STP for supply of tertiary treated sewage in multiples of thousand litres and that non-compliance of the directions attracts penal provisions in accordance with section 15 and section 17 of the Environment (Protection) Act 1986.

- In major metropolitan cities like Delhi, Mumbai, Bangalore and Chennai treated grey water is being used for toilet flushing in some of the major condominiums and high rise apartment complexes on a pilot scale. Care should be taken to ensure that Ultra filtration membranes are used in the treatment process to safeguard against chances of waterborne diseases.

- Secondary treated sewage is purchased and treated for use in cooling water makeup in the industrial sector from as early as 1991 in major industries like Madras Refineries, Madras Fertilizers, GMR Vasavi Power plant in Chennai as also in Rashtriya Chemicals and Fertilizsers in Maharashtra and most recently in the Indira Gandhi International Airport in Delhi and Mumbai International Airport.

- In Kolkata, the Mudiali fish farm occupying an area of 400 hectares is used for growing fish, which is then sold for human consumption.

- The UNDP conducted a detailed study in the 1970s and identified a sand basin on the coast of Bay of Bengal, where secondary treated sewage of the Chennai city can be infiltrated through percolation ponds and extracted for specific industrial use in the nearby petro-chemical complex. However, this project has not been implemented.

- The Bengaluru city is facing a freshwater crisis and it has been considered to study a pilot model of the Singapore NEWater for indirect augmentation of water by advanced treatment of secondary facilities. At present, this project proposal is a statement of capability to formulate a technically feasible and financially viable project and of course the biggest challenge of going through and obtaining public acceptance is understandably a long drawn out process.

2.17.2 Current Practices in the World

The use of treated sewage elsewhere in the world is listed herein.

- **Agriculture:** It is used for irrigation in certain places in Africa, Israel, Mexico and Kuwait.

- **Farm Forestry:** Treated sewage is used for watering urban forests, public gardens, trees, shrubs and grassed areas along roadways in certain places in Egypt, Abu Dhabi, Woodburn in Oregon USA. It is also used for timber plantation in Widebay Water Corporation in Queensland, Australia. It is used for alfalfa plantation in Albirch Palestine.

- **Horticulture:** Certain places in Elpaso in Texas, Durbin Creek in Western California in USA.

- **Toilet Flushing:** Certain locations in Chiba Prefecture, Kobe City, and Fukuoka City and Tokyo Metropolitan in Japan.

- **Industrial and Commercial:** essentially used for cooling purposes in Sakaihama Treated Wastewater Supply Project, Japan, Bethlehem Steel mills, USA. Sewage reclaimed as high quality water is supplied to Mondi Paper Mill and SAPREF Refinery in Durban, South Africa. Landscape and golf course irrigation in Hawai,

- **Fish Culture:** It is used in fish hatcheries / fish ponds in Vietnam and in Bangladesh

- **Groundwater Recharge:** Orlando and Orange County Florida, Orange County California, Phoenix (Arizona), Santa Rosa (California) Recharge Project all in USA.

- **Indirect Recharge of Impoundments**: Restoration of Meguro River in Japan, NEWater project in Singapore, Windhoek in Namibia, Berlin in Germany

- **Other Uses:** Coach cleaning, subway washing and water for building construction is being practised in Jungnang, Nanji, Tancheon, Seonam in Seoul and treated sewage sprinkled on the water retentive pavement that can store water inside paving material at Shiodome Land Readjustment District (Shio Site) in Tokyo and this reduces the surface temperature.

2.17.3 Treatment Options

Appropriate technology unit processes include (but are not limited to) the following:

- Preliminary Treatment by Rotating Micro Screens;

- Vortex Grit Chambers;

- Lagoons Treatment (Anaerobic, Facultative and Polishing), including recentdevelopments in improving lagoons performance;

- Anaerobic Treatment processes of various types, mainly, AnaerobicLagoons, Upflow Anaerobic Sludge Blanket (UASB) Reactors, AnaerobicFilters and Anerobic Piston Reactor (PAR);

- Physicochemical processes of various types such as Chemically EnhancedPrimary Treatment (CEPT); (vi) Constructed Wetlands;

- Stabilization Reservoirs for wastewater reuse and other purposes;

- Overland Flow;

- Infiltration-Percolation;

- Septic Tanks; and

- Phytoremediation technology

- Submarine and Large Rivers Outfalls.

Out of these processes, various combinations can be set up. Combinations can also include some other simple processes such as Sand Filtration and Dissolved Air Floatation (DAF), which are not considered appropriate processes per se but are in fact appropriate processes. One interesting combined process is the generation of effluents suited for reuse in irrigation based on retreatment by one of the mentioned unit processes followed by a stabilization reservoir.

Treatment of wastewater for further reuse

The method of sewage treatment in each specific case depends on the desired product quality and may include the following types of processing:

- Pre-treatment: screens (to remove large solid particles), sand traps, pre-aeration, extraction of oil particles;

- Primary treatment via sedimentation in special tanks to settle a significant part of the deposited particulate matter. The process can be speeded up by use of chemical additives (flocculants);

- Secondary treatment with the use of aerobic bacteria that provides biological destruction of organicload and biological oxidation of organic matter dissolved in the wastewater.

- Treatment processes with suspended biomass (active sludge);

- Removal nutrients (nitrates and phosphates);

- Nitrification, denitrification, dephosphorization: wastewater treatment processes that provide, respectively, the transformation of organic nitrogen into nitrates, the decomposition of nitrates with formation of gas nitrogen and removal from waste water soluble salts of phosphorus;

- Final disinfection is used to provide sanitary and hygienic safety wastewater. The technique involves the use of reagents based on chlorine or ozone, or treatment with ultraviolet irradiation.

- In addition to the above ways there are two technologies of natural wastewater treatment, which may be used as cleaning second or third level – constructed wetlands and bioponds. Both technologies are used primarily in small wastewater treatment plants or in areas where there is the opportunity to use the extensive grounds.

- Essence of constructed wetlands system is that the waste water is gradually passing the channels, where the surface (water depth is 40-60 cm) is directly under the open sky, and the bottom being (all the time under water) is the basis of the roots of particular plant species.

- Biological ponds requires large volumes, which are periodically filled with fecal wastewater. There is a gradual biological degradation of pollution by microbial colonies (due to aerobic or an aerobic metabolism) or algae presented in the pond.

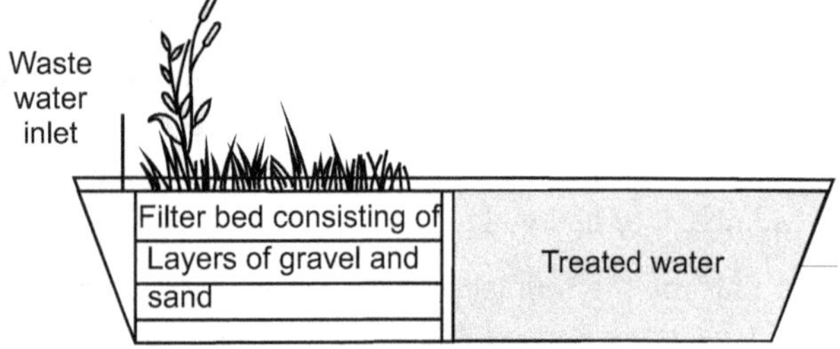

Fig. 2.19 : Constructed wetland

Table : 2.22 Reuse options

Types of Reuse	Treatment
Urban Reuse Landscape irrigation, vehicle washing, toilet flushing, fire protection, commercial air conditioners, and other uses with similar access or exposure to the water.	Secondary Filtration Disinfection
Agriculture Reuse For Non-Food Crops Pasture for milking animals, fodder, fiber and seed crops.	Secondary Disinfection
Indirect Potable Reuse Groundwater recharage by spreading into potable aquifers.	Site specific Secondary and disinfection (min.) May also need filtration and / or advanced wastewater treatment.

2.17.4 Uses of Water Recycled from Water Treatment Plant

1. Secondary Treatment; Biological Oxidation, and Disinfection

- Surface irrigation of orchards and vineyards
- Non-food crop irrigation
- Restricted landscape impoundments
- Groundwater recharge of non-potable aquifer
- Wetlands, wildlife habitat, stream augmentation
- Industrial cooling processes

2. Tertiary and advance treatment

- Landscape and golf course irrigation
- Toilet flushing
- Vehicle washing
- Food crop irrigation
- Unrestricted recreational impoundment
- Indirect potable reuse- Groundwater recharge of potable aquifer and surface water reservoir augmentation

Advantages:

- This technology reduces the demands on potable sources of freshwater.
- It may reduce the need for large wastewater treatment systems, if significant portions of the waste stream are reused or recycled.
- The technology may diminish the volume of wastewater discharged, resulting in a beneficial impact on the aquatic environment.
- Capital costs are low to medium for most systems and are recoverable in a very short time; this excludes systems designed for direct reuse of sewage water.
- Operation and maintenance are relatively simple except in direct reuse systems where more extensive technology and quality control are required.
- Provision of nutrient-rich wastewaters can increase agricultural production in water-poor areas.
- Pollution of rivers and groundwaters may be reduced.
- Lawn maintenance and golf course irrigation is facilitated in resort areas.
- In most cases, the quality of the wastewater, as an irrigation water supply, is superior to that of well water.
Disadvantages
- If implemented on a large scale, revenues to water supply and wastewater utilities may fall as the demand

for potable water for non-potable uses and the discharge of wastewaters is reduced.

- Reuse of wastewater may be seasonal in nature, resulting in the overloading of treatment and disposal facilities during the rainy season; if the wet season is of long duration and/or high intensity, the seasonal discharge of raw wastewaters may occur.
- Health problems, such as water-borne diseases and skin irritations, may occur in people coming into direct contact with reused wastewater.
- Gases, such as sulfuric acid, produced during the treatment process can result in chronic health problems.
- In some cases, reuse of wastewater is not economically feasible because of the requirement for an additional distribution system.
- Application of untreated wastewater as irrigation water or as injected recharge water may result in groundwater contamination.

EXERCISE

1. What are the factors affecting self-purification of polluted streams ? What measures would you recommend to control stream pollution in India ?

2. What is meant by "Environmental Pollution" ?

   Describe what happens when untreated sewage from a town is discharged into a nearby stream.

3. Sewage disposal systems are to be provided for :

   (i) an isolated residential building with ten users

   (ii) a small town of 2000 persons located on the bank of a river

   (iii) a town with 10,000 persons located on the bank of a small river.

   Describe the possible methods of sewage disposal for each case and bring out the advantages and disadvantages of the various methods listed.

4. Write a detailed note on land treatment of sewage dealing with the chemical as well as the engineering aspects of the process.

5. What do you understand by self-purification property of a stream ? Explain the factors affecting this property.

6. Explain clearly the methods, problems, and limitations of land disposal of sewage.

7. Explain how domestic sewage is different from industrial waste.

8. Explain in details the methods of self purification in a stream.

9. Enlist the factors governing the reoxygenation and deoxygenation of stream.

10. Write short notes on :

   (a) Stream standards and effluent standards.

   (b) Oxygen sag curve.

   (c) Disposal of sewage by dilution.

   (d) Dilution factor.

   (e) Sewage farming.

   (f) Irrigation farming.

   (g) Zones of pollution in a stream.

   (h) Sewage sickness.

   (i) Broad irrigation.

   (j) Self purification constant.

   (k) Disposal of sewage in sea water.

   (*l*) Minimum DO content in polluted stream for survival of aquatic life.

11. Explain the importance of effluent standards and stream standards in controlling water pollution.

12. Explain the terms with sketch :

   (i) Self purification of stream.

   (ii) Importance of critical DO deficit.

13. Discuss in details the disposal of sewage on land.

14. Give effluent standards for the following (W/W to be discharged in river).

   (i) BOD, (ii) COD, (iii) Temperature,

   (iv) Total suspended solids.

15. Mention various methods of wastewater disposal. Discuss their merits and demerits. Explain the conditions favourable for their adoption.

16. Explain the difference between dilution process if the wastewater effluents are disposed of in stream water and sea water.

17. A city discharges 2000 litres per second of sewage into a stream whose minimum rate of flow is 7500 litres per second. The temperature of sewage as well as water is 20°C. The BOD_5 at 20°C for sewage is 300 mg/lit and that of the stream is 90% of the saturation DO If the minimum DO to be maintained in the stream is 5 mg/lit, find out the degree of sewage treatment required. Assume the deoxygenation coefficient as 0.1 and reoxygenation coefficient as 0.3.

INDUSTRIAL WASTE WATER TREATMENT MANAGEMENT

3.1 INTRODUCTION

- All industrial wastes affect, in some way, the normal life of a natural stream or river or lake.

- Industrial wastewaters are generally much more polluted than the domestic or commercial wastewaters. If they are discharged directly in the receiving waters, it may result in discolouring, foul smell and killing of aquatic life, apart from making the water unfit for various other purposes.

- The industries are, therefore, generally prevented by legal laws, from discharging their untreated effluents. It, therefore, becomes necessary for the industries to treat their wastewaters in their individual effluent treatment plants (PET) or common effluent treatment plants (CETP) i.e. collect the wastewaters from different industries and give the treatment, before discharging their effluents on land or natural streams or rivers or lakes as per availability of disposal point.

- Industrial wastewater is the result of substances other than water having been dissolved or suspended in water. The objective of industrial wastewater treatment is to remove those dissolved or suspended substances. The best approach to working out an effective and efficient method of industrial wastewater treatment is to examine those properties of water and of the dissolved or suspended substances that enabled or caused the dissolution or suspension, then to deduce plausible chemical or physical actions that would reverse those processes.

3.1.1 Characteristics of Industrial Waste

The following materials can cause pollution:

- **Inorganic Salts:** Inorganic salts, which are present in most industrial wastes as well as in nature itself, cause water to be "hard" and make a stream undesirable for industrial, municipal and agricultural usage. Salt laden waters deposit scale on municipal water- distribution pipelines, increasing resistance to flow and lowering the overall capacity of the lines. Another disadvantage is that, under proper environmental conditions, inorganic salts especially nitrogen and phosphorous induce the growth of microscopic plant life (algae) in surface waters

- **Acids and /or Alkalis:** Acids and Alkalis discharged by chemical and other industrial plants make a stream undesirable not only recreational uses such as swimming and boating, but also for propagation of fish and other aquatic life. High concentrations of sulfuric acid, sufficient to lower the pH below 7.0 when free chlorine is present, have been reported to cause eye irritation to swimmers. A low pH may cause corrosion in air conditioning equipment and a pH greater than 9.5 enhances laundering.

- **Organic Matter:** Organic Matter exhausts the oxygen resources of rivers and creates unpleasant tastes, odours and general septic conditions. It is generally conceded that the critical range for fish survival is 3to 4 mg/l of D.O certain organic chemicals such as phenols, affect the taste of domestic water supplies.

- **Suspended Solids:** Suspended solids settle to the bottom or wash up on the banks and decompose, cause sing odours and depleting oxygen in the river water. Fish often die because of a sudden lowering of the oxygen content of a stream. Visible sludge creates unsightly conditions and destroys the use of a river for recreational purposes. These solids also increase the turbidity of the watercourse.

- **Floating Solids and Liquids:** These includes oils, greases, and other materials which float on the surface, they not only make the river unsightly but also obstruct passage of light through the water, retarding the growth of vital plant food.

  Some specific objections to oil in streams are that it

  (i) Interferes with natural reaeration

  (ii) is toxic to certain species of fish and aquatic life

  (iii) Causes trouble in conventional water treatment processes by imparting tastes and odours to water and coating sand filters with a tenacious film.

- **Heated Water:** An increase in water temperature, brought about by discharging wastes such as condenser waters in to streams, has various adverse effects. Streams waters which vary in temperature from one hour to the next are difficult to process efficiently in Municipal and industrial water treatment plants, and heated stream water are of decreased value for industrial cooling, indeed are industry may so increase the temperature of a stream that a neighbouring industry downstream cannot use the water since there may be less D.O in warm water than in cold, aquatic life suffers and less D.O is available for natural biological degradation of any organic pollution discharged into these warm surface waters. Also bacterial action increases in higher temperatures, resulting in accelerated repletion of the streams oxygen resources.

- **Colour :** Colour is contributed by textile and paper mills, tanneries, slaughterhouses and other industries, is an indicator of pollution. Colour interferes with the transmission of sunlight into the stream and therefore lessens photosynthetic action. Furthermore, municipal and industrial water plants have great difficulty, and scant success in removing colour from raw water.

- **Toxic Chemicals:** Both inorganic and organic chemicals, even in extremely low concentrations, may be poisonous to fresh water fish and other smaller aquatic microorganisms. Many of these compounds are not removed by municipal treatment plants and have a cumulative effect on biological systems.

- **Microorganisms :** A few industries, such as tanneries and slaughterhouses, sometimes discharge wastes containing bacteria. These bacteria are of two significant types: i) bacteria which assist in the degradation of the organic matter as the waste moves down stream. This process may aid in "seeding" a stream and in accelerating the occurrence of oxygen sag in water. ii) bacteria which are pathogenic, not only to other bacteria but also to humans.

- **Radio Active Materials:** Cumulative damaging effects on living cells.

- **Foam Producing Matter:** Foam producing matter such as is discharged by textile mills, paper and pulp mills and chemical plants, gives an undesirable appearance to the receiving streams. It is an indicator of contamination and is often more objectionable in a stream than lack of oxygen.

- **Effects On Sewage Treatment Plants:** The Pollution Characteristics of Wastes having readily definable effects on Sewers and Treatment Plants can be Classified as follows:

- **Biochemical Oxygen Demand:** It is usually exerted by Dissolved and Colloidal Organic Matter and imposes a load on the Biological units of the Treatment Plant. Oxygen must be provided so that Bacteria can grow and oxidise the organic matter. An Added B.O.D load,

caused by an increase in Organic Waste, requires more Bacterial Activity, more oxygen, and greater Biological Unit capacity for its Treatment, which (makes) increases the capital cost and operating cost.

- **Suspended Solids:** Suspended Solids are found in considerable quantity in many Industrial Wastes, such as Paper & Pulp Effluents. Solids removed by settling and separated from the flowing Sewage are called Sludge, which may then undergo an Anaerobic Decomposition known as Digestion and pumped to drying beds or vacuum filters for extraction of additional water. Suspended Solids in Industrial Waste may settle more rapidly or slowly than Sewage Suspended.

- **Matter:** If Industrial Solids settle faster than those of Municipal Sewage, Sludge should be removed at shorter intervals to prevent excessive build up: a Slow Settling one will require a longer detention period and larger basins and increases the likelihood of sludge Decomposition with accompanying nuisances, during Sewage-Flow Periods. Any Increased demands on the System usually require larger Sludge handling devices and may ultimately necessitates an increase in the Plants capacity, with resulting Higher Capital and Operating Expenses.

- **Floating and Coloured Materials**: Floating Materials and Coloured Matter such as Oil, Grease and Dyes From Textile-Finishing Mills, are disagreeable and visible nuisances. A Modern Treatment Plant will remove normal Grease loads in Primary Settling Tanks, but abnormally high loads of predominantly emulsified Greases from Laundries, Slaughterhouses etc Passing through the Primary Units into the Biological Units will clog Flow Distributing Devices and Air Nozzles. Volume: A Sewage Plant can handle any Volume of Flow if its units are sufficiently large. The Hydraulic Capacity of all Units must be analysed, Sewer Lines must be examined for Carrying Capacity, and all other Treatment Units are to be Designed for excessive loading

- **Harmful Constituents:** Toxic Metals, Acids, or Alkalis, Pieces of Fat, Flammable Substances, Detergents and Phenols etc. cause nuisance in Treatment Plants.

3.1.2 Impact of Industrial Waste

Environmental Impacts

- The impacts of environmental degradation may result in decreased levels of dissolved oxygen, physical changes to receiving waters, release of toxic substances, bioaccumulation or biomagnifications in aquatic life, and increased nutrient loads.

- Wastewater is a complex resource, with both advantages and inconveniences for its use. Wastewater and its nutrient contents can be used for crop production, thus providing significant benefits to the farming communities and society in general. However, wastewater use can also impose negative impacts on communities and on ecosystems.

- The wide spread use of wastewater containing toxic wastes and the lack of adequate finances for treatment is likely to cause an increase in the incidence of wastewater-borne diseases as well as more rapid degradation of the environment. Although the harmful effects of using contaminated wastewater effluents could be delayed for several years using intensive and heavy irrigations, it adversely affect groundwater quality when nutrients leach down the soil. Eutrophication due to excessive amounts of nutrients contributes to the depletion of dissolved oxygen. It is important to note that other constituents of wastewater effluents also play an important role in the depletion of DO. The bacterial breakdown of organic solids present in wastewater and the oxidation of chemicals in it can consume much of the dissolved oxygen in the receiving water bodies. These effects may be immediate and short-term or may extend over months or years as a result of the buildup of oxygen consuming material in the bottom sediments

- The impacts of low dissolved oxygen levels include an effect on the survival of fish by increasing their susceptibility to diseases, retardation in growth, hampered swimming ability, alteration in feeding and migration, and, when extreme, lead to rapid death. Long-term reductions in dissolved oxygen concentrations can result in changes in species composition (Welch, 1992; Chambers & Mills, 1996; Environmental Canada, 1997).

- Poorly treated wastewater effluent can also lead to physical changes to receiving water bodies. All aquatic life forms have characteristic temperature preference and tolerance limits. Any increase in the average temperature of a water body can have ecological impacts. Because municipal wastewater effluents are warmer than receiving water bodies, they are sources of thermal enhancement (Welch, 1992; Horner et al., 1994).

- Also, the release of suspended solids into receiving waters can have a number of direct and indirect environmental effects, including reduced sunlight penetration (reduced photosynthesis), physical harm to fish, and toxic effects from contaminants attached to suspended particles

- Another environmental impact of untreated wastewater effluent, which at times can be linked to health, is the phenomenon of bioaccumulation and biomagnifications of contaminants. Due to the phenomenon of bioaccumulation, certain substances which are in low concentrations or barely measurable in water can sometimes be found in high concentrations in the tissues of plants and animals. These substances tend to be stable, live long chemically, and are not easily broken down by digestive processes.

- In some cases, through the process of biomagnification, the concentrations of some of the contaminants may be increased dramatically through passage in the food chain that is prey to predators. Because of the processes of bioaccumulation and biomagnification, very low concentrations of certain substances in wastewater are of concern.

- Examples of such substances include organochlorine pesticides, and heavy metals. Although there are several other sources of persistent bioaccumulatives (such as toxic substances in the environment), including industrial discharges and deposition of atmospheric contaminants, municipal wastewater remains one of the most significant

- Also, the release of toxic substances from wastewater into receiving water bodies has direct toxic impacts onterrestrial plants and animals. The toxic impacts may be acute or cumulative. Acute impacts from wastewater effluents are generally due tohigh levels of ammonia and chlorine, high loads of oxygen demanding materials, or toxic concentrations ofheavy metals and organic contaminants. Cumulative impacts are due to the gradual buildup of pollutants in receiving water, which only become apparent when a certain threshold is exceeded.

- In addition, eutrophication of water sources can lead to nutrient enrichment effects. Nutrient-induced production of aquatic plants in receiving water bodies has the following detrimental consequences:

  - ➢ Algal clumps, odours and decolouration of the water, thus interfering with recreational and aesthetic water use;

  - ➢ extensive growth of rooted aquatic life interferes with navigation, aeration and channel capacity;

- ➤ Dead macrophytes and phytoplankton settle to the bottom of a water body, stimulating microbial breakdown processes that require oxygen, thus causing oxygen depletion;
- ➤ Extreme oxygen depletion can lead to the death of desirable aquatic life;
- ➤ Cilliated diatoms and filamentous algae may clog water treatment plant filters and result in reduced backwashing, and algal blooms may shade and submerge aquatic vegetation, thus reducing or eliminating photosynthesis and productivity.

- In lakes, rivers, streams and coastal waters where large algal blooms are present, the death of the vast numbers of phytoplankton that make up the blooms may smother the lake bottom with organic materials. The decay of these materials can consume most or all of the dissolved oxygen in the surrounding water, thus threatening the survival of many species of fish and other aquatic life forms.

- The net effect of eutrophication on an ecosystem is usually an increase of a few plant types and a decline in the number and variety of other plant and animal species in the system In most surface waters, total ammonia concentrations greater than 2mg/L are toxic to aquatic life, although this varies between species and life stages.

- Studies that have been carried out on the toxicity of ammonia to freshwater vegetation have shown that concentrations greater than 2.4mg/L inhibit photosynthesis. Nitrate is believed to cause a reduction in amphibian populations. Adverse effects are reported to be poor larval growth, reduced body size, and impaired swimming ability.

3.2 SAMPLING   (May 17)

The sampling techniques used in a wastewater survey must assume that representative samples are obtained.

1. Grab Sample :

Samples are taken at a point beneath the surface where the turbulence is thoroughly mixing up the sewage particles.

2. Composite Sample :

Grab samples are collected at regular intervals during a day. These different samples are now mixed together and amount utilised from each specimen is proportional to the rate of flow at the time the specimen was collected.

Sample Preservation :Sample preservation is must to maintain the physical, chemical and biological integrity of the samples during the interim periods between sample collection and sample analysis. Table 3.1 shows preservation of wastewater samples.

Table 3.1 : Preservation of Wastewater Samples

Sr. No.	Parameter	Preservative	Maximum Holding Period
1.	pH	None available	–
2.	BOD	Refrigeration at 4°C	6 hrs
3.	COD	2 ml/l H_2SO_4	7 days
4.	Colour	Refrigeration at 4°C	24 hours
5.	Dissolved oxygen (DO)	Determined on site	No holding
6.	Oil and grease	2 ml/l H_2SO_4, 4°C	24 hours
7.	Solids	None available	–
8.	Turbidity	None available	–
9.	Acidity – Alkalinity	Refrigeration at 4°C	24 hours
10.	Hardness	None required	–
11.	Metals, total	5 ml/l HNO_3	6 months
12.	Metals, dissolved	Filtrate : 3 ml/l 1 : 1 HNO_3	6 months
13.	Nitrogen, ammonia	40 mg/l $HgCl_2$, 4°C	7 days
14.	Nitrogen, Kjedahl	40 mg/l $HgCl_2$, 4°C	Unstable
15.	Calcium	None required	–
16.	Chloride	None required	–
17.	Cyanide	NaOH to pH 10	24 hours
18.	Fluoride	None required	–
19.	Phosphorus	40 mg/l $HgCl_2$, 4°C	7 days
20.	Phenolics	1.0 g $CuSO_4$ + H_3PO_4 to pH 4, 4°C	24 hours
21.	Sulphate	Refrigeration at 4°C	7 days
22.	Sulphide	2 ml/l Zn acetate	7 days

3.3 CHARACTERISTICS OF WASTEWATER

- The characterization of the raw waste is essential in the planning for effective and economical methods of water pollution control. As the varying nature of the industrial wastes, many of the recent installations have designed their treatment units with due consideration to the raw waste characteristics and the effluent characteristics as established by the Indian Standards Institution (ISI), State Pollution Control Boards or Central Pollution Control Boards (CPCB) or by the local administrative authorities.

- Table 3.2 shows pollution characteristics of certain typical Indian Industries.

Table 3.2 : Pollution Characteristics of Certain Typical Indian Industries

Industry	Pollution Characteristics	Suggested Treatment Methods
Paper and pulp	High BOD, strong colour, high COD/BOD ratio, highly alkaline, high sodium content.	Lime treatment for colour, chemicals recovery, biological treatment.
Sugar	High volatile solids, high BOD, low pH.	Biological treatment
Textile (cotton)	High BOD, highly alkaline, high suspended solids.	Chemical and biological treatment
Dairy	High suspended solids, high dissolved solids, high BOD, presence of oil and grease.	Biological treatment.
Distillery and Brewery	High chloride, strong colour, very high BOD, high sulphate.	Biological treatment.

3.4 EFFLUENT STANDARDS

Table 3.3 shows ISI tolerance limits for the sewage and industrial effluents and that of inland surface water.

Table 3.3 : ISI Tolerance Limits for the Sewage and Industrial Effluents and that of Inland Surface Water

Characteristics	Tolerance Limits for Sewage Effluents Discharged into Inland Surface Water. IS : 4764 – 1973	Tolerance Limits for Industrial Effluents Discharged into — Inland Surface Water IS : 2490-1974	Tolerance Limits for Industrial Effluents Discharged into — Public Sewers IS : 3306-1974	Tolerance Limits for Inland Surface Water, When Used as Raw Water for Public Water Supplies and Bathing Ghats IS : 2296-1974
1	2	3	4	5
BOD (5 day, 20°C), mg/lit	20	30	500	3
COD, mg/lit	–	250	–	–
pH	–	5.5 – 9.0	5.5 – 9.0	6.0 – 9.0
Total suspended solids, mg/lit	30	100	600	–
Temperature, °C	–	40	45	–
Oil and Grease, mg/lit	–	10	100	0.1
Phenolic compounds, mg/lit	–	1.0	5	0.005
Cyanides (as CN), mg/lit	–	0.2	2.0	0.01
Sulphides (as S), mg/lit	–	2.0	–	–
Fluorides (as F), mg/lit	–	2.0	–	1.5
Total residual chlorine, mg/lit	–	1.0	–	–
Insecticides, mg/lit	–	zero	–	zero
Arsenic (as As), mg/lit	–	0.2	-	0.2
Cadmium (as Cd), mg/lit	–	2.0	–	–
Chromium, hexavalent (as Cr), mg/lit	–	0.1	2.0	0.05 (Total chromium)
Copper, mg/lit	–	3.0	3.0	–
Lead, mg/lit	–	0.1	1.0	0.1
Mercury, mg/lit	–	0.01	–	–
Nickel, mg/lit	–	3.0	2	–
Selenium, mg/lit	–	0.05	–	0.05
Zinc, mg/lit	–	5.0	15.0	–
Chloride (as Cl), mg/lit	–	–	600	600
Sulphates, mg/lit	–	–	–	1000
% Sodium	–	–	60	–
Ammoniacal Nitrogen, mg/lit	–	50	50	–
Nitrates (as NO_3), mg/lit	–	–	–	50
Radioactive materials :				
α-emitters $\mu cm/l$ β– emitters $\mu cm/l$	– –	10^{-7} 10^{-6}	– –	10^{-9} 10^{-8}

...Conti.

Dissolved Oxygen, mg/lit	–	–	–	40% of the saturation value, or 3 mg/lit, whichever is higher.
Coliform organism (monthly average) – MPN per 100 ml	–	–	–	Should not exceed 5000. (should not exceed 20000 with less than 5% samples, and 5000 with less than 20% samples).

3.5 POLLUTANTS IN INDUSTRIAL WASTEWATER

The following materials can cause pollution :

- **Inorganic Salts :** These are present in most industrial wastes as well as in nature itself, cause water to be hard and make a stream undesirable for municipal, industrial and agricultural usage. These include carbonates, chlorides, nitrogen, phosphorus, etc.

- **Acids and/or Alkalies :** Due to acids and/or alkalies, a stream unsuitable not only for recreational uses such as swimming and boating, but also for propagating of fish and other aquatic life.

- **Organic Matter :** Due to the presence of organic matter this exhausts the oxygen resources of rivers and creates unpleasant tastes, odours and general septic conditions.

- **Suspended Solids :** These are settled to the bottom and decomposed causing odours and depleting oxygen in the river water.

- **Floating Solids :** These include oil, greases and other materials which float on the surface.

  Some of the specific objections to oil in streams are that, it (1) is toxic to aquatic life, (2) interferes with natural reaeration, (3) creates a fire hazard when present in the water surface, (4) renders boiler feed and cooling water unusable.

- **Colour :** It is contributed by tanneries, textile, paper mills etc. Colour producing substances impart objectionable colour in the receiving water bodies. Colour interferes with the transmission of sunlight into the stream and therefore lessens photosynthetic action.

- **Toxic Chemicals :** These include sulphides, cyanides, acetylene, alcohol, phenols, heavy metals (i.e. zinc, cadmium, copper, hexavalent chromium, arsenic etc.) due to which flora and fauna of receiving waters is greatly affected.

Difference between Domestic Wastewater And Industrial Wastewater

Domestic Wastewater	Industrial Wastewater
Domestic sewage is wastewater discharged from sanitary conveniences in residential, office, commercial, factories and various institutional properties.	Industrial (including agro-industrial) wastewaters have very varied compositions depending on the type of industry and materials processed. In other words Industrial wastewater is the aqueous discard that results from the use of water in an industrial manufacturing process or the cleaning activities that take place along with that process.
It is a complex mixture containing primarily water (approximately 99%) together with organic and inorganic constituents.	Some of these wastewaters can be organically very strong, easily biodegradable, largely inorganic, or potentially inhibitory. This means TSS, BOD5 and COD values may be in the tens of thousands mg/L.
These constituents or contaminants comprised suspended, colloidal and dissolved materials.	Industrial wastewater is the result of substances other than water having been dissolved or suspended in water.
Domestic sewage, since it contains human wastes, also contains large numbers of micro-organisms and some of these can be pathogenic. Waterborne bacterial diseases that can be present in sewage include cholera, typhoid and tuberculosis. Viral diseases can include infectious hepatitis.	If wastewater from sanitary units is added in industrial wastewater it may also contain pathogens and organic matter.

...Conti.

Inorganic constituents include chlorides and sulphates, various forms of nitrogen and phosphorous, as well as carbonates and bicarbonates.	Inorganic contents differs from industry to industry. Heavy metals are also found in many industrial wastewaters.
Proteins and carbohydrates constitute about 90% of the organic matter in domestic sewage. These arise from the excreta, urine, food wastes, and wastewater from bathing, washing, and laundering, and because of the latter, soaps, detergents, and other cleaning products can be found as well.	Organic food processing industries may contain carbohydrates and other allied organics
Domestic sewage has a flow pattern which typically shows two peaks — in the morning before the start of working hours and in the evening after the population has returned from work.	The flow pattern of industrial wastewater streams can be very different from that of domestic sewage since the former would be influenced by the nature of the operations within a factory rather than the usual activities encountered in the domestic setting. A significant factor influencing the flow pattern would be the shift nature of work at factories.

3.6 METHODS OF TREATMENT

(Nov. 15, 16, 17, May 16, 17, 18)

Treatment of industrial wastewater generally consists of one or more of the following processes :

1. Equalization and proportioning.
2. Neutralization.
3. Physical treatment.
4. Chemical treatment.
5. Biological treatment.

1. Equalization and Proportioning :

- Equalization is a method of retaining wastes in a mixed basin until the effluent discharged is fairly uniform in its sanitary characteristics (colour, pH, turbidity, BOD etc.). Air is sometimes injected into these basins to provide : (i) chemical oxidation of reduced compounds, (ii) better mixing, (iii) agitation, to keep suspended solids from settling out.

- Only holding of waste is not sufficient to equalize waste. Adequate mixing must take place, so that each unit volume of waste discharged will be mixed with other unit volume of waste discharged many hours ago. The mixing may be brought about in the following ways :

➢ Proper distribution and baffling.

➢ Mechanical agitation.

➢ Aeration.

➢ Combinations of all three.

- Proportioning means discharge of industrial wastes in proportion to the flow of municipal sewage. The objective of proportioning is to keep the percentage of industrial wastes entering the municipal sewage plant constant. This procedure has several purposes :

1. To minimise fluctuations in sanitary standards in the treated effluent,
2. To protect biological treatment devices from receiving shock loads of industrial wastes; which may inactive the bacteria.

- Equalization Tank in any Treatment Plant serves the purpose of maintaining desired flow rate as well as for making mixture homogeneous. It is evident that effluent of any industry is heterogeneous in nature , to bring the best out of any wastewater treatment plant homogeneous mixture is must Homogeneous mixture in Equalization Tank is done via the actions of coarse bubble diffusers, oxygen transfer efficiency of a coarse bubble diffuser is 10%-20% and are capable of delivering 6 - 12 m3 / hour air, typical diameter of coarse bubble diffuser is 150 mm.

Criteria or Precautions to be followed for Designing any Equalization Tank :

- **Easy Accessible :** The tank must be easy accessible for periodic cleaning with safety and comfort for a gang of cleaners to carry out the task.

- **Safe Entry :** The operator has to access the inside of the tank for periodic cleaning and to maintain the diffusers. The tank has relatively low oxygen level and the raw wastewater emits hazardous gases and strong odor. So the operator must be provided with safety equipments such as mask, gloves, full body harness, gum boots . The following features must be present, at minimum: Ventilation to dispel the gases/odor Good lighting that reaches inside the tank Platform that allows easy reach inside the tank.

- **Aeration and Mixing :** Diffusers in sufficient number to cover the entire floor Uniform placement of diffusers The aeration is uniform across the surface. To prevent the solids from settling in dead zones (which in turn avoids the necessity to clean the tank frequently)

- **Use Coarse Bubble Diffusers :** Unlike the fine bubble diffusers, the coarse bubble diffusers are not affected by fluctuating water level.
- **Floor Slope Towards Suction Pit for Pumps :** For complete evacuation of contents by pump. Floor slope is given so that during tank cleaning, all the water is collected in the suction pit of the pumps and the equalization tank is evacuated by pumps alone, with minimum manual cleaning required.
- **Openness of the Tank :** To prevent accumulation of gases, easy access to diffusers for maintenance purposes, and safe and secure entry for periodic cleaning.

2. **Neutralization :**

- Neutralization means neutralizing the excessive acidity or alkalinity of the particular wastewater, by adding alkali or acid respectively to the wastewater. This may be done either in the equalization tank, if the conditions so permit, or else in a separate neutralization tank.
- There are many acceptable methods for neutralizing overacidity or alkalinity of wastewaters.

Some of these methods include :

- Passing acid wastewaters through beds of limestone,
- Mixing wastes so that the net effect is a near-neutral pH,
- Adding the proportions of concentrated solutions of soda ash (Na_2CO_3) or caustic soda ($NaOH$),
- Mixing acid wastes with lime slurries,
- Adding compressed CO_2 to alkaline wastes,
- Adding sulphuric acid to alkaline wastes.

3. **Physical Treatment :**

- It is similar to primary treatment of domestic wastewater. It consists of domestic wastewater. It consists of separating the suspended inorganic matter by physical processes, like screening, sedimentation, floatation and filtration.
- Sedimentation is necessary to remove high percentage of heavy organic solids. Floatation is provided to remove the finer particles. In floatation, create fine air bubbles, by injecting air into the tank from the bottom. The rising air bubbles lift the finer particles to the surface from where these are removed by skimming.

4. **Chemical Treatment :**

- Physico-chemical treatment is applied to the industrial wastewaters in absence of biological treatment or in conjunction with biological treatment.
- The chemical treatment is used to recover the dissolved organic matter from the wastewater. One or more of the following chemical processes, that are used for industrial wastewater treatment, are :

- > chemical oxidation;
- > chemical coagulation or chemical precipitation;
- > hyper filtration or reverse osmosis;
- > electrodialysis;
- > adsorption;
- > deionisation;
- > thermal reduction; and
- > air-stripping.

5. **Biological Treatment :**

- When industrial wastewaters contain large quantities of biodegradable substances, then biological treatment is essential.

Table 3.4 : Type of Treatment depends on the following BOD/COD Ratio

$\dfrac{BOD}{COD}$ Ratio	Type of Treatment
$\dfrac{BOD}{COD}$ > 0.6	Biologically treatable, without acclimatisation.*
$\dfrac{BOD}{COD}$ > 0.3 and < 0.6	Acclimatisation is essential for biological treatment.
$\dfrac{BOD}{COD}$ < 0.3	Biological treatment is not necessary.

* Acclimatisation is a process of seeding or raising initial microbial population under a controlled condition, by gradual exposure of the wastewater in increasing concentration.

- The following various biological treatment methods that can be used are :
- > trickling filter;
- > activated sludge process;
- > oxidation pond;
- > aerated lagoon;
- > oxidation ditch; and
- > oxidation lagoon, followed by aerated lagoon.

3.6.1 Prevention and Control of Industrial Pollution Volume Reduction

- In general, the first step in minimizing the effects of industrial wastes on receiving streams and treatment plants is to reduce the volume of such wastes. This may be accomplished by :
1. Classification of wastes
2. Conservation of wastewater
3. Changing production to decrease wastes
4. Reusing both industrial and municipal effluents as raw water supplies
5. Elimination of batch or slug discharges of process wastes.

1. Classification of Wastes

- If wastes are classified, so that manufacturing process waters are separated from cooling waters, the volume of water required for intensive treatment may be reduced considerably. Sometimes it is possible to classify and separate the process waste themselves, so that only the most polluted ones are treated and the relatively uncontaminated are discharged without treatment.

2. Conservation of Waste Water

- Water conserved is waste saved. For e.g. a paper mill which recycles white water (water passing through a wire screen upon which paper is formed) and thus reduces the volume of wash waters.

- Concentrated recycled wastewaters are often treated at the end. Since usually it is impractical and uneconomical to treat the wastewaters as they complete each cycle. The saving are two folds : both water costs and waste treatment costs are lower.

- Steel mills reuse cooling waters to quench ingots and coal processors reuse water to remove dirt and other non-combustible materials from coal.

3. Changing Production to Decrease Wastes

- This is an effective method of controlling the volume of wastes, but it is difficult to put into practice. It is hard to persuade production men to change their operations just to eliminate wastes. Normally, the operational phase of engineering is planned by the chemical, mechanical or industrial engineer, whose primary objective is cost savings. The sanitary engineer, on the other hand, has the protection of public health and the conservation of a natural resources as his main considerations. Yet there is no reason why both objectives cannot be achieved.

- Waste treatment at the source should be considered an integral part of production. If a chemical engineer argues that it would cost the company money to change its methods of manufacture in order to reduce pollution at the source, the sanitary engineer can do more than simply enter a plea for the improvement of mankind's environment. The reduction in the amount of sodium sulfite used in dyeing ; or sodium cyanide used in plating, and of other chemical used directly in production has resulted in both reduction in quantify of wastewaters and saving of money. The balancing the quantities of acids and alkalis used in a plant often results in a neutral waste, with saving of chemicals, money and time spent in waste treatment.

4. Reusing both Industrial and Municipal effluents for raw water supplies

- Practiced mainly in areas where water is scarce and / or expensive, this is proving a popular and economical method of conservation, of all the sources of water available to industry, sewage plant effluent is the most reliable at all seasons of the year.

5. Elimination of Batch or slug discharges of process wastes

- During manufacturing process, one or two steps are sometimes repeated. which results in production of a significantly higher volume and strength of waste during that period If this waste is discharged in a short period of time ; it is usually referred to as a slug discharge. This type of wastes because of its concentrated contaminants and / or surge in volume, can be troublesome to both treatment plants. and receiving streams. There are at least two methods of reducing the effects of these discharges :
  (i) The manufacturing firm alters its practice so as to increase frequency and lessen the magnitude of batch discharges.
  (ii) Sludge wastes are retained in holding basins from which they are allowed to flow continuously and uniformly over an extended (usually 24 hours) periods

3.6.2 Strength Reduction

Waste strength reduction is the second major objective in industrial waste management. The strength of waste can be reduced by :
  1. Process change
  2. Equipment modifications
  3. Segregation of wastes
  4. Equalization of wastes
  5. By-product recovery
  6. Proportioning wastes and
  7. Monitoring waste streams.
  8. Accidental spills.

1. Process Change

- Many industries have resolved waste problems through process changes e.g. textile and metal-fabricating industries. Textile finishing mills are faced with a problem of disposal of highly polluted wastes from sizing, kiering, desizing and dyeing processes. Starch has been traditionally used as sizing agent. This starch after hydrolysis became a major source of waste pollutants. New substituent of starch; carboxylmethyl cellulose has been developed to solve above problem which has reduced BOD value significantly.

- In the metal-plating industries, several changes in the process have been suggested to eliminate or reduce cyanide strengths :
  - ➢ Change from copper-cyanide plating solution to acid-copper solutions
  - ➢ Replace the C_uCN_2 strike before the copper-plating bath with a nickel strike
  - ➢ Substitute H_3PO_4 for H_2SO_4 in pickling of steel
- Use alkaline derusters instead of acid solutions to remove light rust which occurs during storage (the overall pH will be raised nearer to neutrality by this procedure, which will also alleviate corrosive effects on piping and sewer lines Coal mining company has modified its process to wash raw coal with acid mine water rather than a public or private water supply. In this way. The mine drainage waste is neutralized, while the washed coal is free from any impurities.

2. **Equipment Modifications**

- Changes in equipment can effect a reduction in the strength of the waste. usually by reducing the amounts of contaminants entering the waste stream. For e.g. in pickle factories, screens placed over drain lines in the tanks prevents the escape of seeds and piece of vegetables and fruits which add to the strength and density of the waste.
- Similarly, traps on the discharge pipeline in poultry plants prevents emission of feathers and piece of fats. Waste strength reduction occurred in the dairy industry by redesigning the large milk-cans used to collect farmer's-milk. The new cans were constructed with smooth necks so that they could be drained faster and more completely. This prevented a large amount of milk waste from entering streams and sewage plants.

3. **Segregation of Wastes**

- Segregation of wastes reduces the strength and / or the difficulty of treating the final waste from an industrial plant. For e.g. In metal finishing plants, which produce wastes containing both chromium and cyanide as well as other metals. It is necessary to segregate the cyanide-bearing waste, make them alkaline and oxidize them. The chromium waste, on the other hand, have to be acidified and reduced.
- Two effluents can then be combined and precipitated in an alkaline solution to remove the metals. Without segregation, poisonous hydrogen cyanide gas would develop as a result of acidification. A textile mill manufacturing finished cloth produced the four types of waste such as Grey waste, white waste, dye waste and kier waste. The combined waste are quite strong, difficult and expensive to treat.

- However, when the strong liquid kiering waste was segregated from the other wastes, chemically neutralized, precipitated, and settled, the supernatant could be treated chemically and biologically along with the other three wastes, because the strength of the resulting mixture was considerably less than that of the original combined waste.

4. **Equalization of Wastes**

- Main objectives of equalization are to stabilization of pH, BOD and settling of solids and heavy metals. Plants which have many products, from a diversity of processes, prefer to equalize their waste. This requires holding wastes for a certain period of time, depending on the time taken for the repetitive processes in plant.
- For e.g., if a manufactured item, requires a series of operations that take eight hours, the plant needs an equalization designed to hold the wastes for that eight hour period. The effluent from an equalization basin is much more consistent in its characteristics than in each separate influent to that same basin. Stable effluents are treated more easily and efficiently than unstable, ones by industrial and municipal treatment plants. Sometimes equalization may produce an effluent which warrants no further treatment.

5. **By-Product Recovery**

- This is the Utopian aspect of industrial-waste treatment; the one phase of the entire problem which may lead to economic gains. The industrial management should consider the possibility of building a recovery plant which will produce a marketable by-product and at the same time solve a troublesome waste problem.
- For e.g. (1) The dairy industry treats skim milk with dilute acid to manufacture casein. Casein manufacturers in turn utilize their waste to precipitate albumin. The resulting albumin waste is used in the crystalization of milk sugar, and the residue from this process is utilized as poultry feed ; (2) Yeast factories evaporate a portion of their waste and sell residue for cattle fed; (3) In sewage treatment plant. Methane gas from sewage digesters is commonly utilized for heat and power and some cities digested and dried sewage sludge is used as a manure ; (4) The sulphite waste liquor by products from paper mills are used in fuel, road binder, cattle fodder, fertilizer, insulating compounds and in production of alcohol and artificial vanillin.

- Once a by-product is developed and put into production, it is difficult to identify the new product with a waste-treatment process. For example, when sugar is extracted from sugar cane, a thick slurry known as molasses is left. It has many uses, one of the best known beings in the production of commercial alcohol.

6. Proportioning Waste

- By proportioning its discharge of concentrated wastes into the main sewer, a plant can often reduce the strength of its total waste to the point where it will need a minimum of final treatment or will cause the least damage to the stream or treatment plant.

7. Monitoring Waste Streams

- Remote sensing devices that enable the operator to stop, reduce, or redirect the flow from any process when its concentration of contaminants exceeds certain limits are an excellent method of reducing waste strengths. In fact, accidental spills are often the sole cause of stream pollution or malfunctioning of treatment plants these can be controlled.

8. Accidental Spills

- Accidental discharges of significant process solutions represent one of the most severe pollution hazards. There are some measures that can be taken to reduce the likelihood of accidents and severity when and if they occur. Some suggestions for genera) use include the following :

    (i) Make certain that all pipelines and valves in the plant are clearly identified.

    (ii) Allow only certain knowledgeable persons to operate these valves.

    (iii) Install indicators and warning systems for leaks and spills.

    (iv) Provide for detention of spilled wastewater in holding basins or lagoons until proper waste treatment can be accomplished.

    (v) Monitor all effluents quantity and quality to provide a positive public record if necessary.

3.7 NEUTRALIZATION

- Excessive acid or alkaline wastes should not be discharged withouttreatment into a receiving stream. There are many acceptable methods forneutralizing over-acidity or over-alkalinity of wastewaters such as :

    1. Mixing wastes ;
    2. Limestone treatment for acid waste ;
    3. Lime slurry treatment for acid waste ;
    4. Caustic Soda treatment for acid waste ;
    5. Waste-boiler - Flue-Gas ;
    6. CO_2 treatment for alkaline waste ;
    7. Producing CO_2 in alkaline wastes ; and
    8. Sulfuric acid treatment for alkaline waste.

- The volume, kind, and quantity of acid or alkali to be neutralized are the main factor to be kept in mind to decide, what kind of neutralizingagent is to be used.

- In any lime neutralization treatment, adequate reaction time for anacid effluent should be given so that they can reach minimum pH value.

- Became during storage of alkaline wastes come in contact with air andCO2, which will slowly dissolve in the waste and lower the pH. Biological treatments are more efficient at pH value nearer to neutrality.

1. Mixing Wastes

- Mixing of wastes can be accomplished within a single plant operation or between neighbouring industrial plants. Acid and alkaline wastes may be produced. Individually within one plant and proper mixing of these waste at appropriate time. Results in neutralization.

- For example, if one plant produces an alkaline waste which can be pumped conventionally to an area adjacent to a plant discharging an acid waste, an economical and feasible system of neutralization results for each plant.

2. Lime Stone Treatment for Acid Wastes

- In this method, acid wastes is passed through beds of limestones ($CaCO_3$) at a rate of about one gallon per minute per square feet (1gpm/ft2). Neutralization takes place with following reaction :

$$CaCO_3 + H_2SO_4 \longrightarrow CaSO_4 + H_2CO_3$$

- This reaction will continue as long as excess limestone is availableland in an active state. The first condition can be fulfilled by providing asufficient quantity of limestone, the second condition is difficult to meet, because excess of acid will precipitate the calcium sulfate and causes ubsequent coating and inactivation of the limestone.

- Disposing of the used limestone beds can be a serious drawbackto this method of neutralization. Since the used limestone must bereplaced by fresh at periodic interval, frequency of replacement depending on the quantity and quality of acid wastes being passed through a bed.

- When there are extremely high acid loads, foaming may occur, especiallywhen organic matter is also present in the waste.

3. Lime-Slurry Treatment for Acid Wastes

- Mixing acid wastes with lime slurries is an effective procedure forneutralization. The reaction is similar to that obtained with limestone beds but the advantages of using lime slurry than lime stone beds are :

(i) In this case, lime is used up continuously because it is convened to $CaSO_4$ and which is carried out in the waste ;

(ii) Lime possesses a high neutralizing power and its action can behastened by heating or by oxygenating the mixture.

(iii) It is relatively inexpensive, but in large quantities the cost can bean important item.Neutralization of nitric and sulfuric acid wastes in concentrations upto about 1.5 per cent (in case of H2SO4) was accomplished satisfactorily by using a burned dolomitic stone containing 47.5 percent CaO, 34.3 percent MgO, and 1.8 per cent $CaCO_3$.

4. Caustic-Soda Treatment for Acid Wastes

- Concentrated solutions of caustic soda or sodium carbonate causeneutralization of acid waste. These neutralizers are more powerful than lime or limestone and some agents are required in small volume. Anotheradvantage is that the reaction products are soluble and thus they do not increase the hardness of receiving waters.

- When sodium hydroxide is used as a neutralizing agent for carbonic andsulfuric acid wastes, the following reactions take place :

$$Na_2CO_3 + CO_2 + H_2O \rightarrow 2NaHCO_3$$

Carbonic Acid Waste

$$2NaOH + CO_2 \rightarrow NaHCO_3 + H_2O$$
$$NaOH + H_2SO_4 \rightarrow NaHSO_4 + H_2O$$

Sulfuric Acid Waste

$$NaHSO_4 + NaOH \rightarrow Na_2SO_4 + H_2O$$

- Both these neutralizations take place in two steps and the end products depend on the final pH desired. For example, one treatment mayrequire a final pH of only 6, and the most of end product is $NaHSO_4$.

- Another treatment may require a pH of 8, with most of product being Na_2SO_4.

5. Neutralization of Alkaline Wastes :

- Using Waste-Boiler-Fuel GasIt is a most economical method for neutralization of alkaline waste.

- In this method, when waste boiler-flue gas (containing 14 percent CO_2) is passed through the alkaline waste CO_2 dissolve in waste water and will form carbonic acid (a weak acid) which in turn reacts with caustic wastesto neutralize the excess alkalinity as follows :

(i) $CO_2 + H_2O \rightarrow H_2CO_3$
     Flue gas + Waste Water Carbonic acid

(ii) $H_2CO_3 + 2NaOH \rightarrow Na_2CO_3 + 2H_2O$
Carbonic acid Soda ash

(iii) $H_2CO_3 + Na_2CO_3 \rightarrow 2NaHCO_3 + H_2O$

- Excess Soda ash SodiumCarbonic Acid in waste biocarbonate in waste

6. Carbon - Dioxide Treatment for Alkaline Wastes

- Bottled CO_2 is applied to wastewaters which neutralizes alkalinewastes on the same principle as boiler-flue gas [i.e. it forms a weak acid(carbonic acid) dissolved in water. When the quantity of alkaline waste is large, there is operational difficulty as well as high cost.

7. Producing Carbon-dioxide in Alkaline Wastes

- CO_2 produced by fermentation of an alkaline, organic waste help inneutralization of alkaline waste by decreasing the pH. Similar by fermenting alkaline fat-sugar wastes with yeast, CO_2 is produced which can be used for neutralization.

8. Sulfuric-Acid Treatment for Alkaline Wastes

- The addition of sulfuric acid to alkaline wastes causes neutralization. In this method, storage and feeding equipment requirementare low due to high acidity value of N3804, but it is difficult to handlebecause of its corrosiveness. The neutralization reaction which occurs, when it is added to wastewater is as follows :
$$2NaOH + H_2SO_4 \rightarrow Na_2SO_4 + 2H_2O$$

3.8 EQUALIZATION AND PROPORTIONING

- Equalization is a method of retaining wastes in a basin so that the effluent discharged is :

(i) Uniform in its sanitary characteristics (pH, color, turbidity. Alkalinity & BOD).

(ii) Concentration of effluent contaminants lowered

(iii) Physical, chemical, and biological reactions takes place. Sometimes, air is injected into these basins to provide

    1. Better mixing ;

    2. Chemical oxidation of reduced compounds

    3. Some degree of biological oxidation. And

    4. Agitation to prevent suspended solids from settling.

- The size and shape of the basin vary with the quantity of waste andthe pattern of discharge from the factory. Most basins are rectangular, square or triangular in shape. The capacity of the tank should beadequate to hold all the waste from the plant. Almost all industrial plantsoperate on a cycle basis, thus if the cycle of operations is repeated every two hours, an equalization tank which can hold a two-hour flow will usually be sufficient.

- If the cycle is repeated only each 24 hours, the equalization basin must be high enough to hold a 24 hours flow of waste.The mere holding of waste however, is not sufficient to equalize it.Each unit volume of waste discharged must be adequately mixed with other unit volumes of waste discharged many hours previously. Thismixing may be brought about in the following :

    (i) Proper distribution and baffling

    (ii) Mechanical agitation

    (iii) Aeration

    (iv) Combinations of all three.

3.9 PROPORTIONING

- Proportioning means the discharge of industrial wastes inproportion to the flow of municipal sewage in sewers. In most cases, it ispossible to combine equalization and proportioning in the same basin.

- The objective of proportioning in sewers is to keep constant the percentage of industrial wastes to domestic sewage flow entering themunicipal sewage plant. This procedure have several other purposes also:

    (i) To protect municipal sewage treatment plant using chemical by sudden overdose of chemicals contained in the industrialwaste.

    (ii) To protect biological treatment devices from shock loads ofindustrial wastes, which may inactivate the bacteria

    (iv) To minimize fluctuations of sanitary standards in the treatedeffluents.

- The rate of flow of industrial waste varies from instant to instant, asdoes the flow of domestic sewage, and both empty into the same sewagesystem. Therefore, the industrial waste must be equalized and retained, then proportioned to the sewer or stream according to the volume ofdomestic sewage or stream flow. To facilitate proportioning, an industryshould construct a holding tank with a variable speed pump to control the effluent discharge.

- But in some cases, domestic sewage treatment plant isusually located some distance from an industry, hence signalling the timeand amount of flow is difficult and sometimes quite expensive. To solve this problem, many industries have separate pipelines through which they pump their wastes to the municipal treatment plant. The wastes areequalized separated at the site of the municipal plant and proportioned to the flow of incoming municipal waste water. Separate lines are not ; ofcourse, always possible or even necessary.

3.10 DISCHARGE OF INDUSTRIAL WASTE

There are two general methods of discharging industrial waste inpropomon to the flow of domestic sewage at the municipal plant :

1. Manual control ;

2. Automatic electronic control.

1. Manual Control

- It is lower in initial cost but less accurate. It involves determiningthe flow pattern of domestic sewage for each day in the week, over a period of months. Usually one does this by examining the flow records ofthe sewage plant or by studying the hourly water-consumption figures forthe city.

2. Automatic Control

- It involves placing a metering device, which registers the amount offlow at the most convenient main sewer connection. This device translatesthe rate of flow in the sewer to a recorder which is located near the industrial plant's holding tank. The pen on the recorder accelerates eithera mechanical or a pneumatic (air) control system for opening or closingthe diaphragm of the pump.

- The initial cost of this automatic control system is higher than that of manual control, they will usually return theinvestment many times by saving in labour costs.Some industrial and municipal sewage-plant operations think thatthe best time to release a high proportion of industrial waste to the seweris at night ; when the domestic sewage flow is low. Whether night releaseis a good idea depends on the type of treatment used and the character ofthe industrial waste.

- If the treatment is primarily biological and theindustrial wastes contain readily decomposable organic matter and notoxic elements discharging the largest part of the industrial waste to thetreatment plant at night is indeed advisable, since, this ensures arelatively constant organic load delivered to the plant day and night.

3.10.1 Removal of Oil By Air Floatation

- To remove oil from wastewater, the wastewater is normally pumped to a gravity oil/water separator. There, with time and quiescence, most of the free oil droplets rise to the surface where they are skimmed off, but emulsified and dissolved oils remain. Normally, further oil removal is accomplished by chemically breaking the emulsion followed by treatment of the wastewater in an air flotation unit.

- The oil/water emulsion leaving the American Petroleum Institute (API) separator normally, according to Sylvester and Byseda,138 has oil droplets less than 30 |i.e. in diameter and oil concentrations less than 200 mg/L Air flotation has been used for many years in the beneficiation of ores. Its first application in the wastewater-treatment field was in the flotation of suspended solids, fibres, and other lowdensity solids. Flotation also was used for the thickening of activated sludge and flocculated chemical sludges.

- More recently, air flotation has been applied to the removal of oils and greases from wastewater because it is a practical, reliable, and efficient treatment process. Air flotation is widely used to treat oil-bearing effluents from a wide variety of sources: refineries, ship's bilge and ballast waste, de-inking operations, metal plating, meat processing, laundries, iron and steel plants, soap manufacturing, chemical processing and manufacturing plants, barrel and drum cleaning, wash rack and equipment maintenance, glass plants, soybean processing, mill waste, and aluminium forming. The process of flotation consists of four basic steps:

  1. Bubble generation in the oily wastewater.
  2. Contact between the gas bubble and the oil droplet suspended in the water
  3. Attachment of the oil droplet to the gas bubble.
  4. Rise of the air/oil combination to the surface where the oil (and normally attendant suspended solids) is skimmed off.

- Flotation utilizes the differential density between the bubbles to which the oil droplets and small solid particles become attached and the water to effect separation. Since the agglomerates have a lower density than the medium in which they are immersed, they rise to the surface where they are removed. There are essentially five different types of flotation systems, their classification being based on the method of bubble formation:

1. **Dissolved Air :** The gas precipitates from a supersaturated solution as a result of the reduction of pressure.
2. **Induced (Dispersed) Air :** The gas and liquid are mechanically mixed to induce bubble formation in the liquid.
3. **Froth :** The gas is directly injected into the fluid by means of a sparger.
4. **Electrolytic :** The bubbles are generated by electrolysis of the water.
5. **Vacuum :** The air is extracted from a saturated solution by a negative pressure.

- However, of the five systems, only the first four are utilized industrially to any extent for wastewater treatment.
- Following table 3.5 shows methods those can be adopted in prevention of Industrial pollution

Table 3.5 : Industrial pollution prevention methods.

Practice	Method
Temporary Construction Sediment Control	Implement and encourage practices to retain sediment within construction project area; see Temporary Construction Erosion and Sediment Control Factsheets for additional information
Wind Erosion Control	Institute a local program for wetting of open construction surfaces and other sources for windblown pollutants.
Emission Regulation 1	Follow local, state and federal regulatory requirements for control of air emissions.
Material Storage Control 1	Reduce or eliminate spill and leakage loss by properly inspecting, containing, and storing hazardous materials and having a cleanup plan that can be quickly and efficiently implemented.
Dumpster and Landfill Management	Ensure that contaminated material is contained to prevent solid and/or liquid waste from being washed into storm sewer systems or water bodies.
Better Turf Management 1	Ensure that mowing, fertilization, pesticide application, and irrigation are completed in ways that will prevent or reduce grass clippings, sediment, and chemicals from entering storm sewer systems; use native vegetation where possible.
Better Parking Lot Cleaning 1	Maintain streets and parking lots frequently and especially in the spring by sweeping, picking up litter, and repairing deterioration; pressure wash pavement only as needed and avoid using cleaning agents.

...Conti.

Better Impervious Surface Deicing1	Properly store and conservatively apply salt, sand, or other deicing substances in order to prevent excessive and / or unnecessary contamination; implement anti-icing and prewet salt techniques for increased deicing efficiency.
Proper Vehicle Management 1	Ensure that vehicles are fueled, maintained, washed and stored in a manner that prevents the release of harmful fluids, including oil, antifreeze, gasoline, battery acid, hydraulic and transmission fluids, and cleaning solutions.
Storm Sewer System Maintenance	Regularly clean debris from storm sewer inlets, remove sediment from catch basin sumps, and remove any illicit connections to storm sewer systems.
Sanitary Sewer System Maintenance 2	Regularly inspect and flush sanitary pipes to ensure that there are no leaks in the system and that the system is properly functioning.

3.10.2 Control of Industrial Pollution

- Till now, there are about 17 industries which are declared to be most polluting. These include the caustic soda, cement, distillery, dyes and dye intermediaries, fertilisers, iron and steel, oil refineries, paper and pulp, pesticides and pharmaceuticals, sugar, textiles, thermal power plants, tanneries and so on. The table 3.6 enlists few of the industries, their wastes (important) and the type of pollution these induce in the environment.

- The wide variety of pollutants as shown above enter the environment and disturb the natural eco-system affecting the biota. Due to industrial activities, a variety of poisonous gases like NO, SO_2, NO_2, SO_3, Cl_2, CO, CO_2, H_2SO_4 etc. volatile chemicals, dusts etc., are liberated into the atmosphere causing acute pollution problem. Besides, the accidental leakage of poisonous gases can cause havoc.

- For example, Methyl Isocyanate gas leakage from Union Carbide factory at Bhopal caused mass killing which is known as Bhopal gas tragedy. In addition to accidents, many of the above poisonous gases induce

depletion of ozone layer, creation of ozone hole. Green House effect, Global warming. Acid rain, destruction of monument and killing of living organisms disturbing the natural eco-systems.

Table 3.6 : Major Industries causing pollution

Industry	Wastes Produced	Type of Pollution
Caustic Soda	Mercury, Chlorine gas	Air, water and land
Cement dust, smoke	Particulate matter	–
Distillery	Organic waste	Land and water
Fertiliser	Ammonia, cyanide, oxides of nitrogen, oxides of sulphur	Air and water
Dye	Inorganic waste pigment	Land and water
Iron and steel	Smoke, gases, coal dust, fly ash, fluorine	Air, water and land
Pesticides	Organic and inorganic waste	Water and land
Oil Refineries	Smoke, toxic gases, organic waste	Air and water
Paper and Pulp	Smoke, organic waste	Air and water
Sugar	Organic waste, molasses	Land and water
Textiles	Smoke, particulate matter	Land and water
Tanneries	Organic waste	Water
Thermal power	Fly ash, SO_2 gas	Air and water
Nuclear power station	Radioactive wastes	Water and land
Food processing	Alkalies, phenols chromates, organic wastes	Water and land

- The nature of the industrial waste depends upon the industrial process in which these originate and the raw materials the use.

- Broadly the industrial wastes may be divided into two groups:
  1. Process waste;
  2. Chemical waste.

1. Process Waste:

- The waste generated in an industry during washing and processing of raw materials is known as process waste. The process waste may be organic or inorganic in nature depending upon the raw materials used and nature of the industry.

- The organic process wastes are liberated from food processing units, distilleries, breweries, paper and pulp industry, sugar mills etc. The inorganic process wastes may be the effluents of chemical industries; caustic soda industry, paint industry, petroleum industry, pesticide industry etc. Both organic and inorganic process wastes are toxic to living organisms.

- The solid wastes released by different industries can be divided in to two different groups i.e.

  (a) process wastes, and

  (b) packing wastes.

- Since different industries use different raw materials, the quality and quantity of solid wastes differ from industry to industry. Industries releasing the solid wastes in the form of fly ash is dumped on the ground which leads to soil pollution.

- Some amount of fly ash also contaminate atmospheric tract causing respiratory tract disorders. Metallic industries produce a lot of solid metallic waste and large quantities of slag. In addition to the release of hazardous chemical pollutants, the industries may also cause thermal pollution and noise pollution. The thermal pollution is due to release of hot water from industries into aquatic bodies. The noise pollution is due to running of heavy machinery producing a lot of noise.

2. Chemical Wastes:

- The chemical substance generated as a by-product during the preparation of a product is known as chemical waste product. The chemical waste include heavy metals and their ions, detergents, acids and alkalies and various other toxic substances.

- These are usually produced by the industries like fertiliser factories, paper and pulp industries, iron and steel industries, distilleries, sugar mills etc. These are usually liberated into nearby water bodies like rivers, lakes and seas and sometimes into lands. The entry of these chemicals into bodies may alter the pH, BOD (Biological Oxygen Demand) and COD (Chemical Oxygen Demand).

- The loading of suspended solids (ss), heavy metals and their ions brings about a drastic change in physiochemical nature of the water. The aquatic animals and plants absorb, accumulate and bio-concentrate the chemical wastes leading to bio magnifications and finally destroying the trophic levels and food chains of the eco-system. Hence these disturb the eco-system dynamics and eco-system balance of the nature.

3.10.3 Control of Industrial Pollution

- The ultimate object behind the measures to control pollution to maintain safety of Man, Material and Machinery (Three Ms). The implementation of control measures should be based on the principle of recovery or recycling of the pollutants and must be taken as an integral part of production i.e. never as a liability but always an asset.

Some Important Control Measures are:

1. **Control at Source:** It involves suitable alterations in the choice of raw materials and process in treatment of exhaust gases before finally discharged and increasing stock height upto 38 metres in order to ensure proper mixing of the discharged pollutants.

2. **Selection of Industry Site:** The industrial site should be properly examined considering the climatic and topographical characteristics before setting of the industry.

3. **Treatment of Industrial Waste:** The industrial wastes should be subjected to proper treatment before their discharge.

4. **Plantation:** Intensive plantation in the region, considerably reduces the dust, smoke and other pollutants.

5. **Stringent Government Action:** Government should take stringent action against industries which discharge higher amount of pollutants into the environment than the level prescribed by Pollution Control Board.

6. **Assessment of the Environmental Impacts:** Environmental impact assessment should be carried out regularly which intends to identify and evaluate the potential and harmful impacts of the industries on natural eco-system.

7. **Strict Implementation of Environmental Protection Act:** Water/Air/Environment Protection Act should be strictly followed and the destroyer of the environment should be strictly punished.

EXERCISE

1. What is characteristics of Industrial waste water treatments ?

2. Write a difference between domestic waste water and industrial waste water.

3. Explain the methods of treatment.

4. Explain prevention and control of Industrial pollution.

5. Explain Removal of oil by air floatation.

6. Define industrial waste water and type.

7. Explain sampling.

8. Explain strength of reduction.

◇ ◇ ◇

WASTE WATER TREATMENT METHODS

4.1 INTRODUCTION

Types of Dissolved Solids

- The dissolved solids are of both organic and inorganic types. A number of methods have been investigated for the removal of inorganic constituents from waste water. Methods which are finding wide application in advanced waste treatment are :
    1. Ion-exchange,
    2. Electrodialysis,
    3. Reverse osmosis,
    4. Evaporation,
    5. Air stripping,
    6. Adsorption.
- For the removal of soluble organics from waste water the most commonly used method is adsorption on activated carbon. Solvent extraction is also used to recover certain organic chemicals like phenol and amines from industrial waste waters.

Removal of Inorganic Solids

4.2 EVAPORATION

- "Evaporation" is a process of bringing wastewater to its boiling point and vaporizing pure water.
- The vapor is either used for power production, condensed and used for heating, or simply wasted to the surrounding atmosphere.

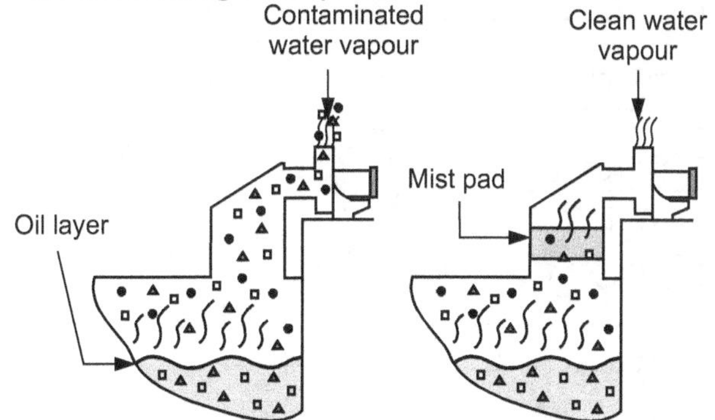

- Oil entrained water molecules
- Soap entrained water molecules
- Metal entrained water molecules

Fig. 4.1 : Evaporation Process

- The mineral solids concentrate in the residue may be sufficiently concentrated for the solids either to be reusable in the production cycle or to be disposed easily.
- This method of disposal is used for radioactive wastes, and paper mills have for a long time been evaporating their sulfate cooking liquors to a degree where they may be returned to the cookers for reuse.
- Wastewater evaporation is a time-tested method for reducing the water portion of water-based wastes. In it's simplest form, the evaporator converts the water portion of water-based wastes to water vapor, while leaving the higher boiling contaminants behind.
- This greatly minimizes the amount of waste that needs to be hauled off-site.

4.3 ION EXCHANGE

- This technique has been used extensively to remove hardness, and iron and manganese salts in drinking water supplies.
- It has also been used selectively to remove specific impurities and to recover valuable trace metals like Chromium, Nickel, Copper, Lead and Cadmium from industrial waste discharges.

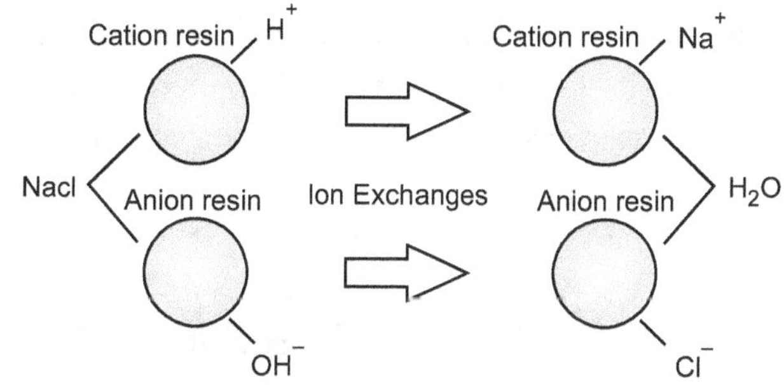

Fig. 4.2 : Ion Exchange

2R-NA	+	Ca(HCO₃)₂	->	R₂Ca	+	2NaHCO₃
Sodium Ion Exchange Resin	+	Calcium Bicarbonate in Water		Calcium Ion Exchange Resin	+	Sodium Bicarbonate in Water

4.3.1 Fully Charged Resin

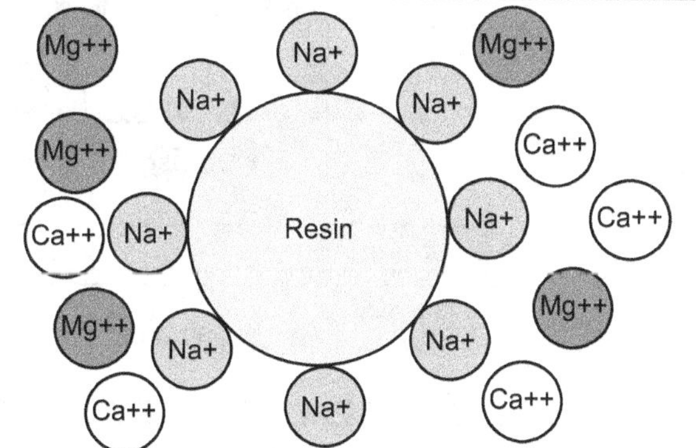

Fig. 4.3 : Fully charged Resin

4.3.2 ION Exchanged Resin

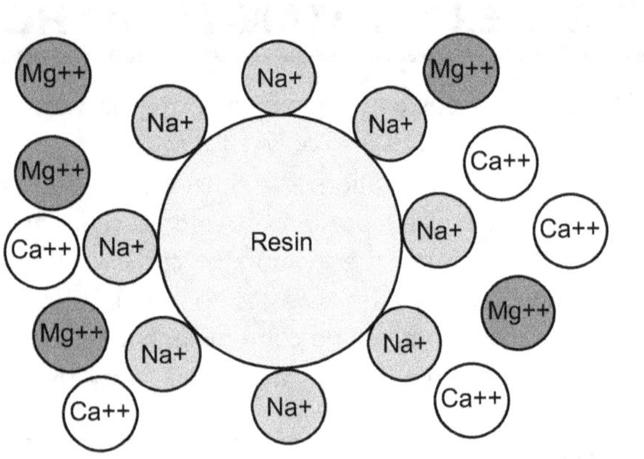

Fig. 4.4 : Ion exchanged resin

- The process takes advantage of the ability of certain natural and synthetic materials to exchange one of their ions. A number of naturally occurring minerals have ion exchange properties. Among them the notable ones are aluminum silicate minerals, which are called zeolites. Synthetic zeolites have been prepared using solutions of sodium silicate and sodium aluminate.

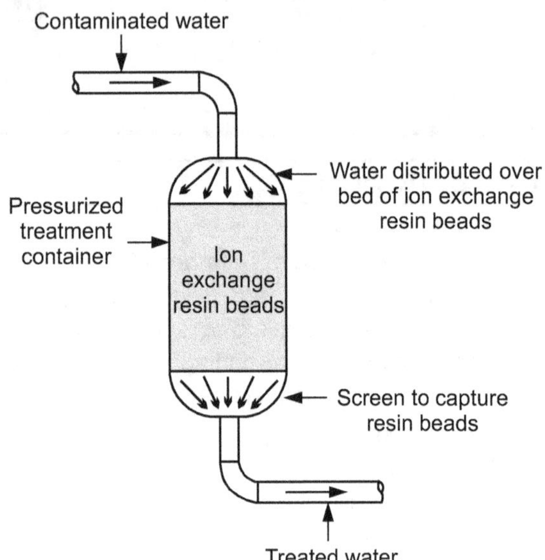

Fig. 4.5 : Ion exchange Treatment Process

4.4 MEMBRANE FILTRATION

- A membrane is a thin layer of semi-permeable material that separates substances when a driving force is applied across the membrane. Membrane processes are increasingly used for removal of bacteria, microorganisms, particulates and natural organic material, which can impart color, tastes, and odours to water and react with disinfectants to form disinfection by products.

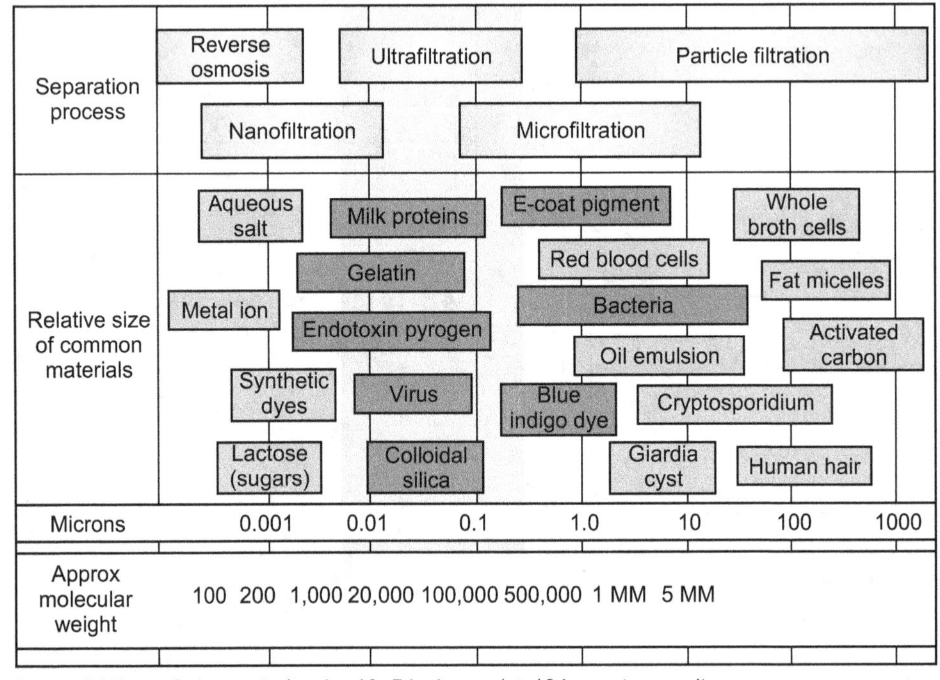

Note : 1 Micron (micrometer) = 4 × 10–5 inches = 1 × 104 angstrom units

Fig. 4.6 : Size of materials that are removed by various separation processes

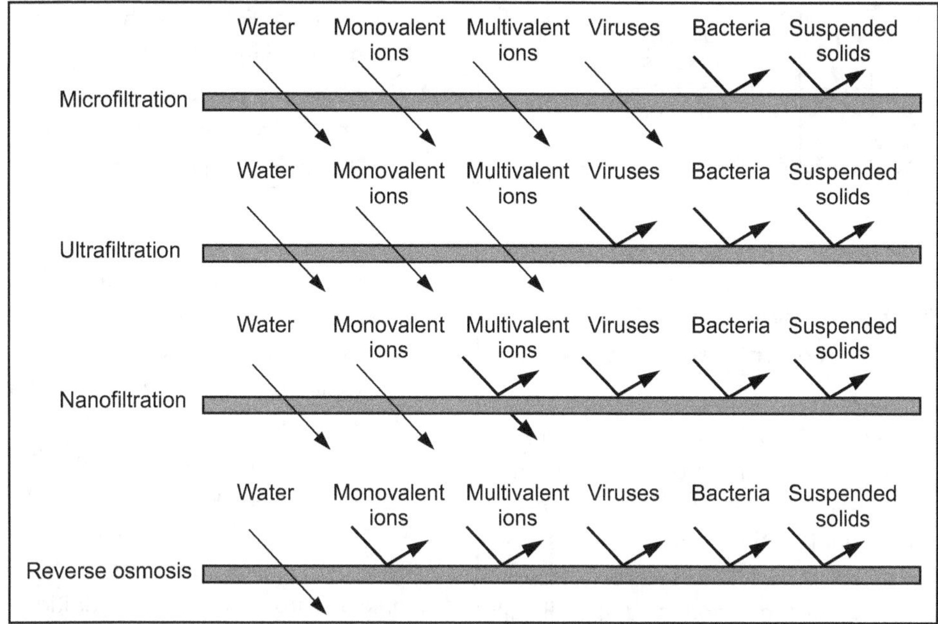

Fig. 4.7 : Substances removed from water by membrane filtration processes

- As advancements are made in membrane production and module design, capital and operating costs continue to decline. The membrane processes discussed here are Microfiltration (MF), Ultrafiltration (UF), Nanofiltration (NF), and Reverse Osmosis (RO).

4.4.1 Ultrafiltration

- An ultrafiltration filter has a pore size around 0.01 micron. A microfiltration filter has a pore size around 0.1 micron, so when water undergoes microfiltration, many microorganisms are removed, but viruses remain in the water. Ultrafiltration would remove these larger particles, and may remove some viruses. Neither microfiltration nor ultrafiltration can remove dissolved substances unless they are first adsorbed (with activated carbon) or coagulated (with alum or iron salts).

4.4.2 Nanofiltration

- A nanofiltration filter has a pore size around 0.001 micron. Nanofiltration removes most organic molecules, nearly all viruses, most of the natural organic matter and a range of salts. Nanofiltration removes divalent ions, which make water hard, so nanofiltration is often used to soften hard water.

4.4.3 Reverse Osmosis

- Reverse osmosis filters have a pore size around 0.0001 micron. After water passes through a reverse osmosis filter, it is essentially pure water. In addition to removing all organic molecules and viruses, reverse osmosis also removes most minerals that are present in the water. Reverse osmosis removes monovalent ions, which means that it desalinates the water. To understand how reverse osmosis works, it is helpful to understand osmosis.

- Osmosis occurs when a semi-permeable membrane separates two salt solutions of different concentrations. The water will migrate from the weaker solution to the stronger solution, until the two solutions are of the same concentration, because the semi-permeable membrane allows the water to pass through, but not the salt. In the following Fig. 4.8 (a) and (b) illustrate the process of osmosis.

- In reverse osmosis, the two solutions are still separated by a semi-permeable membrane, but pressure is applied to reverse the natural flow of the water. This forces the water to move from the more concentrated solution to the weaker. Thus, the contaminants end up on one side of the semi-permeable membrane and the pure water is on the other side. In the Fig. 4.8 below, Reverse Osmosis is represented in (c).

4.4.4 Difference between Osmosis and RO

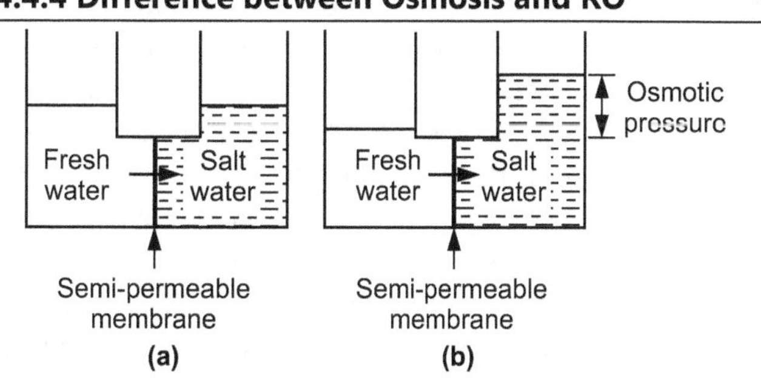

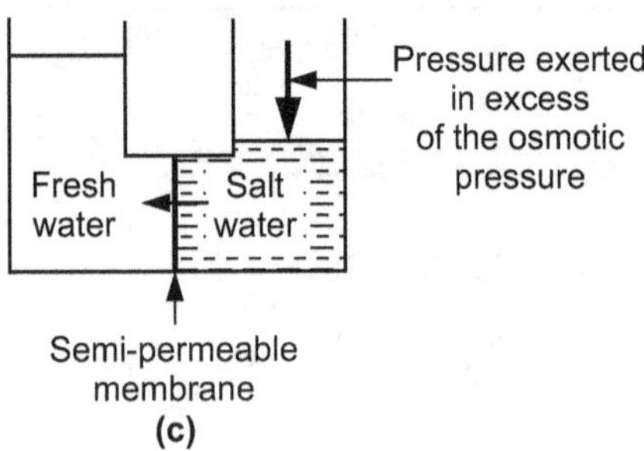

Fig. 4.8 : Osmosis / Reverse osmosis

4.4.5 What do These Three Processes Remove?

- Ultrafiltration removes bacteria, protozoa and some viruses from the water. Nanofiltration removes these microbes, as well as most natural organic matter and some natural minerals, especially divalent ions which cause hard water. Nanofiltration, however, does not remove dissolved compounds. Reverse osmosis removes turbidity, including microbes and virtually all dissolved substances.

- However, while reverse osmosis removes many harmful minerals, such as salt and lead, it also removes some healthy minerals, such as calcium and magnesium. This is why water that is treated by reverse osmosis benefits by going through a magnesium and calcium mineral bed.

- This adds calcium and magnesium to the water, while also increasing the pH and decreasing the corrosive potential of the water. Corrosive water may leach lead and copper from distribution systems and household water pipes.

4.4.6 Advantages of using Ultrafiltration, Nanofiltration or Reverse Osmosis to Treat Water

- All three of these membrane filtration processes are effective methods of treating water that cannot be treated using conventional treatment methods. Reverse osmosis, in particular, has been responsible for ending several nearly decade long Boil Water Advisories.

- The water in the First Nations community, which is located in Saskatchewan, contained high levels of organic matter, iron, manganese, ammonium and arsenic, to name a few. Besides the obvious benefit of providing safe drinking water to a community which had been under a Boil Water Advisory for approximately nine years, the reverse osmosis system (together with the biological treatment) allowed the community to treat their water using small quantities

of chemicals. For more information about the water treatment facility at Yellow

4.4.7 What are the Disadvantages of Using Ultrafiltration, Nanofiltration or Reverse Osmosis to Treat Water?

- Compared with the benefits of using membrane filtration to treat water, there are very few disadvantages. If conventional water treatment processes can effectively treat the water, then constructing a reverse osmosis water treatment facility would be an unnecessary cost. But for the First Nations communities that have been on Boil Water Advisories for many years, a reverse osmosis treatment system can be a valuable investment that can provide safe drinking water for the residents.

- Reverse osmosis removes a number of healthy minerals from water, in addition to the harmful minerals and particles. The removal of these minerals, including calcium and magnesium, can actually make water unhealthy, especially for people with inadequate diets and people who live in hot climates, as water can provide these necessary minerals. The addition of calcium and magnesium, as described above, can resolve these concerns.

- Reverse Osmosis (RO) is a membrane-technology filtration method that removes many types of large molecules and ions from solutions by applying pressure to the solution when it is on one side of a selective membrane.

- The result is that the solute is retained on the pressurized side of the membrane and the pure solvent is allowed to pass to the other side.

- To be "selective," this membrane should not allow large molecules or ions through the pores (holes), but should allow smaller components of the solution (such as the solvent) to pass freely.

- Commonly used membrane is Cellulose Acetate in RO systems. Pre-treatment of wastewater is needed to avoid membrane fouling. High COD and BOD can also affect membrane. Flushes away impurities and does not collect them. It can remove almost everything. It allows only water (H_2O) to pass very efficient but cost is high.

4.4.8 Applications of RO

- Applications include treatment and recycle of wastewaters generated from metal finishing and plating operations; printed circuit board and semiconductor manufacturing (treatment and recycle of rinse waters used in electroplating processes); automotive manufacturing (treatment and recycle of water used for cleaning and painting); food and beverage (concentration of wastewater for reuse and reduction of BOD prior to discharge); groundwater and landfill leachate (removal of salts and heavy metals prior to discharge).

4.5 ELECTRODIALYSIS

- Electrodialysis (ED) is a membrane process, during which ions are transported through semi permeable membrane, under the influence of an electric potential. The membranes are cation- or anion-selective, which basically means that either positive ions or negative ions will flow through. Cation-selective membranes are polyelectrolytes with negatively charged matter, which rejects negatively charged ions and allows positively charged ions to flow through.

- By placing multiple membranes in a row, which alternately allow positively or negatively charged ions to flow through, the ions can be removed from wastewater.

- In some columns concentration of ions will take place and in other columns ions will be removed. The concentrated saltwater flow is circulated until it has reached a value that enables precipitation. At this point the flow is discharged.

- This technique can be applied to remove ions from water. Particles that do not carry an electrical charge are not removed.

- Cation-selective membranes consist of sulphonated polystyrene, while anion-selective membranes consist of polystyrene with quaternary ammonia. Sometimes pre-treatment is necessary before the electro dialysis can take place. Suspended solids with a diameter that exceeds 10 µm need to be removed, or else they will plug the membrane pores. There are also substances that are able to neutralize a membrane, such as large organic anions, colloids, iron oxides and manganese oxide. These disturb the selective effect of the membrane.

- Pre-treatment methods, which aid the prevention of these effects are active carbon filtration (for organic matter), flocculation (for colloids) and filtration techniques.

4.5.1 Applications

- Nitrogen removal from drinking water (nitrate, ammonium)
- Desalination of organic substances
- Concentration of salts, acids and bases

4.6 ADSORPTION

- Organic pollution is the term used when large quantities of organic compounds. It originates from domestic sewage, urban run-off, industrial effluents and agriculture wastewater. Sewage treatment plants and industry including food processing, pulp and paper making, agriculture and aquaculture.

- During the decomposition process of organic pollutants the dissolved oxygen in the receiving water may be consumed at a greater rate than it can be replenished, causing oxygen depletion and having severe consequences for the stream biota. Wastewater with organic pollutants contains large quantities of suspended solids which reduce the light available to photosynthetic organisms and, on settling out, alter the characteristics of the river bed, rendering it an unsuitable habitat for many invertebrates. Organic pollutants include pesticides, fertilizers, hydrocarbons, phenols, plasticizers, biphenyls, detergents, oils, greases, pharmaceuticals, proteins and carbohydrates

- Among the possible techniques for water treatments, the adsorption process by solid adsorbents shows potential as one of the most efficient methods for the treatment and removal of organic contaminants in wastewater treatment. Adsorption has advantages over the other methods because of simple design and can involve low investment in term of both initial cost and land required. The adsorption process is widely used for treatment of industrial wastewater from organic and inorganic pollutants and meet the great attention from the researchers.

- In recent years, the search for low-cost adsorbents that have pollutant –binding capacities has intensified. Materials locally available such as natural materials, agricultural wastes and industrial wastes can be utilized as low-cost adsorbents. Activated carbon produced from these materials can be used as adsorbent for water and wastewater treatment.

- One of the most commonly used techniques for removing organics involves the process of adsorption, which is the physical adhesion of chemicals on to the surface of the solid.

- The effectiveness of the adsorbent is directly related to the amount of surface area available to attract the particles of contaminant.

- The most commonly used adsorbent is a very porous matrix of granular activated carbon, which has an enormous surface area (@1000 m^2/g).

- Adsorption on activated carbon is perhaps the most economical and technically attractive method available for removing soluble organics such as phenols, chlorinated hydrocarbons, surfactants, and colour (dyes) and odour producing substances from waste water.

  Adsorbate: material being adsorbed.

  Adsorbent: material doing the adsorbing. (examples are activated carbon or ion exchange resin).

- Generally some combination of physical and chemical adsorption is responsible for activated carbon adsorption in water and wastewater.

4.6.1 Physical Adsorption

- Van der Waals attraction between adsorbate and adsorbent.

- The attraction is not fixed to a specific site and the adsorbate is relatively free to move on the surface.

- This is relatively weak, reversible, adsorption capable of multilayer adsorption.

4.6.2 Chemical Adsorption

- Some degree of chemical bonding between adsorbate and adsorbent characterized by strong attractiveness. Adsorbed molecules are not free to move on the surface.

- There is a high degree of specificity and typically a monolayer is formed.

- The process is rarely reversible.

- Granular Activated Carbon (GAC) adsorption is a process used as tertiary treatment of municipal and industrial wastewater (physical-chemical treatment, followed by secondary treatment) or as a step in the physical-chemical treatment (coagulation, sedimentation, filtration, GAC adsorption) instead of the secondary treatment.

- In case of being applied as a tertiary treatment, GAC is fundamentally used for organic molecules adsorption incorporated to effluent from the biological treatment. It normally requires a pretreatment, such as lime precipitation followed by rapid filtration.

- In industrial water treatment, the GAC adsorption can be used to meet pretreatment discharge standards to municipal wastewater treatment plants or to meet discharge standards to environment.

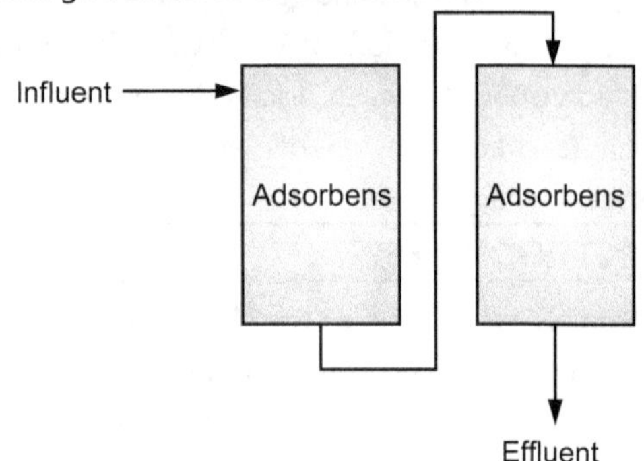

Fig. 4.9 : Adsorption tower

4.6.3 Factors Influencing Adsorption

- Number of factors influencing the adsorption process in AC can be listed: -
  1. Surface area of adsorbent
  2. Physical and Chemical characteristics of the adsorbate: as molecular weight and solubility in the water media.

- **Adsorbate Polarity:** A polar solute is preferably adsorbed by a polar adsorbent, whereas a nonpolar solute is more easily adsorbed by a nonpolar adsorbent. Activated carbon adsorbs nonpolar molecules better than polar molecules.

- **pH:** Adsorption of most organic compounds is higher at neutral pH conditions (around pH 7).

- **Temperature:** Temperature influences on solvent affinities and compounds diffusion in water.

- **Porosity of the Adsorbent:** The total porosity is usually classified into three groups: micropores less than 2 nm, mesopores pores between 2 and 50 nm and macropores above 50 nm pores. Most of the total surface area is found in the micropores and the contribution of the macropores to the total surface area is very low.

- **Surface Chemical Characteristics:** Adsorption capacities of thermally activated carbons have been found to be relatively higher than those of chemically activated carbons, although it varies depending on the adsorbate.

4.6.4 Transport Mechanisms

- **Bulk Solution Transport (Advection):** Adsorbates must first be transported from the bulk solution to the boundary layer of water (liquid film) surrounding the activated carbon particle.

- **External Diffusion:** Transport of adsorbate from the bulk of solution across the stationary layer of water, called the hydrodynamic boundary layer, liquid film or external film that surrounds the adsorbent particles. It occurs by molecular diffusion.

- **Intraparticle (Internal) Diffusion :** Intraparticle diffusion involves the transfer of adsorbate from the activated carbon surface to sites within the particle. In modeling of the adsorption of pollutants from water or wastewater, the surface diffusion is usually assumed to be the dominant intraparticle transport mechanism.

- **Adsorption :** After the transport of the adsorbate to an available site, an adsorption bond is formed. In the case of physical adsorption, the actual physical attachment of adsorbate onto adsorbent is regarded as taking place very rapidly. Therefore, the slowest step among the preceding diffusion steps (advection, external diffusion or internal diffusion), called the rate-limiting step, will control the overall rate at which the adsorbate is removed from solution. However, if adsorption is accompanied by a chemical reaction that changes the nature of the molecule, the chemical reaction may be slower than the diffusion step and may thereby control the rate of removal.

4.6.5 Activated Carbon Regeneration and Activation

- The carbon adsorption capacity gradually deteriorates with use. When the quality of the effluent reaches the minimum level set in quality standards, the spent carbon should be regenerated, reactivated or removed. Activated carbon is a high capital investment and operating cost to operate in both batch and continuous, mainly due to system regeneration need.

- The regeneration of the spent adsorbent is the most difficult and expensive part of the adsorption technology. It carries about 75% of the cost of operation and maintenance of a filtration unit in fixed bed GAC.

- Regeneration represents carbon contaminant removal without destroying contaminants.

- Reactivation involves the destruction of contained matter and reactivation of coal, which normally occurs at high temperatures.

- If regeneration of spent carbon is not feasible or it is irreversibly contaminated by the adsorbed substance, used carbon used will be removed.

- The most used techniques for regeneration are thermal, chemical and electrochemical regeneration.

4.7 AIR STRIPPING

- Air stripping removes a range of volatile and semi-volatile contaminants from water. And the performance of various types and sizes of tray-type air strippers for treating contaminated water now is highly predictable because of laboratory studies. Air stripping can be a fast, efficient and economical approach to treating industrial wastewater.

- However, since every industrial wastewater stream is unique, each must be evaluated to determine its constituents, its potentially adverse effects on treatability, and any pretreatment steps necessary to ensure desired results.

4.7.1 How Air Strippers Work

- The general principles of air stripping are simple. In an air stripper, the surface area of a film of contaminated water is maximized while air is directed across it. Contaminants at the aid water interface volatilize and are discharged to the atmosphere or to an off-gas treatment system.

- The two main types of air strippers are packed towers and low profile, tray-type strippers. In a packed tower air stripping system, water trickling over a packing media creates a water film. As water flows downward, it spreads thinner, creating more surface area. These thin films of water are met by a counterflow of air blown in from the bottom of the tower.

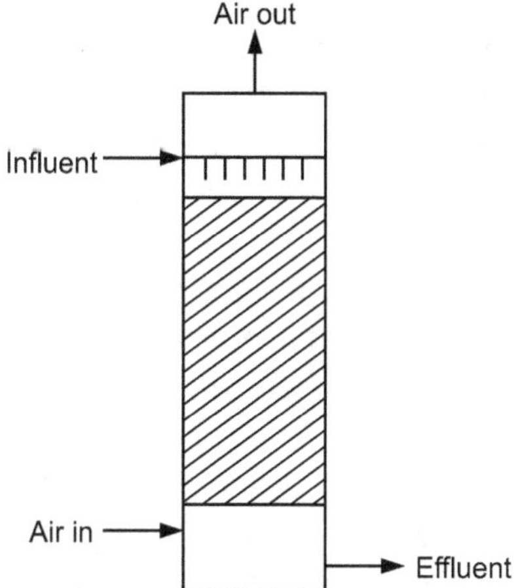

Fig. 4.10 : Air Stripping Column

- The amount of contamination that can be stripped from water depends on many factors, including:
  - ➤ The strippability of the compound(s);
  - ➤ Flow rate of the influent;
  - ➤ Residence time in the system;
  - ➤ Air flow;

- ➤ Water temperature;
- ➤ Air temperature; and
- ➤ The effects of solids, minerals and other chemicals in the water.

- The air stripper is a powerful tool because most of these variables can be controlled to obtain an optimized, site specific solution for contaminant removal. Air stripping is effective in removing contaminants found in a wide cross section of industrial waste streams, including:
  - ➤ Food processing (Ammonia);
  - ➤ Electronics manufacturing (chlorinated solvents);
  - ➤ Petroleum products (BTEX and MTBE);
  - ➤ Laundries (chlorinated solvents);
  - ➤ Printing plants (solvents);
  - ➤ Plating solutions (solvents);
  - ➤ Synthetic fiber processing (acetone).

- The stripping process is cheap and reliable, and provides relatively good substance transfer. One of the disadvantages of this process is that it is susceptible to pollution if sufficient attention is paid to pre-purification, stripping is a robust technique that can be easily implemented in a wide variety if sectors and in a wide volume range. Here are a few examples:
  - ➤ In both inorganic and organic chemistry, stripping is used for the removal of volatile organic substances, sulphur compounds (H_2S, phosphine) and NH_3. Stripping is normally carried out on the concentrated partial flow;
  - ➤ Air stripping is used in the pharmaceutical sector for the removal of chlorinated solvents from wastewater;
  - ➤ In viscose production, air stripping is the standard technique for the removal of CS_2 from wastewater;
  - ➤ In glass engraving with ammonium-based solvents, pH supplementation and air stripping can be used for the removal of nitrogen from wastewater;
  - ➤ In the graphics sector, stripping is used for the removal of Toluene from condensate discharged by recuperation systems;
  - ➤ In a variety of sectors, air stripping is implemented for the removal of chlorinated solvents from wastewater:
  - ➤ PER in the dry cleaning sector;
  - ➤ Methylene chloride in the removal of paint layers from wood.

- Air stripping is currently being used for cleaning groundwater in soil remediation, possibly followed by an air-based active carbon filter or bio-filter. The following compounds are primarily removed when stripping groundwater: Aromatic compounds (BTEX) and/or volatile chlorinated hydrocarbons (VOCs, incl. trichloro ethene, perchlore ethene, tetrachloro methane, chloroform).

- In order to implement the technique, initially the main factor is the ratio between the partial vaporisation of pollution in the gas phase and the concentration of the pollution in the water phase. This ratio is a fixed substance property and is referred to as the Henry coefficient. In fact, this coefficient indicates the "stripability" of the compound. The Henry coefficient is determined by the temperature.

- Further, a major role is played by the speed of substance transport from the water phase to the air phase. This substance transfer is determined by:
  - ➤ The interacting surface between the air phase and water phase;
  - ➤ The retention time in the system or the time that a substance is in contact with the surface (contact time);
  - ➤ The concentration difference of the substance between the gas phase and the liquid phase (the driving force or the deviation from the thermo-dynamic balance).
  - ➤ The technique is used for volatile compounds that (generally) have a Henry coefficient greater than 0.001 atm.m^3/mol. The concentration range in which the technique can be implemented varies from around ten micrograms per litre to a few dozen milligrams per litre.
  - ➤ Iron-removal is needed prior to air stripping for iron contents above 2 mg/l in a stripping column and 5 mg/l in a plate stripper. If iron is not removed, iron deposits could result in blockages which, in time, would require the package to be replaced or cleaned with acid.
  - ➤ The pH must be corrected if the water leaves lime deposits.
  - ➤ The water must be free of suspended matter, in order to prevent blockages. The wastewater must be pre-treated so that all suspended matter is removed (for example, using a sand filter or by using coagulation / flocculation followed by sedimentation).

- ➤ Air stripping is primarily implemented for the removal of volatile organic matter (incl. chlorinated hydrocarbons, VOX and BTEX) from wastewater.

- ➤ If the established condition for the Henry coefficient is met, air stripping can be used to realize high removal yield (e.g. >99% for VOX) per placed unit. If necessary, multiple units can be placed in a series to further increase total performance.

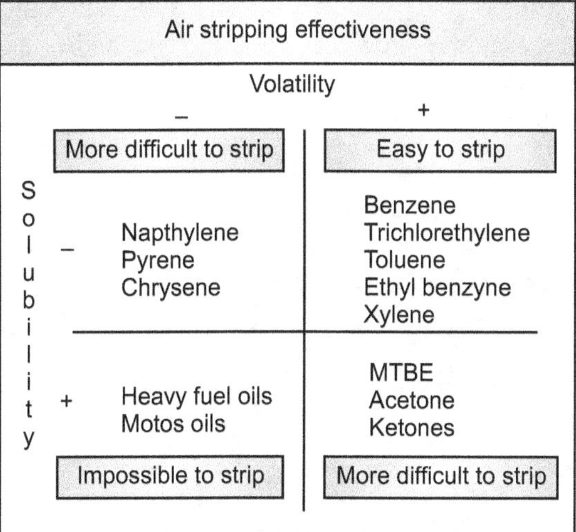

Fig. 4.11 : Air stripping effectiveness

- The ability to air strip a chemical compound depends on the compounds solubility and volatility. Those easiest to strip exhibit high volatility and low solubility.

4.8 CHEMICAL OXIDATION

4.8.1 Chemical Oxidation Techniques

Method Diagram

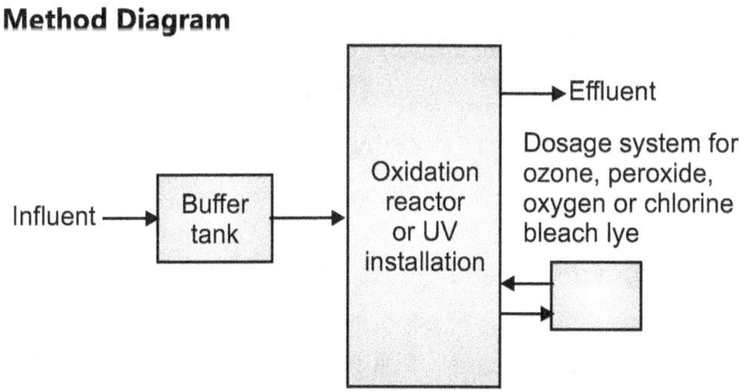

Fig. 4.12 : Chemical Oxidation

4.8.2 Method and Theoretical Description

- The aim of chemical oxidation is to oxidise organic pollutants to less dangerous or harmless substances. In the best case scenario, complete oxidisation of organic substances will result in CO_2 and H_2O. This technique can also be used to remove inorganic components (e.g. oxidisation of cyanide). Chemical oxidisation can also be used in combination with biological purification. In this case, we refer to partial oxidisation. The purpose of chemical oxidisation as a pre-treatment technique is to either break down difficult to degrade components and make them suitable for biological degradation or to limit sludge production by partly oxidising the sludge.

- Chemical oxidisation involves adding or generating oxidants in the wastewater. A few currently used oxidants include ozone (O_3), hydrogen peroxide (H_2O_2), natrium hypochlorite or bleaching liquor (NaOCl), chlorine dioxide (ClO_2), chlorine gas (Cl_2), peroxy acetic acid ($C_2H_4O_3$) and pure oxygen (O_2). Combinations of oxidants are also possible. The most active oxidant is hydroxyl radical (OH°). This can be formed from ozone or hydrogen peroxide after activation with a catalyst (e.g. Fe^{2+} in a Fenton reaction) or via UV light.

- The installation for chemical oxidation consists of a buffer tank, a reactor and a dosage unit for the oxidant. This could be supplemented by a UV installation. Most oxidants are not selective, whereby prior purification (e.g. filtration step) of wastewater is often necessary.

4.8.3 Specific Advantages and Disadvantages

- Each oxidant has its own advantages and disadvantages. In general, this technique requires little space. The potential risks of over-dosage must be taken into account (e.g. potential to kill a later biological purification technique). Because most chemicals are not selective, (partial) oxidisation may create products that are actually more toxic than the initial pollutants.

- The disadvantage of the Fenon reaction is the pH sensitivity and the increased sludge production.

4.8.4 Applications

- Below is a brief description of some examples of chemical oxidisation:

  - ➤ Treatment of groundwater for the removal of cyanides, PAH, BTEX, phenols and other organic micro-pollutants.

  - ➤ Chemical oxidisation of percolation water as post-biological purification. The aim of this is to further oxidise certain residual pollutants (persistent COD or AOX).

  - ➤ Increasing biodegradability by treating the influents and effluents of a biological purification system. Besides increasing the BOD/COD ratio, potential toxic substances can be converted into molecules that are easy to degrade biologically.

- ➤ The removal of colour components (e.g. textile sector or paper industry).
- ➤ Detoxification of galvano-effluents.
- ➤ Treatment of cooling water (AOX reduction and biological growth).

4.8.5 Boundary Conditions

- There are few, or no requirements at all, for the wastewater that will be treated. A cheap pre-treatment (e.g. removal of Fe from groundwater) will sometimes be carried out to restrict the costs of chemical oxidisation.

- An important point of note in the use of UV light is the turbidity and colour of the wastewater. The turbidity reduces the permeability of the UV light and the colour may adsorb the UV light. Therefore, in order to achieve the required purification yield, a more intensive light may be necessary. It is thus important for suspended components, amongst other things, to be removed from the wastewater in advance. This can be done easily using sand filtration, for example.

- Testing is used to determine the most optimum process conditions. The parameters that play a role in this are the type of oxidant, the required dosage, the acidity and the retention time in the reactor.

4.8.6 Effectiveness

- Chemical oxidisation is primarily implemented for the removal of persistent organic substances (e.g. dioxins, pesticides and biocides), organic compounds (e.g. BOD and COD, AOX, EOX, TOC, TOX, BTEX (Benzene, Toluene, Ethylbenzene and Xylene, MAH, Phenols and PAHs), Nutrients (Nitrogen and Organophosphorous compounds) and inorganic salts (e.g. CN^-, S^{-2} and SO_3^{-2}).

- In the case of recalcitrant COD and colour components, there are no known limits for the treatment of input concentrations. Both parameters can be removed up to 100%.

- In general, the yield from chemical oxidisation is good to excellent. Screening should indicate whether this technology can be used for a particular case, possibly with prior treatment steps. The intended yield can also be realised by possibly increasing the oxidant dosage. However, when doing so, a trade-off must be made with the overall cost price.

4.9.1 Significance of Nitrogen Removal

- In its various forms, nitrogen can deplete dissolved oxygen in receiving waters, stimulate aquatic plant growth (Fig. 4.13), exhibit toxicity toward aquatic life, present a public health hazard, and affect the suitability of wastewater for reuse purposes.

- Wastewater effluents containing nutrients such as nitrogen and phosphorus can cause eutrophication, the excessive growth of aquatic plants and/or algae in lakes, streams, rivers, wetlands or any surface water subject to runoff.

Fig. 4.13 : Algae growing in a small creek

4.9.2 Forms of Nitrogen Found in Wastewater

- Nitrogenous gasses (N_2O, N_2)
- Organic nitrogen (urea, amino acids, fecal material)
- Ammonia (NH_3 as a dissolved gas)
- Ammonium (NH_4^+ ions in solution)
- Nitrate (NO_3^- and nitrite (NO_2^- ions in solution
- Total Kjeldahl Nitrogen (TKN) is the sum of the ammonia and organic nitrogen Total Nitrogen (TN) is the sum of TKN and nitrate and nitrite.

Nitrate (NO_3^- is a primary contaminant in drinking water and can cause a human health condition called methemoglobinemia (blue baby syndrome). This is due to the conversion of nitrate to nitrite (NO_2^-) by nitrate reducing bacteria in the human gastrointestinal tract.

4.9.3 Nitrification

- Nitrification is an Oxidation Process (loss of electrons or gain of the oxidation state by an atom or compound takes place). This process starts with the ammonium which gets oxidized into nitrite (NO_2^-), this action is performed by the bacteria Nitrosomonas sp. Later on, this nitrite (NO_2^-) gets oxidized into nitrate (NO_3^-), and this action is performed by the *Nitrobacter sp.*

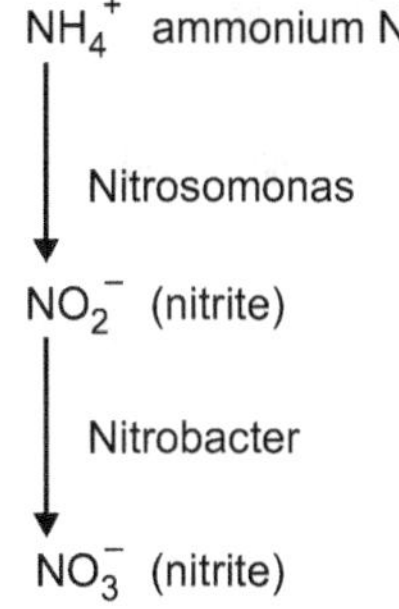

Fig. 4.14

- The bacteria are autotrophic, and the reaction is performed under aerobic condition. The importance of this step in the nitrogen cycle is the conversion of ammonia into nitrate, as nitrate is the primary nitrogen source present in the soil, for the plant. Though nitrate is toxic to the plants.

- The activity of nitrifying bacteria gets slower in acidic solution, and are best at pH between 6.5 to 8.5 and temperature vary from 16 to 35°C.

- Nitrification is the biological conversion of the ammonium to nitrate nitrogen, and is a two-step process. In the first step, aerobic bacteria known as *Nitrosomonas* convert ammonium to nitrite. Another group of aerobic bacteria called *Nitrobacter* finish the conversion of nitrite to nitrate.

- The reactions are generally coupled and proceed rapidly to the nitrate form; therefore nitrite levels at any given time are usually low. These bacteria known as 'nitrifiers' are strict aerobes, which means they must have free dissolved oxygen (O_2) to perform their work, and are active only under aerobic conditions. Complete nitrification requires approximately 4.6 pounds of oxygen for every pound of ammonium converted to nitrate. In comparison, CBOD can be consumed with only about 1.5 pounds of oxygen.

- The growth rate of nitrifiers is affected by the concentration of dissolved oxygen (DO), and at DO less than 0.5 mg/*l* the growth rate is minimal. Typical operational guidelines call for a minimum DO concentration of 1.0 mg/*l* at peak flow and an average daily DO concentration of 2.0 mg/*l*. For nitrification to proceed the oxygen should be well distributed throughout the aeration tank and its level should not be below 1.0 mg/*l*. Similar to humans, activated sludge organisms need nutrients to survive and reproduce. Nitrifying bacteria are no different, and need calcium in their diet. Luckily, there is usually enough calcium in the raw wastewater in the form of calcium carbonate ($CaCO_3$) to allow nitrifiers to survive nicely. Later in this manual, we will discuss the consequences of insufficient calcium (alkalinity) in the wastewater. As the nitrifiers use the ammonium as an energy source, they consume the calcium carbonate as a carbonsource.

- The process of nitrification produces acids. This acid formation, along with the calcium carbonate reduction, can lower the pH of the MLSS and cause a decline in the growth rate of nitrifying bacteria. The optimum pH for *Nitrosomonas* and *Nitrobacter* organisms is between 7.5 and 8.5 and nitrification stops at pH levels at or below 6.0. Approximately 7.14 pounds of alkalinity (as $CaCO_3$) are consumed per pound of ammonia oxidized to nitrate.

- Water temperature also affects the rate of nitrification. Nitrification reaches a maximum rate at liquid temperatures between 30 and 35 degrees C (86°F and 95°F). At temperatures of 40°C (104°F) and higher, nitrification rates fall to near zero. Conversely, cold water temperatures can also affect the rate of nitrification. As water temperature decreases, the nitrifiers slow down.

- Many wastewater treatment plants increase the MLVSS amounts during the winter months to increase the amount of aerobic autotrophic bacteria available to perform nitrification efficiently.

- Nitrifying bacteria are sensitive organisms, and react quickly to environmental changes. Rapid changes in liquid temperature can shock nitrifiers and other organisms we rely upon to treat our wastewater. As seasons change, cold fronts can bring swift changes in air temperatures. Water temperatures change fast as well, and treatment plants with mechanical aeration equipment where the MLSS is thrown into the air can see temperature decreases much faster than facilities with submerged, compressed air diffusers. We can see the effects of a MLSS temperature drop of one degree or more per day as cloudiness in secondary clarifiers, and an increase in effluent ammonium levels.

- Sludge Age and Mixed Liquor amounts are also integral components in the nitrification process. When performing sludge age calculations (MCRT or SRT) to determine the detention time required for nitrification, the capacity of the oxic (aerated) portion of the plant should be used. Since anoxic or fermentation basins are not aerated and nitrifying organisms are strict aerobes, the capacity of these basins should not be included in calculations for oxic SRT.

Fig. 4.15 : Surface aerator in Aeration Tank

- Extended aeration (package type) wastewater plants are more capable of nitrification than contact-stabilization and some other activated sludge modifications due to the high sludge age and long periods of aeration. Toxicity and sources of inhibition to microorganisms present problems to operators and nitrifying organisms. Some of the most toxic compounds to nitrifiers include cyanide, thiourea, phenol and heavy metals such as silver, mercury, nickel, chromium, copper and zinc.

- Some of these compounds can enter a wastewater treatment plant from landfill leachate and septage. Nitrifying bacteria can also be inhibited by high concentrations of free forms of their own substrate.

- Nitrite oxidizing bacteria are sensitive to free nitrous acid, and ammonium oxidizing bacteria are sensitive to free ammonia (NH_3). Increased levels of free ammonia can decrease nitrifier growth rates. Some treatment plants that may have increased influent organic nitrogen and ammonia levels include plants that serve highway rest areas and wastewater plants serving schools.

4.9.4 Denitrification

- Denitrification is the reduction process, where the nitrate is removed in the form of nitrogen and is converted to nitrogen gas. The action is performed by bacteria like *Bacillus, Aerobacter, Lactobacillus, Spirillum, Pseudomonas*.

- The bacteria are heterotrophs, and the action is completed under anaerobic condition. Even the small amount of oxygen may hamper the process, but there is a need of organic carbon. Denitrification is useful for wastewater treatment, aquatic habitats.

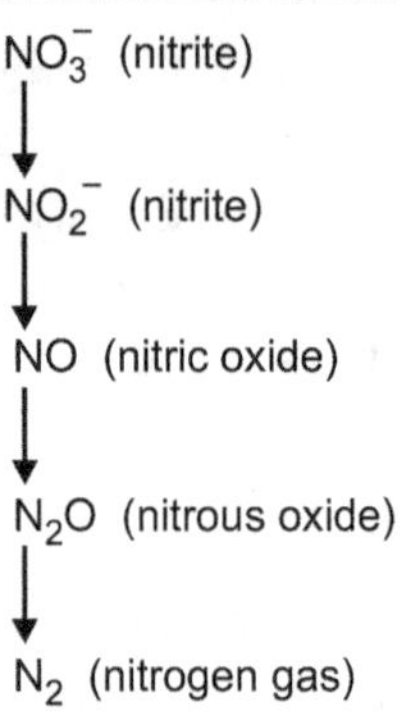

Fig. 4.16

- The denitrification is performed best at pH between 7.0 to 8.5 and at the temperature between 26 to 38°C.

- Denitrification is an anaerobic respiration process in which nitrate serves as the electron acceptor. In simpler terms, denitrification occurs when free dissolved oxygen levels are depleted and nitrate becomes the primary bound oxygen source for facultative heterotrophic microorganisms.

- Heterotrophic organisms can obtain their carbon and energy from the same source – by biodegrading influent wastewater containing carbon. We measure the influent carbon by sampling for and running a carbonaceous biochemical oxygen demand (CBOD) test.

- When bacteria break apart nitrate (NO_3^-) to gain the oxygen (O_2), the nitrate is reduced to nitrite (NO_2^-) then quickly to nitrous oxide (N_2O), and finally nitrogen gas (N_2) as the oxygen is stripped away by the microorganisms. Since nitrogen gas has low water solubility, it tends to escape as gas bubbles once the surrounding liquid is saturated with nitrogen. These gas bubbles can become bound in the settled sludge floc in clarifiers and cause the sludge to rise to the surface.

4.9.5 Conditions Required for Effective Denitrification

- Conditions that affect the efficiency of denitrification include nitrate concentration, anoxic conditions, presence of organic matter, pH, temperature, alkalinity and the effects of trace metals.

- Since denitrifying bacteria are facultative organisms, they can use either dissolved oxygen or nitrate or sulfate as an oxygen source for metabolism and oxidation of organic matter. If dissolved oxygen and nitrate are present, facultative bacteria prefer to use the dissolved oxygen. This will occur since dissolved oxygen is readily available and yields more energy to the organisms. Therefore it is imperative to keep dissolved oxygen levels as close to zero as possible in anoxic basins or timed anoxic cycles.

- Excessive dissolved oxygen in basins designed as anoxic is possible through aerated return sludge (air lift type RAS), excessive splashing of liquid streams into anoxic basins, and air diffusion being used for tank mixing instead of using mixing pumps or mixing devices.

- Another important aspect of denitrification is the presence of organic matter to drive the denitrification reaction. Organic matter may be in the form of raw wastewater, food processing wastes, or chemical sources such as methanol, ethanol, acetic or citric acid.

- When these sources are not present, bacteria may depend on internal (endogenous) carbon reserves as the source of organic matter. This material is released during the death phase of organisms and may not be a consistent enough source of carbon to drive denitrification to completion.

- Whatever organic source is used to drive the denitrification reaction, it should be fed consistently and at a rate to keep denitrification levels maximized. Conversely, it is important to avoid raising effluent CBOD values and avoid spending excessive money on organic sources such as methanol.

- An advantage of denitrification is the production of alkalinity and an increase of pH. Approximately 3.0 to 3.6 mg of alkalinity (as $CaCO_3$) is produced per milligram of nitrate reduced to nitrogen gas. Optimum pH values for denitrification are 7.0 to 8.5. Temperature affects the growth rate of denitrifying organisms, with increased growth rate at higher temperatures. Denitrification can occur between 5 to 30°C (41°F to 86°F), and these rates increase with temperature and type of organic source present. The highest growth rate can be found when using methanol or acetic acid. A slightly lower rate using raw wastewater will occur, and the lowest growth rates are found when relying on endogenous carbon sources at low water temperatures. Denitrifying organisms are generally less sensitive to toxic chemicals than nitrifiers and recover from toxic shock loads faster than nitrifiers.

4.9.6 Nitrification, an Oxidation Process

- Once the ammonium reaches the aeration tank, the aerobic autotrophic bacteria (nitrifiers) get to work. As long as there is enough DO left over after CBOD removal by aerobic and facultative bacteria, the nitrifiers will consume calcium carbonate while oxidizing the ammonium as an energy source.

- *Nitrosomonas* bacteria take the lead by converting the ammonium (NH_4^+) to nitrite (NO_2^-), then the *Nitrobacter* step in to finish off the conversion of nitrite to a final fully oxidized form called nitrate (NO_3^-). While the calcium carbonate is being consumed, nitrous acid is being formed and the total alkalinity of the liquid begins to decline. By the time the wastewater has reached the end of the aeration tank system,

  - ➤ The CBOD has been oxidized to low levels by aerobic and facultative bacteria as a carbon food source

  - ➤ The incoming ammonia/ammonium has been oxidized to nitrite, then to nitrate

  - ➤ Ammonia (NH_3) and ammonium (NH_4^+) levels have decreased, nitrate (NO_3^-) levels have increased

  - ➤ Dissolved oxygen demand has decreased, residual DO has increased

  - ➤ Total alkalinity (as calcium carbonate) has decreased

  - ➤ pH may have decreased

  - ➤ A total of 6.1 pounds of oxygen has been consumed by aerobic and facultative bacteria to oxidize CBOD and ammonia

  - ➤ Significant bacterial floc has been formed, trapping suspended and colloidal solids, including particulate organic nitrogen

4.9.7 Denitrification, a Reduction Process

- Here's where the process gets a little tricky. After the aeration tank, the MLSS normally enters a secondary clarifier for settling and separation of the solids from the treated water.

- If there is a carbon source available for bacteria to consume, little or no dissolved oxygen and nitrate is present, denitrification can take place.

- In the settled activated sludge that becomes the sludge blanket in the clarifier, the free available dissolved oxygen is rapidly used up by aerobic and facultative bacteria. Once free DO is gone, nitrification basically stops. If there is sufficient carbon present in the form of CBOD or carbon from within the bacteria itself, facultative organisms will continue to thrive. Since there is little to no free DO present, they use a different internal enzyme to allow the use of a combined oxygen source instead of free oxygen. The most readily available and easily reduced oxygen is found within the nitrate compound, NO_3^-.

- Facultative bacteria are able to consume carbon in a oxygen depleted environment, using nitrate as an alternative oxygen supply. A bacterium reduces the nitrate molecule by taking one oxygen atom off the nitrate (NO_3^-), reducing it to nitrite (NO_2^-). Another bacterium takes one more oxygen atom from the nitrite, reducing it to nitrous oxide (N_2O). Finally, another bacterium takes the last oxygen reducing it to nitrogen gas (N_2).

- Where does the nitrogen gas go? Some of it dissolves into the liquid solution, however once the water is saturated with nitrogen gas, no more nitrogen can be dissolved in the water. Microscopic bubbles of nitrogen form, becoming larger bubbles as they bump into each other. Some of the bubbles rise directly to the tank surface, some of the bubbles remain trapped in the MLSS floc they formed inside of. Once enough of these bubbles accumulate inside the floc particle, the bubbles make the floc particle positively buoyant causing the particle to rise to the surface.

- Some of the floc may be neutrally buoyant, where it remains suspended mid-water, neither rising nor settling. If enough of these nitrogen bubbles accumulate inside a dense floc particle, or inside a clump of sludge, the whole clump of sludge may rise to the surface. This is commonly referred to as sludge 'pop-ups' or clumping.

- It has been noted that nitrification and denitrification can take place simultaneously within an aeration tank (Hope, GRU, 1999). Since the outer layer of bacteria making up a floc particle are exposed directly to oxygen from aeration, this layer stays mostly aerobic. Oxygen may not penetrate deep within the floc particle, causing this area to become oxygen deficient or anoxic.

4.9.8 Comparison between Nitrification and Denitrification

Basis for Comparison	Nitrification	Denitrification
Meaning	The part of nitrogen cycle where ammonium (NH_4^+) is converted into nitrate (NO_3^-) is called nitrification.	Denitrification is the level where reduction of nitrate (NO_3^-) is made into nitrogen gas (N_2).
The process involves	Nitrifying bacteria like Nitrobacter, Nitrosomonas.	Denitrifying bacteria like Spirillum, Lactobacillus, Pseudomonas, Thiobacillus.
	Grows slowly.	Grows rapidly.
	Requires aerobic condition.	Requires anaerobic condition.
The microbes	Autotrophic.	Heterotrophic.
Precursor	Ammonium .	Nitrate.
End product	Nitrate.	Nitrogen.
pH and Temperature	The process occurs at the pH between 6.5 to 8.5 and temperature between 16 to 35 degree C.	The process occurs at the pH between 7.0 to 8.5 and temperature between 26 to 38 degree C.
Importance	Provides nitrate to the plant, which acts as the important nitrogen source.	Denitrification is used in wastewater treatment and is beneficial for aquatic habitats.

4.10 WASTEWATER DISPOSAL OPTIONS

Once wastewater has been treated, it is *disposed* of by reintroducing it to the environment. Three methods for disposing of treated wastewater effluent are surface water discharge, subsurface discharge, and land application for beneficial use.

1. **Discharge to Surface Waters**
   - Pollution Control board prohibit the disposal of untreated wastewater into storm drains or surface waters. In some cases, with a permit, a facility may discharge treated wastewater into surface waters.

2. **Subsurface Discharge**
   - Subsurface discharge occurs on site where wastewater is treated by discharging septic tank effluent underground where it leaches through a drain field. Dispersion trenches etc.

3. **Land Application (Reuse)**
   - In some situations, treated wastewater can be applied to land for irrigation. The small amount of pollutants remaining in the wastewater after treatment is absorbed by the crop or are assimilated into the soil structure.
   - Depending on the contaminants, the water may require pretreatment before discharge to meet water quality standards. State authorities issues permits for facilities treating wastewater and sludge through this method.

4.11 COMMON EFFLUENT TREATMENT PLANT (CETP)

Common Effluent Treatment Plant is the concept of treating effluents by means of a collective effort mainly for a cluster of small scale industrial units. This concept is similar to the concept of Municipal Corporation treating sewage of all the individual houses. The main objective of CETP is to reduce the treatment cost for individual units while protecting the environment.

   - ➢ To achieve 'Economics of scale' in waste treatment, thereby reducing the cost of pollution abatement for individual factory.
   - ➢ To minimize the problem of lack of technical assistance and trained personnel as fewer plants require fewer people.
   - ➢ To solve the problem of lack of space as the centralized facility can be planned to ensure that adequate space is available.
   - ➢ To reduce the problems of monitoring for the pollution control boards.
   - ➢ To organize the disposal of treated wastes and sludge and to improve the recycling and reuse possibilities.

- It is difficult for each industrial unit to provide and operate individual wastewater treatment plant because of the scale of operations or lack of space or technical manpower.

- However, the quantum of pollutants emitted by SSIs clusters may be more than an equivalent large-scale industry, since the specific rate of generation of pollutants is generally higher because of the inefficient production technologies adopted by SSIs.

- Hence the desirable option is of the shared or combined treatment, wherein, managerial and operational aspects are collectively addressed and the cost of treatment, becomes affordable as enunciated in the scheme of the common effluent treatment plants, which are proving to be a boon especially for small entrepreneurs, given the methodical planning, regular operation and equitable contribution of member units. Such common facilities also facilitate proper management of effluent and compliance of the effluent quality standards.

Advantages of Common Treatment

- Saving in Capital and operating cost of treatment plant. Combined treatment is always cheaper than small scattered treatment units.

- Availability of land which is difficult to be ensured by all individual units in the event they go for individual treatment plants. This is particularly important in case of existing old industries which simply do not have any space.

- Contribution of nutrient and diluting potential, making the complex industrial waste more amenable to degradation.

- The neutralization and equalization of heterogeneous waste makes its treatment techno-economically viable.

- Professional and trained staff can be made available for operation of CETP which is not possible in case of individual plants.

- Disposal of treated wastewater & sludge becomes more organized.

- Reduced burden of various regulatory authorities in ensuring pollution control requirement.

EXERCISE

1. Differentiate between Nitrification and Denitrification.
2. Explain detailed working with neat sketch for
   1. Ion exchange
   2. Reverse osmosis
   3. Air stripping
   4. Adsorption
   5. Chemical Oxidation
3. Enlist methods for solids removal and explain any one in detail.
4. Explain Ion exchange.
5. What is membrane filtration.
6. Explain ultra filtration and non filtration.
7. Explain reverse osmosis.

8. Write short note on :
   (i)   Reverse osmosis
   (ii) Electrodialysis
   (iii) Ultrafiltration
   (iv) Nanofiltration
9. What is electrodialysis process, Explain in brief with its application.
10. Explain adsorption with its types.
11. Explain Air stripping with schematic diagram.

12. Write notes on :
    (i) Adsorption
    (ii) Physical adsorption
13. What are the factors affecting / influencing adsorption.
14. How air strippers work?
15. What do you mean by common effluent treatment plant (CEPT), Explain in brief.

5.1 INTRODUCTION

Communicable Disease :

- A communicable disease is one that is spread from include: contact with blood and bodily fluids; one person to another through a variety of ways that breathing in an airborne virus; or by being bitten by an insect.

- Reporting of cases of communicable disease is important in the planning and evaluation of disease prevention and control programs, in the assurance of appropriate medical therapy, and in the detection of common-source outbreaks. California law mandates healthcare providers and laboratories to report over 80 diseases or conditions to their local health department. Some examples of the reportable communicable diseases include Hepatitis A, B and C, influenza, measles, and salmonella and other food borne illnesses.

5.1.1 Spread of Communicable Disease

How these diseases spread depends on the specific disease or infectious agent. Some ways in which communicable diseases spread are by:

- Physical contact with an infected person, such as through touch (staphylococcus), sexual intercourse (gonorrhea, HIV), fecal/oral transmission (hepatitis A), or droplets (influenza, TB)
- Contact with a contaminated surface or object (Norwalk virus), food (salmonella, E. coli), blood (HIV, hepatitis B), or water (cholera);
- Bites from insects or animals capable of transmitting the disease (mosquito: malaria and yellow fever; flea: plague); and
- Travel through the air, such as tuberculosis or measles.

5.1.2 List of Communicable Diseases

- Local health departments are required to report some 80 communicable diseases to the department of Public Health as part of the disease surveillance and investigation efforts.
- The diseases below are among them.
  - CRE
  - Ebola
  - Enterovirus D68
  - Flu
  - Hantavirus
  - Hepatitis A
  - Hepatitis B
  - HIV/AIDS
  - Measles
  - MRSA
  - Pertussis
  - Rabies
  - Sexually Transmitted Disease (STD)
  - Shigellosis
  - Tuberculosis
  - West Nile Virus
  - Zika

5.1.3 Protection Against Communicable Diseases

1. **Handle and Prepare Food Safely**
   - Food can carry germs. Wash hands, utensils, and surfaces often when preparing any food, especially raw meat. Always wash fruits and vegetables. Cook and keep foods at proper temperatures. Don't leave food out - refrigerate promptly.

2. **Wash Hands Often**

3. **Clean and Disinfect Commonly Used Surfaces**
   - Germs can live on surfaces. Cleaning with soap and water is usually enough. However, you should disinfect your bathroom and kitchen regularly. Disinfect other areas if someone in the house is ill.

4. **Cough and Sneeze Into Your Sleeve**

5. **Don't Share Personal Items**
   - Avoid sharing personal items that can't be disinfected, like toothbrushes and razors, or sharing towels between washes. Needles should never be shared, should only be used once, and then thrown away properly.

6. **Get Vaccinated**
   - Vaccines can prevent many infectious diseases. There are vaccines for children and adults designed to provide protection against many communicable diseases. There are also vaccines that are recommended or required for travel to certain parts of the world. Our Immunization Program can advise you on immunizations and clinics where you to get needed shots.

7. Avoid Touching Wild Animals

- Be cautious around wild animals as they can spread infectious diseases to you and your pets.

8. Stay Home When Sick

I Whack Germs Concept

Try this way of remembering the most important steps to staying well.

I	Immunizations are important to protect you from diseases
W	Wash your hands often with soap and water.
H	Home is where you stay when you are sick.
A	Avoid touching your eyes, nose, and mouth – especially when you are sick.
C	Cover your coughs and sneezes so you do not spread germs to others.
K	Keep your distance from sick people so you don't get sick too.

5.1.4 Chain of Disease Transmission

This refers to a logical sequence of factors or links of a chain that are essential to the development of the infectious agent and propagation of disease. The six factors involved in the chain of disease transmission are:

1. Infectious agent (etiology or causative agent)
2. Reservoir
3. Portal of exit
4. Mode of transmission
5. Portal of entry
6. Susceptible host

1. Infectious Agent: An organism that is capable of producing infection or infectious disease. On the basis of their size, etiological agents are generally classified into:

- Metazoa (multicellular organisms). (e.g. Helminths).
- Protozoa (Unicellular organisms) (e.g. Ameobae)
- Bacteria (e.g. Treponema pallidum, Mycobacterium tuberculosis, etc.)
- Fungus (e.g. Candida albicans)
- Virus (e.g. Chickenpox, polio, etc.)

2. Reservoir of Infection: Any person, animal, arthropod, plant, soil or substance (or combination of these) in which an infectious agent normally lives and multiplies, on which it depends primarily for survival and where it reproduces itself in such a manner that it can be transmitted to a susceptible host.

- **Types of Reservoirs**

(i) Man: There are a number of important pathogens that are specifically adapted to man, such as: measles, smallpox, typhoid, meningococcal meningitis, gonorrhea and syphilis. The cycle of transmission is from human to human.

(ii) Animals: Some infective agents that affect man have their reservoir in animals. The term "zoonosis" is applied to disease transmission from animals to man under natural conditions.

Example:

(a) Bovine tuberculosis - cow to man

(b) Brucellosis - Cows, pigs and goats to man

(c) Anthrax - Cattle, sheep, goats, horses to man

(d) Rabies - Dogs, foxes and other wild animals to man Man is not an essential part (usual reservoir) of the life cycle of the agent.

(iii) Non-Living Things as Reservoir: Many of the agents are basically saprophytes living in soil and fully adapted to live freely in nature. Biologically, they are usually equipped to withstand marked environmental changes in temperature and humidity.

Example : Clostridium botulinum etiologic agent of Botulism Clostridium tetani etiologic agent of Tetanus Clostridium welchi etiologic agent of gas gangrene

3. Portal of Exit (Mode of Escape from the Reservoir): This is the site through which the agent escapes from the reservoir.

Examples include:

GIT: typhoid fever, bacillary dysentery, amoebic dysentery, cholera, ascariasis, etc.

Respiratory: tuberculosis, common cold, etc.

Skin and Mucus Membranes: Syphilis

4. Mode of Transmission (Mechanism of Transmission of Infection): Refers to the mechanisms by which an infectious agent is transferred from one person to another or from a reservoir to a new host. Transmission may be direct or indirect.

(i) Direct Transmission: Consists of essentially immediate transfer of infectious agents from an infected host or reservoir to an appropriate portal of entry. This could be:

- Direct Vertical Such as: transplacental transmission of syphilis, HIV, etc. b. Direct horizontal Direct touching, biting, kissing, sexual intercourse, droplet spread onto the conjunctiva or onto mucus membrane of eye, nose or mouth during sneezing coughing, spitting or talking; Usually limited to a distance of about one meter or less.

(ii) **Indirect Transmission a. Vehicle-borne Transmission:** Indirect contact through contaminated inanimate objects (fomites) like:

- Bedding, toys, handkerchiefs, soiled clothes, cooking or eating utensils, surgical instruments.

Contaminated Food and Water

- Biological products like blood, serum, plasma or IV-fluids or any substance serving as intermediate means by which an infectious agent is transported and introduced into a susceptible host through a suitable portal of entry. The agent may or may not multiply or develop in the vehicle before it is introduced into man.

- **Vector-Borne Transmission:** Occurs when the infectious agent is conveyed by an arthropod (insect) to a susceptible host.

- **Mechanical Transmission:** The arthropod transports the agent by soiling its feet or proboscis, in which case multiplication of the agent in the vector does not occur. (e.g. common house fly.)

- **Biological Transmission:** This is when the agent multiplies in the arthropod before it is transmitted, such as the transmission of malaria by mosquito.

- **Air-Borne Transmission:** Dissemination of microbial agent by air to a suitable portal of entry, usually the respiratory tract. Two types of particles are implicated in this kind of spread: dusts and droplet nuclei.

  (i) **Dust:** small infectious particles of widely varying size that may arise from soil, clothes, bedding or contaminated floors and be resuspended by air currents.

  (ii) **Droplet Nuclei :** Small residues resulting from evaporation of fluid (droplets emitted by an infected host). They usually remain suspended in the air for long periods of time.

5. **Portal of Entry:** The site in which the infectious agent enters to the susceptible host. For example:

- Mucus membrane
- Skin
- Respiratory tract
- GIT
- Blood

6. **Susceptible Host (Host Factors):** A person or animal lacking sufficient resistance to a particular pathogenic agent to prevent disease if or when exposed. Occurrence of infection and its outcome are in part determined by host factors. The term "immunity" is used to describe the ability of the host to resist infection.

Resistance to infection is determined by non-specific and specific factors:

Non-Specific Factors

- Skin and mucus membrane
- Mucus, tears, gastric secretion
- Reflex responses such as coughing and sneezing.

Specific Factors

- **Genetic :** Hemoglobin resistant to Plasmodium falciparum Naturally acquired or artificially induced immunity. Acquired immunity may be active or passive.

- **Active Immunity :** Acquired following actual infection or immunization.

- **Passive Immunity :** pre-formed antibodies given to the host.

5.2 DISEASES COMMUNICATED BY DISCHARGES OF INTESTINES, NOSE AND THROAT

5.2.1 Oral-Fecal Transmitted Diseases

- What the diseases in this group have in common is that the causative organisms are excreted in the stools of infected persons (or, rarely, animals). The portal of entry for these diseases is the mouth.

- Therefore, the causative organisms have to pass through the environment from the feces of an infected person to the gastro-intestinal tract of a susceptible person. This is known as the fece-oral transmission route. Oral-oral transmission occurs mostly through unapparent fecal contamination of food, water and hands.

- As indicated in the schematic diagram below, food takes a central position; it can be directly or indirectly contaminated via polluted water, dirty hands, contaminated soil, or flies.

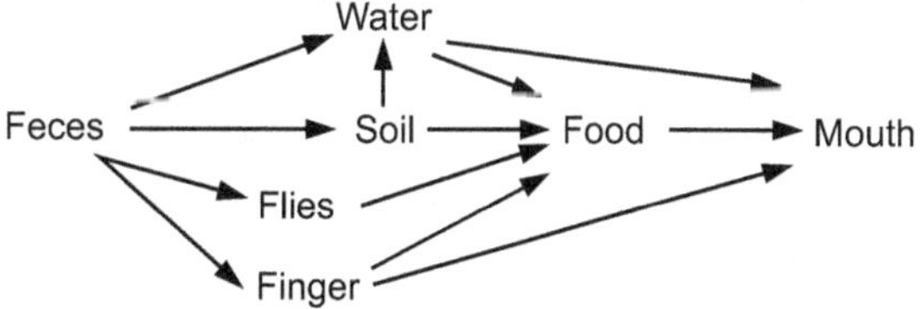

Fig. 5.1 : The five 'Fs' which play an important role in fecal oral diseases transmission (finger, flies, food, fomites and fluid).

1. Typhoid Fever

Definition

A systemic infectious disease characterized by high continuous fever, malaise and involvement of lymphoid tissues.

Infectious Agent

- Salmonella typhi
- Salmonella enteritidis (rare cause)

Occurrence : It occurs worldwide, particularly in poor socioeconomic areas. Annual incidence is estimated at about 17 million cases with approximately 600,000 deaths worldwide. In endemic areas the disease is most common in preschool and school aged children (5-19 years of age).

Reservoir : Humans

Mode of Transmission : By water and food contaminated byfeces and urine of patients and carriers. Flies may infectfoods in which the organisms then multiply to achieve aninfective dose.

Incubation Period : 1-3 weeks

Treatment

- Ampicillin or co-trimoxazole for carriers and mild cases.
- Chloramphenicol or ciprofloxacin or ceftriaxone for seriously ill patients.

Prevention and Control

- Treatment of patients and carriers
- Education on handwashing, particularly food handlers, patients and childcare givers
- Sanitary disposal of feces and control of flies.
- Provision of safe and adequate water
- Safe handling of food.
- Exclusion of typhoid carriers and patients from handling of food and patients
- Immunization for people at special risk (e.g. Travelers to endemic areas)
- Regular check-up of food handlers in food and drinking establishments

2. Dysentery

Definition

An acute bacterial disease involving the large and distal small intestine, caused by the bacteria of the genus shigella.

Infectious Agent

- Shigella is comprised of four species or serotypes.

Group A = Shigella dysentraie (most common cause)

Group B = Shigella flexneri

Group C = Shigella boydii

Group D = Shigella sonnei

Occurrence : It occurs worldwide, and is endemic in both tropical and temperate climates. Outbreaks commonly occur under conditions of crowding and where personal hygiene is poor, such as in jails, institutions for children, day carecenters, mental hospitals and refugee camps. It is estimated that the disease causes 600,000 deaths per year in the world. Two-thirds of the cases, and most of the deaths, are in children under 10 years of age.

Reservoir : Humans

Mode of Transmission : Mainly by direct or indirect fecal-oral transmission from a patient or carrier. Transmission through water and milk may occur as a result of direct fecal contamination. Flies can transfer organisms from latrines to a non-refrigerated food item in which organisms can survive and multiply.

Incubation Period : 12 hours-4 days (usually 1-3 days).

Treatment

- Fluid and electrolyte replacement
- Co-trimoxazole in severee cases or Nalidixic acid in the case of resistance.

Prevention and Control

- Detection of carriers and treatment of the sick will interrupt an epidemic.
- Handwashing after toilet and before handling or eating food.
- Proper excreta disposal especially from patients, convalescent and carriers.
- Adequate and safe water supply.
- Control of flies.
- Cleanliness in food handling and preparation.

3. Amoebiasis (Amoebic Dysentery)

Definition

An infection due to a protozoan parasite that causes intestinal or extra-intestinal disease.

Infectious Agent

- Entamoeba histolytica

Occurrence- worldwide but most common in the tropics and sub-tropics. Prevalent in areas with poor sanitation, in mental institutions and homosexuals.

Invasive amoebiasis is mostly a disease of young people (adults). Rare below 5 years of age, especially below 2 years.

Mode of Transmission : Fecal-oral transmission by ingestion of food or water contaminated by feces containing the cyst. Acute amoebic dysentery poses limited danger.

Incubation Period : Variable from few days to several months or years; commonly 2-4 weeks.

Treatment

- Metronidazole or Tinidazole

Prevention and Control

- Adequate treatment of cases
- Provision of safe drinking water
- Proper disposal of human excreta (feces) and handwashing following defecation.
- Cleaning and cooking of local foods (e.g. raw vegetables) to avoid eating food contaminated with feces.

4. **Cholera**

Definition

An acute illness caused by an enterotoxin elaborated by vibrio cholerae.

Infectious Agent

- Vibrio cholerae

Occurrence : has made periodic outbreaks in different parts of the world and given rise to pandemics. Endemic predominantly in children.

Reservoir : Humans

Mode of Transmission : by ingestion of food or water directly or indirectly contaminated with feces or vomitus of infected person.

Incubation Period : From a few hours to 5 days, usually 2-3 days.

Treatment

- Prompt replacement of fluids and electrolytes Rapid IV infusions of large amounts Isotonic saline solutions alternating with isotonic sodium bicarbonate or sodium lactate.
- Antibiotics like tetracycline dramatically reduce the duration and volume of diarrhea resulting in early eradication of vibrio cholerae.

Prevention and Control

- Case treatment.
- Safe disposal of human excreta and control of flies.
- Safe public water supply.
- Handwashing and sanitary handling of food.
- Control and management of contact cases.

5. **Infectious Hepatitis**

Definition

An acute viral disease characterized by abrupt onset of fever, malaise, anorexia, nausea and abdominal discomfort followed within a few days by jaundice.

Infectious Agent

- Hepatitis A virus

Occurrence : Worldwide distribution in sporadic and epidemic forms. In developing countries, adults are usually immune and epidemics of HA are uncommon. Infection is common where environmental sanitation is poor and occurs at an early age.

Reservoir : Humans.

Mode of Transmission : Person to person by fecal-oral route. Through contaminated water and food contaminated by infected food handlers.

Incubation Period : 15-55 days, average 28-30 days.

Treatment

- **Symptomatic:** Rest, high carbohydrate diet with low fat and protein.

Prevention and Control

- Public education about good sanitation and personal hygiene, with special emphasis on careful handwashing and sanitary disposal of feces.
- Proper water treatment and distribution systems and sewage disposal.
- Proper management of day care centers to minimize possibility of fecal-oral transmission.
- HA vaccine for all travelers to intermediate or highly endemic areas.
- Protection of day care centers' employees by vaccine.

5.2.2 Air-Borne Diseases

- The organisms causing the diseases in the air-borne group enter the body via the respiratory tract. When a patient or carrier of pathogens talks, coughs, laughs, or sneezes, he/she discharges fluid droplets. The smallest of these remain up in the air for some time and may be inhaled by a new host.
- Droplets with a size of 1-5 microns are quite easily drawn in to the lungs and retained there.
- Droplets that are bigger in size will not remain air-borne for long but will fall to the ground. Here, however, they dry and mix with dust. When they

contain pathogens that are able to survive drying, these may become air-borne again by wind or something stirring up the dust, and they can then be inhaled.

- Air-borne diseases, obviously, will spread more easily when there is overcrowding, as in overcrowded class rooms, public transport, canteens, dance halls, and cinemas. Good ventilation can do much to counteract the effects of overcrowding. Air-borne diseases are mostly acquired through the respiratory tract.

1. Common Cold

Definition

An acute catarrhal infection of the upper respiratory tract.

Infectious Agent

- Rhino viruses (100 serotypes) are the major causes in adults. Parainfluenza viruses, respiratory syncytial viruses (RSV), Influenza, and Adeno viruses cause common cold-like illnesses in infants and children.

Occurrence : Worldwide both in endemic and epidemic forms. Many people have one to six colds per year. Greater incidence in the highlands. Incidence is high in children under 5 years and gradually declines with increasing age.

Reservoir : Humans

Mode of Transmission : By direct contact or inhalation of airborne droplets. Indirectly by hands and articles freshly soiled by discharges of nose and throat of an infected person.

Incubation Period : Between 12 hours and 5 days, usually 48 hours, varying with the agent.

Treatment

No effective treatment but supportive measures like:

- Bed rest
- Steam inhalation
- High fluid intake
- Anti pain
- Balanced diet intake

Prevention and Control

Educate the public about the importance of:

- Handwashing
- Covering the mouth when coughing and sneezing
- Sanitary disposal of nasal and oral discharges

Avoid crowding in living and sleeping quarters especially in institutions

Provide adequate ventilation

2. Measles (Rubella)

Definition

An acute highly communicable viral disease

Infectious Agent

- Measles virus

Occurrence : Prior to widespread immunization, measles was common in childhood so that more than 90% of people had been infected by age 20; few went through life without any attack.

Reservoir : Humans

Mode of Transmission : Airborne by droplet spread, direct contact with nasal or throat secretions of infected persons and less commonly by articles freshly solid with nose and throat secretion. Greater than 94% herd immunity may be needed to interrupt community transmission.

Incubation Period : 7-18 days from exposure to onset of fever.

Treatment

- No specific treatment
- Treatment of complications
- Vitamin A provision

Prevention and Control

- Educate the public about measles immunization.
- Immunization of all children (less than 5 years of age) who had contact with infected children.
- Provision of measles vaccine at nine months of age.
- Initiate measles vaccination at 6 months of age during epidemic and repeat at 9 months of age.

3. Influenza

Definition

An acute viral disease of the respiratory tract

Infectious Agent

- Three types of influenza virus (A,B and C)

Occurrence : In pandemics, epidemics and localized outbreaks.

Reservoir : Humans are the primary reservoirs for human infection.

Mode of Transmission : Airborne spread predominates among crowded populations in closed places such as school buses.

Incubation Period : Short, usually 1-3 days.

Treatment

Same as common cold, namely:

- Anti-pain and antipyretic
- High fluid intake
- Bed rest
- Balanced diet intake

Prevention and control

- Educate the public in basic personal hygiene, especially the danger of unprotected coughs and sneezes and hand to mucus membrane transmission.
- Immunization with available killed virus vaccines may provide 70-80% protection.
- Amantadize hydrochloride is effective in the chemprophylaxis of type A virus but not others.

4. **Tuberculosis**

Definition

A chronic and infectious mycobacterial disease important as a major cause of illness and death in many parts of the world.

Infectious Agent.

- Mycobacterium tuberculosis- human tubercle bacilli (commonest cause)
- Mycobacterium bovis- cattle and man infection
- Mycobacterium avium- infection in birds and man.

Occurrence : Worldwide, however underdeveloped areas are more affected. Affects all ages and both sexes. Age groups between 15-45 years are mainly affected. According to the WHO 1995 report, 9 million cases and 3 million deaths have occurred. According to the Ministry of Health report in 1993

E.C, tuberculosis was a leading cause of outpatient morbidity (ranked 8th with 2.2%), leading cause of hospitalization (ranked 3rd with 7.8%) and leading cause of hospital death (ranked 1st with 10.1%). Tuberculosis has two major clinical forms. Pulmonary (80%) primarily occurs during childhood and secondarily 15-45 years or later. The other is extra pulmonary, which affects all parts of the body. Most common sites are lymph nodes, pleura, Genitourinary tract, bone and joints, meninges and peritoneum.

Mode of Transmission : Through aerosolized droplets mainly from persons with active ulcerative lesion of lung expelled during talking, sneezing, singing, or coughing directly. Untreated pulmonary uberculosis positive (PTB+) cases are the source of infection. Most

important is the length of time of contact an individual shares volume of air with an infectious case. That is intimate, prolonged or frequent contact is required. Transmission through contaminated fomites (clothes, personal articles) is rare. Ingestion of unpasteurized milk transmits bovine tuberculosis. Overcrowding and poor housing conditions favor the disease transmission.

Incubation Period : 4-12 weeks

Treatment

The following drugs are being used for treatment of TB in Ethiopia.

- Streptomycin (s) daily IM injection
- Ethambutol (E)
- Rifampin (R)
- Thiacetazone (T)
- Isoniazid (H)
- Pyrazinamide (Z)

All drugs, except streptomycin, which is administered daily through in route) are to be taken orally as a single daily dose preferably on an empty stomach.

Prevention and Control

- Chemotherapy of cases
- Chemoprophylaxis for contacts
- INH (Isoniazid) for adults and children who have close contact with the source of infection
- Immunization of infants with BCG
- Educate patients with TB about the mode of disease transmission and how to dispose their sputum and cover their mouth while coughing, sneezing, etc.
- Public health education about the modes of disease transmission and methods of control
- Improved standard of living
- Adequate nutrition
- Health housing
- Environmental sanitation
- Personal hygiene; etc.
- Active case finding and treatment

5. **Measles**

Definition

Measles is a highly contagious virus that lives in the nose and throat mucus of an infected person. The symptoms of measles generally appear about seven to 14 days after a person is infected.

Symptoms (Occurrence)

Measles typically begins with:

- high fever,
- cough,
- runny nose (coryza), and
- red, watery eyes (conjunctivitis)
- Two or three days after symptoms begin, tiny white spots (Koplik spots) may appear inside the mouth.
- Three to five days after symptoms begin, a rash breaks out. It usually begins as flat red spots that appear on the face at the hairline and spread downward to the neck, trunk, arms, legs, and feet. Small raised bumps may also appear on top of the flat red spots. The spots may become joined together as they spread from the head to the rest of the body. When the rash appears, a person's fever may spike to more than 104° Fahrenheit.
- After a few days, the fever subsides and the rash fades.

Mode of Transmission

Measles can spread to others through coughing and sneezing. Also, measles virus can live for up to two hours in an airspace where the infected person coughed or sneezed. If other people breathe the contaminated air or touch the infected surface, then touch their eyes, noses, or mouths, they can become infected. Measles is so contagious that if one person has it, 90 percent of the people close to that person who are not immune will also become infected.

Infected people can spread measles to others from four days before through four days after the rash appears.

Prevention and Control

Measles can be prevented with the MMR (measles, mumps, and rubella) vaccine. One dose of MMR vaccine is about 93 percent effective at preventing measles if exposed to the virus, and two doses are about 97 percent effective.

Treatment

- No treatment can get rid of an established measles infection. However, some measures can be taken to protect vulnerable individuals who have been exposed to the virus.
- **Post-Exposure Vaccination :** Nonimmunized people, including infants, may be given the measles vaccination within 72 hours of exposure to the measles virus to provide protection against the disease. If measles still develops, the illness usually has milder symptoms and lasts for a shorter time.
- **Immune Serum Globulin :** Pregnant women, infants and people with weakened immune systems who are exposed to the virus may receive an injection of proteins (antibodies) called immune serum globulin. When given within six days of exposure to the virus, these antibodies can prevent measles or make symptoms less severe.
- **Fever Reducers :** You or your child may also take over-the-counter medications such as acetaminophen (Tylenol, others), ibuprofen (Advil, Motrin, others) or naproxen (Aleve) to help relieve the fever that accompanies measles. Use caution when giving aspirin to children or teenagers. Though aspirin is approved for use in children older than age 3, children and teenagers recovering from chickenpox or flu-like symptoms should never take aspirin. This is because aspirin has been linked to Reye's syndrome, a rare but potentially life-threatening condition, in such children.
- **Antibiotics :** If a bacterial infection, such as pneumonia or an ear infection, develops while you or your child has measles, your doctor may prescribe an antibiotic.
- **Vitamin A :** People with low levels of vitamin A are more likely to have a more severe case of measles. Giving vitamin A may lessen the severity of the measles. It's generally given as a large dose of 200,000 international units (IU) for two days.

6. Mumps

Definition

Mumps is a contagious disease caused by a virus. An outbreak is considered three confirmed cases and can occur at any time of the year, but often occur in the winter and spring.

Occurrence

Symptoms of mumps may include:

- Swelling and tenderness in front of and below one or both ears and along the jaw
- Pain along the jaw and in front of and below one or both ears
- Fever
- Tiredness
- Muscle aches
- Loss of appetite

Mode of Transmission

Most people recover completely in a few weeks. People who do not have swelling may still spread the virus to others.

Mumps is spread through indirect or direct contact with an infected person's nose or throat droplets.

- It can be spread when an infected person coughs or sneezes or shares drinks or eating utensils.

- People with mumps can spread it for up to 2 days before and 5 days after the start of symptoms. Anyone with mumps should stay home during that time to prevent giving the illness to others.

- Symptoms typically appear 16-18 days after infection, but this period can range from 12-25 days after infection.

Treatment :

- Since mumps is caused by a virus, antibiotics cannot cure or treat mumps. Most treatment is to alleviate symptoms. Bed rest, a soft diet to reduce pain when chewing, and pain and fever relievers are often recommended.

- Complications of mumps are rare but may include orchitis (painful swelling of the testicles), meningitis (in 1-10 percent of cases), encephalitis (swelling of the brain; less than 1 percent of cases), and/or hearing loss (very rare). There may be an increased risk of miscarriage with mumps in the first trimester.

- Anyone who has not had two doses of mumps vaccine (usually via the MMR or measles-mumps-rubella vaccine) is at risk for mumps. Even those who have had two MMR vaccines can get mumps because the vaccine does not produce 100 percent immunity.

- Another risk for getting mumps is being on a college campus during a mumps outbreak (three or more confirmed cases). The risk is greatest for international travelers or people who are in contact with international travelers.

5.2.3 Vector Borne Diseases

Mosquito-Borne Diseases

1. Malaria

Definition

An acute infection of the blood caused by protozoa of the genus plasmodium.

Infectious Agent.

- **Plasmodium Falciparum / Malignant Tertian:** Invades allages of red blood cells. Red blood cell cycle is 48 hours

- **Plasmodium Vivax / Benign Tertian:** Invades reticulocytes only. Red blood cell cycle is 48 hours.

- Plamodiumovale/tertian: Invades reticulocytes only. Red blood cell cycle is 48 hours.

- **Plasmodium Malariae / Quartan Malaria:** Invadesreticulocytes only. Red blood cell cycle is 72 hours.

Epidemiology

Occurrence : Endemic in tropical and sub-tropical countries of the world. Affects 40% of the world population. Children less 5years of age, pregnant women and travelers to endemic areas are risk groups.

Predisposing factors are:

- Environment- physical environment for the propagation

- Patient source

- Susceptible recipients

- Anopheles capable to transmit the parasite

- Socio-economic factors like immigration, war, poverty, ignorance, agricultural irrigation farms, etc.

Reservoir : Humans

Mode of Transmission : By the bite of an infective female anopheles mosquito, which sucks blood for egg maturation. Blood transfusion, hypodermic needles, organ transplantation and mother to fetus transmission is possible. Since there is nopre-erythrocytic (tissue) cycle, the incubation period is short.

Incubation Period : Varies with species

- Plasmodium falciparum 7-14 days

- Plasmodium virvax 8-14 days

- Plasmodium ovale 8-14 days

- Plasmodium malariae 7-30 days

Treatment

- Plasmodium vivax, ovale and sensitive plasmodiumfalciparum

- Chloroquine or

- Fansidar

- Chloroquine resistant falciparum and when sensitivity pattern is not known.

- Quinine or

- Fansidar

Prevention and Control

(i) Chemoprophylaxis : For those who go to endemic areas but not for those who live in the endemic area (travellers and newcomers); for under-five children and pregnant mothers who have not enough immunity.

(ii) Vector Control

- Avoiding mosquito breeding sites
- Residual DDT spray or other chemicals
- Personal protection against mosquito bite (use of bednets, etc.)

Chemotherapy of Cases

2. **Plague**

Definition

A highly infectious bacterial disease which can kill many people within a short time.

Infectious Agent

- Yersinia pestis, the plague bacillus.

Epidemiology

Occurrence : Endemic in wild rodents living in forests in the highlands. Wild rodent plague exists in western USA, large are as of South America, North, Central, Eastern and Southern Africa, Central and Southeast Asia. However, urban plague is controlled in most of the world.

Reservoir : Wild rodents (especially ground squirrels) are the natural vertebrate reservoir of plague. Wild carnivores and domestic cats may also be a source of infection to people.

Mode of Transmission : Through the bite of infected fleas. Handling of tissues of infected animals.

Incubation Period : 1-7 days.

Treatment

- Early treatment with antibiotics like streptomycin ortetracycline or sulfa groups.

Prevention and Control

- Chemotherapy of patient
- Chemoprophylaxis of all contacts with Sulfa drugs
- The area where disease occurs must be quarantined (isolated from outer world)
- Insecticides to kill fleas
- Encourage people to kill rats
- Notify the disease to the concerned health authority

5.3 COMMUNICABLE DISEASE CONTROL

This refers to the reduction of the incidence and prevalence of communicable disease to a level where it cannot be a major public health problem.

5.3.1 Methods of Communicable Disease Control

There are three main methods of controlling communicable diseases:

1. **Elimination of the Reservoir**

- **Man as Reservoir:** When man is the reservoir, eradication of an infected host is not a viable option. Instead, the following options are considered:

- **Detection and Adequate Treatment of Cases :** Arrests the communicability of the disease (e.g. Treatment of active pulmonary tuberculosis).

- **Isolation:** Separation of infected persons for a period of communicability of the disease. Isolation is indicated for infectious disease with the following features:
  - ➤ High morbidity and mortality
  - ➤ High infectivity

- **Quarantine:** Limitation of the movement of apparently well person or animal who has been exposed to the infectious disease for a duration of the maximum incubation period of the disease.

- **Animals as Reservoir:** Action will be determined by the usefulness of the animals, how intimately they are associated to man and the feasibility of protecting susceptible animals.

Example:

(i) **Plague:** The rat is regarded as a pest and the objective would be to destroy the rat and exclude it from human habitation.

(ii) **Rabies:** Pet dogs can be protected by vaccination but stray dogs are destroyed.

Infected animals used for food are examined and destroyed.

- **Reservoir in Non-Living Things:** Possible to limit man's exposure to the affected area (e.g. Soil, water, forest, etc.).

2. **Interruption of Transmission**

This involves the control of the modes of transmission from the reservoir to the potential new host through:

- Improvement of environmental sanitation and personal hygiene
- Control of Vectors
- Disinfections and sterilization

3. **Protection of Susceptible Host:**

   This can be achieved through:

   - Immunization: Active or Passive
   - Chemo-prophylaxis- (e.g. Malaria, meningococcal meningitis, etc.)
   - Better nutrition
   - Personal protection. (e.g. wearing of shoes, use of mosquito bed net, insect repellents, etc.)

Chemical Control for Vector Borne Diseases

- Use of Indoor Residual Spray (IRS) with insecticides recommended under the program.
- Use of chemical larvicides like Abate in potable water Aerosol space spray during day time
- Malathion fogging during outbreaks.

Biological Control for Vector Borne Diseases

- Use of larvivorous fish in ornamental tanks, fountains etc.
- Use of biocides.
- Personal Prophylactic Measures that individuals / communities can take up
- Use of mosquito repellent creams, liquids, coils, mats etc.
- Screening of the houses with wire mesh
- Use of bed nets treated with insecticide
- Wearing clothes that cover maximum surface area of the body

5.3.2 Definition of Vector

In ancient times, insects were very important in the transmission of communicable diseases. The definition of **vector** was then related mostly to insects. Later on the term vector has been used more widely to include other non-human animals including snails, dogs and rats. Alternative definitions are found. For example, vectors can be defined as:

- Arthropods and other invertebrates which transmit infection by inoculation into or through the skin or mucous membrane by biting or by deposit of infective materials on the skin or on food or other objects.

- This classical definition considers mainly the **arthropods** (which include insects and other organisms such as mites). It shows the mechanisms of transmission as inoculation (biting) and depositing infective materials (pathogenic organisms such as bacteria) on skin and food.

- Vectors can also be defined as any non-human carriers of pathogenic organisms that can transmit these organisms directly to humans. Vertebrates, such as dogs and rodents, and invertebrates, such as insects, can all be vectors of disease.

- This second definition focuses on the range of living things involved. Knowing this definition is helpful in the design of preventive measures for controlling living organisms such as insects and rats which carry the disease agent (bacteria, virus) from an infected person to a healthy person.

5.3.3 Public Health Importance of Vectors

- A number of diarrhoeal diseases (acute watery diarrhoea, dysentery, typhoid fever) can also be transmitted by vectors and are commonly observed among children in areas where sanitation is very poor. Diarrhoea alone kills many children before they get to their fifth year.

- **Vector-Borne Diseases** not only cause illness, they also act as a barrier to development. Irrigation and dam workers will not be productive if they get malaria or schistosomiasis (bilharzia or snail fever). A person with malaria will need healthcare and will lose productive days at work. Some diseases like onchocerciasis (river blindness) have a devastating health impact. If onchocerciasis is left untreated the person could go blind. Additionally, vectors like rats destroy food and household materials and weevils damage cereals.

- The public health importance of vectors can be summarised as follows:

   ➢ They cause illness that could be fatal or restrict working capacity.

   ➢ They damage food and household goods.

   ➢ They are a barrier to development.

   ➢ Vector-borne disease transmission mechanisms

There are Two Ways that Vector-Borne Diseases are Transmitted:

1. Mechanical transmission takes place when a vector simply carries pathogenic microorganisms on their body and transfers them to food, which we then consume. Flies and cockroaches are in this category. Flies like to rest on faecal matter and then may move on to fresh food. They can carry infectious agents through their mouth and on their legs and other body parts. They deposit these agents on ready-to-eat foods and the recipient gets infected if they consume the contaminated food.

2. Biological transmission involves the multiplication and growth of a disease-causing agent inside the vector's body.

- Malaria is a good example of biological transmission. The female mosquitoes take the malaria infectious agent (Plasmodium) from an infected person with a blood meal.

- After sexual reproduction in the gut of the mosquito, the infectious agent migrates into the salivary gland of the insect, where it grows in size, matures and becomes ready to infect humans. When the mosquito next bites a human the saliva is injected into the skin and transfers the infection in doing so. An infectious agent may be passed from generation to generation of vector — this happens mostly in ticks and mites.

5.4 INSECT AND RODENT CONTROL

- Almost everyone has had the unpleasant experience of being bitten by a mosquito. Mosquito bites can cause skin irritation through an allergic reaction to the mosquito's saliva - this is what causes the red bump and itching. But a more serious consequence of some mosquito bites may be transmission of serious diseases and viruses such as malaria, dengue virus, Zika and West Nile virus, which can lead to disabling and potentially deadly effects (such as encephalitis, meningitis and microcephaly). Read more about diseases carried by mosquitoes.

- Not only can mosquitoes carry diseases that afflict humans, but they also can transmit several diseases and parasites that dogs and horses are very susceptible to. These include dog heart worms, eastern equine encephalitis and West Nile virus.

- There are about 200 different species of mosquitoes in the United States, which live in specific habitats, exhibit unique behaviors and bite different types of animals. Despite these differences, all mosquitoes share some common traits, such as a four-stage life cycle (egg, larva, pupa, adult).

- Different species of mosquitoes prefer different types of standing water in which to lay their eggs. The presence of beneficial predators such as fish and dragonfly nymphs in permanent ponds, lakes and streams help keep these bodies of water relatively free of mosquito larvae. However, portions of marshes, swamps, clogged ditches and temporary pools and puddles are all prolific mosquito breeding sites. Other sites in which some species lay their eggs include:

- tree holes,
- old tires,
- buckets,
- toys,
- potted plant trays and saucers,
- plastic covers or tarpaulins and even
- places as small as bottle caps!

- Some of the most annoying and potentially dangerous mosquito species, such as the Asian tiger mosquito, come from these sites.

- State and local government agencies play a critical role in protecting public health from mosquito-borne diseases. They serve on the front line, providing information through their outreach programs to the medical and environmental surveillance networks that first identify possible outbreaks.

- They also manage the mosquito control programs that carry out prevention, public education and vector population management. These agencies determine if the use of pesticides for mosquito control is appropriate for their area.

5.5 MOSQUITO LIFE CYCLE

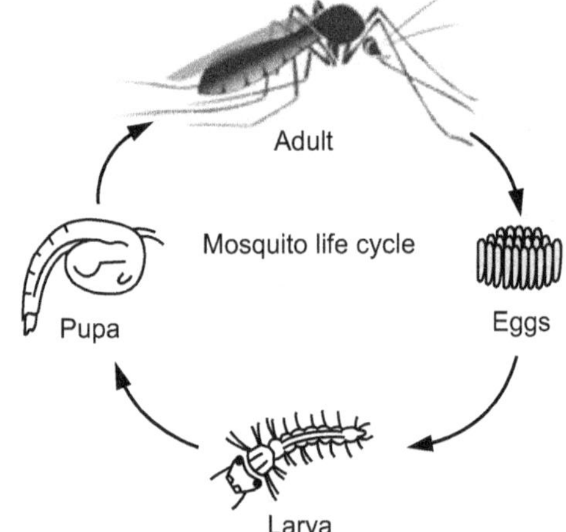

Fig. 5.2 : Life Cycle of Mosquito

- Knowing the different stages of the mosquito's life will help you prevent mosquitoes around your home and also help you choose the right pesticides for your needs, if you decide to use them. All mosquito species go through four distinct stages during their life cycle:

1. **Egg :** hatches when exposed to water.
2. **Larva :** (plural: larvae) "wriggler" lives in water; molts several times; most species surface to breathe air.
3. **Pupa :** (plural: pupae) "tumbler" does not feed; stage just before emerging as adult.
4. **Adult :** flies short time after emerging and after its body parts have hardened.

- The first three stages occur in water, but the adult is an active flying insect. Only the female mosquito bites and feeds on the blood of humans or other animals.
- After she obtains a blood meal, the female mosquito lays the eggs directly on or near water, soil and at the base of some plants in places that may fill with water. The eggs can survive dry conditions for a few months.
- The eggs hatch in water and a mosquito larva or "wriggler" emerges. The length of time to hatch depends on water temperature, food and type of mosquito.
- The larva lives in the water, feeds and develops into the third stage of the life cycle called, a pupa or "tumbler." The pupa also lives in the water but no longer feeds.
- Finally, the mosquito emerges from the pupal case after two days to a week in the pupal stage.
- The life cycle typically takes up two weeks, but depending on conditions, it can range from 4 days to as long as a month.

5.6 DISEASES CAUSED BY MOSQUITO

1. Malaria

- Malaria or paludism is caused by parasites of genus *Plasmodium*, which is transmitted to humans by the bite of female *Anopheles* mosquitoes. *Plasmodium falciporum* is responsible for most severe cases, usually causing coma or anemia in patients, which flows into death. Meanwhile, *Plasmodium vivax* causes recurring fevers and lesions in the brain and liver, but it rarely causes death. Within the measures applied to control, *Anopheles* vectors, for a long time, was based on the application of DDT (1,1,1-trichloro-2,2-bis [p-chorophenyl] ethane), but currently it is beginning to use pyrethroids in outbreaks and transmission foci.
- While the disease appeared to be under control in the 1950s, the infection again reappeared in many countries due to the resistance generated by vectors to insecticides of plasmodia and chloroquine. This disease is responsible for the deaths of between 700,000 and 2.7 million people. Moreover, malaria causes between 400 and 900 million cases of acute fever per year in children fewer than five years in these areas. Therefore, malaria is the disease with the highest prevalence in areas with limited economic resources, causing the largest number of cases in the warm and rainy seasons. The solution to eradicate this disease would be the application of vaccine.

2. Yellow Fever

- This disease is caused by the yellow fever virus, an arbovirus, belonging to the *Flavivirus* genus is present in tropical areas of Africa and South America. *Aedes aegypti* mosquito is the most important vector in the transmission of the yellow fever disease in America. Yellow fever virus infects both humans and monkeys, being monkeys the main reservoir of infection and transmission from monkey to monkey in woodlands and jungle. *Haemagogus jantinomys* and *Sabethes choropterus* mosquitoes are the vectors responsible for the transovarially virus transmission among the primate species. On the other hand, the infection is transmitted to humans through *A. aegypti* mosquito bites.
- Yellow fever virus causes 200,000 clinical cases of disease and 30,000 deaths each year, of which 90% of the cases correspond to the African continent. Unfortunately, the most of the cases and deaths are not recognized because it occurs in rural areas where surveillance and reporting are inadequate.
- Yellow fever distribution in America ranged from Philadelphia, in the United States, until the line connecting Bahía Blanca and Mendoza in Argentina. On the other hand, in Africa, it is located in the sub-Saharan Africa. This disease can be fatal and acute or mild and in apparent. Because there is no specific antiviral treatment against it, the best strategy to prevent its spread is the prevention of infection. In this context, vaccination is the best preventive measure against yellow fever.

3. Dengue

- Dengue is a viral disease caused by infection of four viruses, known as dengue 1, 2, 3, and 4, which is endemic in more than 100 countries in Africa, America, the Eastern Mediterranean, Southeast Asia, and the Western Pacific, the latter two being the most severely affected. These viruses belong to the genus *Flavivirus*, Flaviciridae family.
- The most important mosquito vector is *A. aegypti* and to a lesser degree *A. albopictus*. Once an infected mosquito bites a human, the virus goes through an incubation period of between 3 and 14 days before disease symptoms appear. Furthermore, passive dispersion through means of transport is one of the most important factors that favor the spread of these mosquitoes and dengue virus from one region to another.

- This disease is of major interest to public health because of its great impact on morbidity and mortality in the world since it is the viral disease transmitted by mosquito vectors most common and important worldwide. The World Health Organization estimated that there may be 50 to 100 million dengue infections, a half-million hospitalizations, and 22,000 deaths worldwide every year. Moreover, because of the absence of a vaccine to protect the population at risk, vector control is the most important method for the prevention and interruption of the transmission of the disease.

- The use of chemical insecticides is a key component in the control of larvae and adult mosquito vector populations. However, derived from overuse for over five decades of these insecticides to interrupt the transmission of the virus, it has generated resistance to different molecules of insecticides by part of mosquito vectors.

4. **Chikungunya Fever**

- Chikungunya fever is a viral disease, manifested by fever and severe arthralgia, prevalent in Africa, Asia, and Europe, and now emerging and little studied in the Americas and the Caribbean Islands since 2013. Chikungunya virus is transmitted to humans by *Aedes* vectors. There are two main vectors *A. aegypti* and *A. albopictus*, which are present in the tropics and temperate zones. Although *A. aegypti* has always been the main vector transmitter of the disease, in most recent outbreaks, *A. albopictus* has become the main vector.

- Chikungunya word comes from the makonde dialect spoken by an ethnic group of southeast Tanzania and northern Mozambique, which means the man who walks hunched over, due to the appearance shown by patients due to the severity of joint pain that they suffer.

- The Chikungunya virus is an enveloped positive-strand RNA virus, belonging to the *Alphavirus* genus, group arbovirus A, of the Togaviridae family. The virus affects all humans without distinction, being susceptible to contracting the disease individuals not previously infected with the virus, after individuals are infected, immunity is prolonged. Chikungunya fever epidemics spread rapidly within infected community until the development of immunity in the population affected stops transmission. In general, about 3–28% of infected people are asymptomatic, although they contribute to the spread of the disease. On the other hand, in the symptomatic forms, the clinical manifestations may be acute, subacute, or chronic. After the bite, the incubation period of the disease is 1–12 days characterized by fever, severe arthralgias, back pain, incapacitating myalgia, and conjunctivitis.

- Subsequently, after 2 or 3 days, maculopapular exanthem is described (sometimes only macular) in half of the cases, distributed on the trunk and extremities. Fever usually takes between 2 and 3 days, and leukopenia is the rule. Since there is no specific antiviral treatment or vaccine, treatment against the disease should be symptomatic against the symptoms presented by the patient and can be applied antipyretics, analgesics, anti-inflammatory, and ribavirin.

5.6.1 Mosquito Control

- The prevention and control of mosquito-borne diseases globally is conducted through a comprehensive and thorough method of pest management. Where programs are not intended to completely eliminate mosquito populations but rather are aimed to reduce their number and therefore minimize the risk of disease transmission.

- Methods used to mosquito control include the elimination of breeding sites and the control of mosquito larvae and adults. Larvicides, by applying chemical insecticides in the breeding sites, are the best strategy to kill larvae and pupae of mosquitoes in the water. Larvicides are present in several forms ranging from powder, tablets, or liquids and include methoprene, monomolecular surface films, larvicidal oils, chemical insecticides, neurotoxic insecticides, plant-derived products, and larvicidal bacteria.

- Adulticides technique is usually less efficient for mosquito control. However, it is the only way to kill adult mosquitoes and is the last line of defense in reducing mosquito populations. Some of the adulticide used for mosquito control include products derived from microorganisms, plants or minerals, synthetic molecules, organophosphates, some natural pyrethrins, or synthetic pyrethroids.

5.7 CHEMICAL INSECTICIDES

- Since its discovery, chemical insecticides have represented the most widely method used to control mosquito-borne vectors. However, the effects of chemical insecticides on mosquito vector populations are usually transitory because vectors can rapidly develop resistance against them. On the other hand, the environmental problems caused by the excessive use of chemical insecticides are a matter of current concern because it is estimated that about 2.5 million tons of pesticides are used annually, generating worldwide damage amounting to $100 billion annually. Some of the disadvantages that generates when using only chemical products are:

- ➤ The selection of new insecticide resistance in pest populations;
- ➤ The resurgence of already treated populations;
- ➤ The generation of waste, risks, and legal complications;
- ➤ The destruction of beneficial species; and
- ➤ The high costs in equipment, labor, and material.

- In addition, the highly toxic and nonbiodegradable properties of insecticides and waste generated in soil, water, food, and crops that affect public health are additional reasons to search new methods to help solve the problems caused by chemical insecticides. Consequently, the concept of integrated control arises, a method in which pest and diseases control is performed using chemicals, useful organisms, and cultural practices.

- The progress of science and the chemical industry in the nineteenth century, with the discovery of DDT, made possible the development and emergence of new conventional insecticides or so-called of synthesis. The most used of these insecticides of synthesis are modulators of sodium channels (organochlorines, pyrethroids, and pyrethrins), acetylcholinesterase inhibitors (carbamates and organophosphates), and the chloride channel antagonists regulated by the gamma-aminobutyric acid or also known as GABA (organochlorine cyclodiene and phenylpyrazoles).

- Using these conventional insecticides gave positive results against insects disease vectors at first. However, due to its massive use, insects soon began to develop resistance to them. Thus, an insecticide that initially was effective, just being useless in the long term. In response to this problem, new-generation insecticides also called biorational insecticides have been developed, whose research strategy is based on a good understanding of the physiological processes or mechanisms specific communication of insects, and in obtaining agents that are able to affect them. These products are divided into the following: those who are analogs of juvenile hormone and molting, inhibitors tissue formation, pheromones, insecticides that prevent hatching, and biological insecticides.

5.7.1 Organophosphates and Carbamates

- Organophosphate insecticides are phosphoric acid derivatives, having activity against a wide spectrum of invertebrate. It interferes with the action of enzymes called cholinesterases that regulate the neurotransmitter acetylcholine, resulting in first instance to muscle cramps, paralysis, and eventually death. Therefore, these insecticides have a toxic action that blocks an enzyme acetylcholinesterase of central and peripheral nervous system of insects, in synaptic junctions.

- The enzyme rapidly hydrolyzed acetylcholine, resulting in the repolarization of the membrane or the basal plate in neuromuscular connections, preparing for the arrival of a new impulse. By forming strong covalent bonds between insecticide and acetylcholinesterase, the enzyme is inhibited, causing the accumulation of acetylcholine in the synaptic junction and the interruption of normal transmission of nerve impulses.

- However, due to the generation of resistance in vector insects to these chemical products, the use of many of these organophosphate and carbamates insecticides is no longer effective. Furthermore, because cholinesterases and neurotransmitters acetylcholine also form part of vertebrate nervous system, organophosphate pesticides are highly or moderately toxic to vertebrates. In this regard, temephos are the only organophosphate pesticide that is still used to control mosquito larvae.

- Although temephos are not persistent in the environment being that last 7–10 days, it has been shown in many studies the adverse effects of temephos on a wide range of no target aquatic taxa. Furthermore, carbamate pesticides, just like organophosphates, act by inhibiting the cholinesterase enzyme. Therefore, the symptoms experienced by insects per carbamate poisoning are similar to those experienced with organophosphates. However, carbamate pesticides block acetylcholinesterase enzyme hydrolyzing acetylcholine in muscle by carbamylation, which is a reversible reaction. Therefore, the recovery of carbamate poisoning in humans is faster than with organophosphate intoxication since the acetylcholinesterase enzyme is able to break apart of the carbamate.

5.7.2 Organochlorines, Pyrethroids, and Pyrethrins

- Organochlorine insecticides are chlorinated hydrocarbons, which are known to be effective to control mosquito populations. Its mode of action is by inhibiting GABA receptor in the nervous system through the interruption of nerve impulses due to the closure of chloride channels. Therefore, when an organochlorine binds to a GABA receptor, the receptor is unable to close GABA chloride channel, which results in stimulation of the nervous system and similar symptoms to poisoning with carbamates or organophosphates. However, with the Stockholm Convention on Persistent Organic Pollutants, which entered into force on May 17, 2004, the use of 12 chemicals including DDT, aldrin, dieldrin, heptachlor, mirex, chlordecone, and chlordane was prohibited because of its long average life and toxicity. However, an extension clause allows countries where malaria is endemic to use DDT to control vectors that transmit the disease. Taking into account the negative effects that DDT has for the environment, malaria programs without the use of insecticides have been developed with the assistance of the Pan American Health Organization.

- On the other hand, pyrethroids and pyrethrins used to control mosquitoes break down faster in the sunlight as opposed to chemical or microbial breakdown. However, pyrethroids are considered axonic poisons, composed of more stable substances, or degrade slower in the presence of sunlight than pyrethrins and are generally effective against most of the insect pests of agriculture. Furthermore, pyrethroids can be combined with other active ingredients, such as piperonyl butoxide, to retard its degradation and prevents the insect's system from detoxifying the pyrethroid, making it more effective.

- Delay that allows the chemical product persists longer in the environment, requiring smaller and less frequent doses to kill pests. This type of insecticidal affects the central and peripheral nervous system of insects and have a rapid knock-down effect, by interfering with the sodium channels of nerve membrane causing the interruption of the transfer of ions and transmission of impulses between nerve cells. Moreover, it stimulates nerve cells to produce repetitive discharges and eventually cause paralysis and death. Furthermore, because pyrethroids act on the nervous system of insects through a different pathway from the organophosphate pesticides, they generally have low toxicity in mammals and birds; however, they are toxic to fish and tadpoles.

5.8 BIORATIONAL INSECTICIDES

- Biorational insecticides are those that have relatively low toxicity to humans and have few environmental effects. Among which, methoprene is an insect growth regulator insecticide with a broad spectrum of action that interferes with the insect life cycle preventing maturity or reproductive stage. Meanwhile, the juvenile hormone analogue is a biorational insecticide that causes deformations in larval stage, death in the pupal stage, and sterility effect in adults.

- Spinosad is another biorational insecticide that comes from a *Saccharopolyspora spinosa* neurotoxin, made by a mixture of spinosyns A and D. Spinosad act on the postsynaptic nicotinic acetylcholine receptors and GABA receptors and has proven its usefulness in the dipterans control. Pyriproxyfene is another new-generation insecticide that has been tested in adult and larval mosquitoes causing a reduction in the number of sperm, egg production, blood feeding, and mating activity.

Plants and Their Derivates

- For centuries, nature has created several active substances that, when applied correctly, can control insect pests such as mosquito in an efficient manner. The use of plants by man with insecticide purposes dates back to early human history. Due to their environmental advantages, the use of insecticides of vegetable origin in pest management has been increasing. Among plants with potential activity against mosquitoes, Nim or Neem (*Azadirachta indica*) causes stunted growth, loss of appetite, reduction of fertility, molting disorders, morphological defects, and behavioral changes.

- Moreover, it has been demonstrated that raw or partially purified plant extracts are most effective for mosquito control in place of the purified compounds or extracts. The snuff (*Nicotiana tabacum*) is used, thanks to its insecticide and insect repellent action, where nicotine acts on the nervous system of insects through breathing, ingestion, and contact. Other plants from which oils are extracted are garlic (*Ocimum basilicum*) and cinnamon (*Cinnamomum osmophloeum*), which have been shown to have insecticidal properties against larvae and adults of *A. albopictus*, *Culex quinquefasciatus*, and *Armigeres subalbatus*.

Biological Agents

- Among biological agents used for mosquito control can be mentioned derivatives of viruses, bacteria, and fungi. Entomopathogenic virus spreads from one insect generation to the next causing paralysis and eventually death on mosquito larvae being more effective in the first stage of development. Within bacteria, only reports of *Bacillus thuringiensis*, *B. sphericus*, and *B. popilliae* with possibilities to exercise control over dipterans insects currently exist. These bacteria, during the sporulation process, produce protein crystals with insecticidal effect and/or some toxins with the same effect.

- *Bacillus* initially causes diarrhea and intestinal paralysis in mosquito, giving rise to a decrease of body movements, convulsions, and general paralysis. Internally, within the mosquito stomach, *B. thuringiensis* releases toxic crystals that paralyze the insect gut stopping peristalsis, causing that the insect stop feeding and die by starvation. Within the gut, bacteria multiply until they break the epithelium and invade the rest of the insect body. However, its use for mosquito control is scarce and presents some drawbacks as its duration in the environment is limited, its dispersion is rather inefficient, and the susceptibility to bacterial infection in the pest population is very heterogeneous. There are very sensitive individuals and other highly resistant.

- Fungi are other microorganisms that may be used to control mosquito vectors, of which 400 species are known with insecticide potential. About 20 of them have been given more attention, including those in the *Lagenidium*, *Entomophaga*, *Neozygites*, *Entomophtora*, *Erynia*, *Aschersonia*, *Verticillium*, *Nomuraea*, *Hirsutella*, *Metarhizium*, *Beauveria*, and *Paecilomyces* genera. Although, entomopathogenic fungi are not as specific as bacteria or viruses, spores persist and infect insect successive generations, so that when the infection is established, its effects can last several years.

- Infection occurs by adhesion of the spores on the insect cuticle, where these germinate and penetrate the cuticle leading to insect colonization by mycelium. Cuticle penetration occurs through the use of an enzyme complex that the fungi use to feed. The entomopathogenic fungus most used in controlling mosquito infestations is *Beauveria bassiana*, which produces various active ingredients such as beauvericin.

- The biological control of mosquito larvae with predators and other biological control agents could be a more effective and environmentally friendly strategy, thus avoiding the use of synthetic chemicals and the consequent environmental damage. Among them, some insects and vertebrates such as fish, amphibians, and some mammals have the potential to control mosquito disease vector populations. Within vertebrates, amphibians, bats, and fish have been used to control populations of mosquito.

Example : Using larvivorous fish species, control of mosquito larvae in deposits used to store water has been achieved. Moreover, bats are responsible for capturing flying insects such as mosquitoes at night; similarly, toads and frogs consume large numbers of insects, slugs, worms, and other invertebrates. However, the use of frogs and tadpoles for disease vector control is still largely unexplored.

5.9 FLIES

5.9.1 Life Cycle

- There are four distinct stages in the life of a fly: egg, larva or maggot, pupa and adult (Fig. 5.4). Depending on the temperature, it takes from 6 to 42 days for the egg to develop into the adult fly. The length of life is usually 2–3 weeks but in cooler conditions it may be as long as three months. Eggs are usually laid in masses on organic material such as manure and garbage. Hatching occurs within a few hours.

- The young larvae burrow into the breeding material; they must obtain oxygen from the atmosphere and can, therefore, survive only where sufficient fresh air is available. When the breeding medium is very wet they can live on its surface only, whereas in drier materials they may penetrate to a depth of several centimetres. The larvae of most species are slender, white, legless maggots that develop rapidly, passing through three instars.

- The time required for development varies from a minimum of three days to several weeks, depending on the species as well as the temperature and type and quantity of food available. After the feeding stage is completed the larvae migrate to a drier place and burrow into the soil or hide under objects offering protection. They form a capsule-like case, the puparium, within which the transformation from larva to adult takes place.

- This usually takes 2–10 days, at the end of which the fly pushes open the top of the case and works its way out and up to the surface. Soon after emergence the fly spreads its wings and the body dries and hardens. The adult fly is grey, 6–9mm long and has four dark stripes running lengthwise on the back. A few days elapse before the adult is capable of reproduction. Under natural conditions an adult female rarely lays eggs more than five times, and seldom lays more than 120–130 eggs on each occasion.

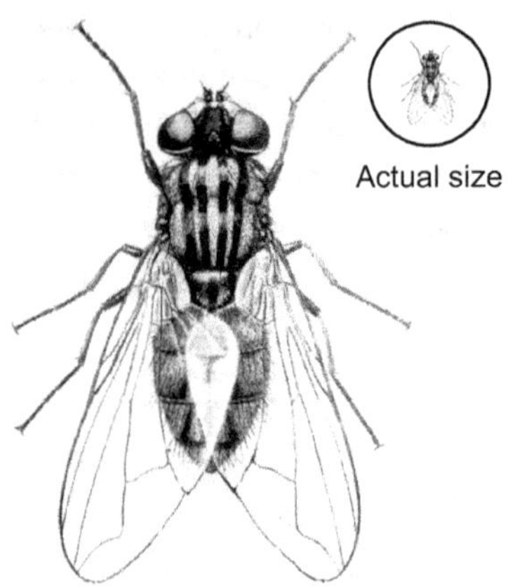

Fig. 5.3 : The housefly (Musca domestica) (by courtesy of the natural history Museum, London)

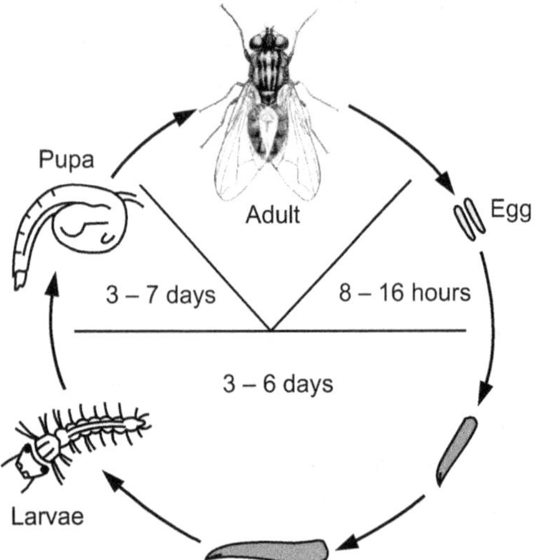

Fig. 5.4 : The life cycle of the fly

5.9.2 Control Measures

- Flies can be killed directly by insecticides or physical means such as traps, sticky tapes, fly swats and electrocuting grids. However, they should preferably be controlled by improving environmental sanitation and hygiene. This approach provides longer-lasting results, is more cost-effective and usually has other benefits.

- Improvement of environmental sanitation and hygiene
- Four strategies can be employed:
  1. reduction or elimination of fly breeding sites;
  2. reduction of sources that attract flies from other areas;
  3. prevention of contact between flies and disease-causing germs;
  4. protection of food, eating utensils and people from contact with flies.
- Reduction or elimination of fly breeding sites

Animal Sheds, Stables, Pens and Feed Lots

- Solid concrete floors with drains should be constructed; dung should be cleaned out and floors should be flushed daily.

Poultry Houses

- Where birds are kept in cages and dung accumulates below them, fans should be used to dry it; leaking water pipes should be repaired, dung should be removed and the floors should be flushed at frequent intervals.

Dung Heaps

- Dung should be stacked to reduce the surface area and the zone in which the temperature is suitable for fly breeding. It should be covered with plastic sheets or other fly-proof material. This prevents egg-laying and kills larvae and pupae as the heat produced in the composting process can no longer escape. It is preferable to stack the dung on a concrete base, surrounded by gutters to prevent the migration of larvae to pupate in soil around the heap. In hot climates, dung may be spread on the ground and dried before the flies have time to develop.

Human Excreta

- Breeding in open pit latrines can be prevented by the installation of slabs with a water seal and a fly screen over the vent pipe. If a water seal is not feasible, a tightly fitting lid may be placed over the drop hole. Installing a ventilated pit latrine can also reduce fly breeding.

- Defecation in the field, other than in latrines and toilets, may provide breeding places for filth flies (Musca sorbens). This is a common problem where large groups of people, e.g. refugees, stay together in temporary camps. Installation of proper latrines should be given priority.

- In the absence of proper facilities, people could be asked to defecate in a special field at least 500m downwind of the nearest habitation or food store and at least 30 m from a water supply. This reduces the numbers of flies in the camp and makes it easier to remove exposed faeces. Covering the faeces with a thin layer of soil may increase breeding since the faeces are then likely to dry out more slowly.

Garbage and Other Organic Refuse

- This breeding medium can be eliminated by proper collection, storage, transportation and disposal. In the absence of a system for collection and transportation, garbage can be burnt or disposed of in a specially dug pit. At least once a week the garbage in the pit has to be covered with a fresh layer of soil to stop breeding by flies. Flies are likely to breed in garbage containers even if they are tightly closed.

- In warm climates the larvae may leave the containers for pupation after only 3–4 days. In such places, garbage has to be collected at least twice a week. In temperate climates once a week is sufficient.

- When emptying a container it is important to remove any residue left in the bottom. In most countries, garbage is transported to refuse dumps, where, to reduce breeding, it is necessary to compact the refuse and cover it daily with a solid layer of soil (15–30 cm). Such dumps should be at least several kilometres away from residential areas.

- As discussed in Chapter 1, refuse can be used for filling mosquito breeding places in borrow-pits, marshy areas and other low-lying sites. If properly covered with soil, the sites are called sanitary landfills.

- In some cities, large quantities of refuse are burned in incinerators. In dry areas, simple small incinerators can be installed.

5.10 RODENTS

- Rodents are relatively small mammals with a single pair of constantly growing incisor teeth specialised for gnawing. The group includes rats and mice. Rodents are abundant in both rural and urban areas. They are found inside houses, in fields and around heaps of waste.

Types of Rodent

Three types of rodent are commonly associated with public health problems.

1. Norway Rats (Rattus Norvegicus)

- Also known as the brown rat or sewer rat, Norway rats are most numerous in urban areas. They burrow and live in the ground, and in woodpiles, debris, sewers and rubbish. Norway rats are omnivorous, which means they eat a wide variety of foods, but they mostly prefer cereal grains, meat, fish, nuts and some fruits. They do not travel more than 100 metres in search of water and food. When Norway rats invade buildings, they usually remain in the basement or ground floor. They reproduce rapidly (four to seven times a year producing eight to twelve young per litter with a gestation period of 22 days). The adult is relatively large in size, with a short tail and small ears. Their lifespan is 9–12 months.

2. Roof Rats (Rattus Rattus)

- Also know as the black or grey rat, roof rats are more numerous in rural areas. They live in roofs, and eat mainly grains. They are smaller than Norway rats with longer tails and ears. They are excellent climbers and usually live and nest above ground in shrubs, trees and dense vegetation.

- In buildings, they are most often found in enclosed or elevated spaces in attics, walls, false ceilings, roofs and cabinets. They usually nest in buildings and have a range of 30–45 metres. They can often be seen at night running along overhead utility lines or fence tops, using their long tails for balance. The average number of litters a female roof rat has per year depends on many factors but generally is between three and five, with five to eight young in each litter.

3. Mice

- Mice are smaller in size than rats and generally prefer cereals to eat. They are excellent climbers and can run up any rough vertical surface. They will run horizontally along wire cables or ropes and can jump up to 30 cm from the floor on to a flat surface.

- Mice can squeeze through openings slightly larger than 1 cm across. In a single year, a female may have five to ten litters of about five to six young. Young are born 19–21 days after mating, and they reach reproductive maturity in 6–10 weeks. The life span of a mouse is about 9–12 months.

Behaviour of Rats

- Rats are active at night. Although the vision of rats is poor, they have keen senses of smell and hearing, and a well-developed sense of touch via their nose, whiskers and hair. They like the same food as people and prefer it fresh, although they will eat almost anything. Rats constantly explore and learn about their environment, memorising the locations of pathways, obstacles, food and water, shelter, and other elements in their domain.

- They quickly detect, and tend to avoid, new objects placed in a familiar environment. Thus, objects such as traps and baits are often avoided for several days or more following their initial placement. While both species exhibit this avoidance of new objects, it is usually more pronounced in roof rats than in Norway rats.

Rodents Cause a Number of Problems:

- **Disease Transmission:** rats are the natural hosts of fleas that may carry bubonic plague and murine typhus or endemic typhus from an infected rat to a human.

- **Food Damage:** mice and rats will eat stored food, mainly grains, and will spoil food by leaving their droppings. One rat can consume 15 kilograms of food per year. Rats are estimated to destroy 20% of the world's crop production.

- **Material Damage:** gnawing by front teeth to doors, windows, wood, boxes, bags, clothes, etc.

5.11 VECTOR MANAGEMENT AND CONTROL

Vectors can be controlled using various methods. Here we describe the basic methods.

Basic Sanitation

- This approach targets the elimination or reduction of that part of the environment that facilitates breeding and **harbourage** (places where vectors find refuge or shelter). It includes the elimination of all possible breeding places for insects, the prevention of stagnation of water to limit the breeding of mosquitoes, and proper solid waste management and use of a latrine to control the breeding of houseflies. The use of clean water from protected sources for drinking prevents the transmission of guinea worm. Rats are controlled by starving them and eliminating their breeding places. Personal hygiene contributes to the control of lice.

- Generally, a clean home and environment will prevent the breeding of insects. The use of ventilation, latrines and adequate water supply play a significant role in the control of insects.

Physical Measures

- These include methods that stop vectors from getting into close contact with humans, and methods that are used to kill vectors. They include bed nets for mosquitoes and wire mesh for flies and mosquitoes Mosquito larvae can be controlled in some water containers by putting a thin layer of used oil on the surface of the water. This acts as a barrier between the water and the air so the larvae cannot access oxygen, and suffocate. Physical methods also include traps

such as adhesives to control flies and traps for rats and mice (Fig. 5.5). Delousing by boiling or steaming infested clothes are physical methods for controlling lice.

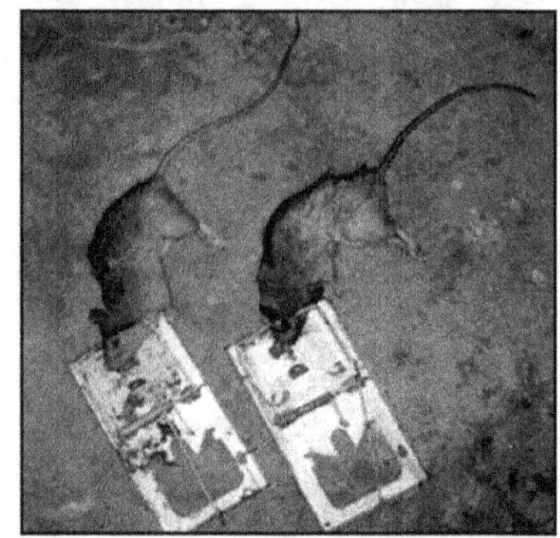

Fig. 5.5 : Rat trapping (urban roof rat)

Use of Chemicals

- Chemical insecticides can be used for the destruction of adults and larvae of insects. Commonly used chemicals are DDT, malathion and pyrethrums. Pyrethrum-containing aerosols are used for the destruction of cockroaches and flies in our homes (Fig. 5.6). Rodenticides can be used to kill rats and mice. The indiscriminate use of these chemicals, however, could have undesired health effects on users and domestic animals. Extreme care should be taken during the application and storage of chemicals. It is always important to look at the instructions for using the chemical. Environmental health workers and veterinary technicians may be able to assist in the use of chemicals against vectors.

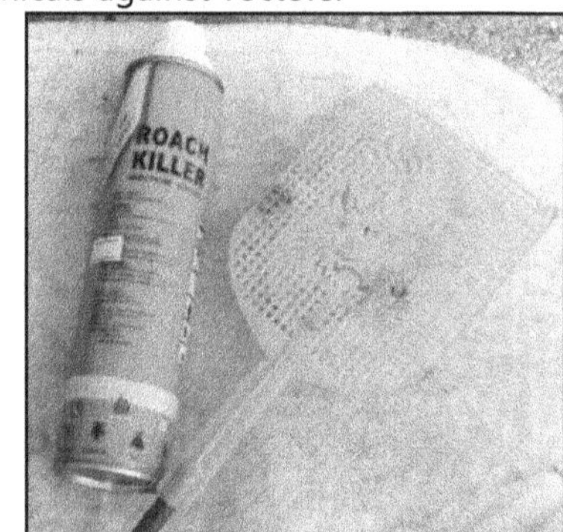

Fig. 5.6 : Insect killer chemical insecticide and fly swat.

Biological Methods

- These include several very advanced methods that prevent the successful reproduction of pest species. They include the sterilisation of males (tsetse fly, mosquito), sex distortion or replacement of genes. All of these methods are expensive and often complex to monitor. Other biological methods involve introducing or encouraging predators of the vector species. For example, small fish can be used to feed on larvae of mosquitoes. Reptiles, birds and frogs feed on adult insects and cats will prey on rats.

Integrated Approach

- **Integrated Vector Management** includes a combination of two or more of the above methods. This is often more effective than using a single method of control. For example, the rat population may be significantly reduced by combining starving with trapping. Sanitation can be combined with other cheap methods in order to be both sustainable and effective.

Planning for the Improvement of Vector Control

- The community may seek your advice on vector management. There are situations where epidemics could be possible because of vectors such as lice and fleas. The following activities are required in order to have good planning in vector management.

Knowing the Scope of Vectors

- You cannot tackle all types of vectors. However, you can be involved in the control of flies, lice, fleas, bedbugs and rats, which are the most important public health vectors. You will probably also be involved in mosquito control.

Identifying the Extent of the Problem

- Knowing the depth of the problem is important in order to mobilise the necessary resources to deal with it. This will also help you in setting priorities for vector control. You need to visit a few dwellings and ask which vectors disturb the family. You should find out how common each vector is in the community.

Identifying Control Methods

- Vector control methods vary depending on the species and you will need to use appropriate methods of intervention according to the above descriptions. Pay attention to breeding site control through the provision of basic sanitation. The use of sanitation, with one or more other methods, is the preferred tool of intervention.

Identifying Partners in Vector Management

- You will probably need to liaise with other people and offices to tackle vector problems. These may include local government institutions (for example, the police office for prison lice management; the school office for nits and lice management among students), local NGOs, and community institutions (idir, traditional leaders). They could provide resources and advice, and help mobilise the people.

Designing the Plan of Action

- This requires the preparation of activities under a specified timeframe based on the identified problems. Such activities include: visiting houses, advocacy, public and individual education, and conferences. Your approach to preparing a plan of action for vector management should be similar to other action plans you have learned about in previous study sessions of this Module.

5.12 PLAGUE

- Plague is an infectious disease caused by the bacteria Yersinia pestis, a zoonotic bacteria, usually found in small mammals and their fleas. It is transmitted between animals through fleas. Humans can be infected through:
  - ➢ The bite of infected vector fleas
  - ➢ Unprotected contact with infectious bodily fluids or contaminated materials
  - ➢ The inhalation of respiratory droplets/small particles from a patient with pneumonic plague.
- Plague is a very severe disease in people, particularly in its septicaemic (systemic infection caused by circulating bacteria in bloodstream) and pneumonic forms, with a case-fatality ratio of 30% to 100% if left untreated. The pneumonic form is invariably fatal unless treated early. It is especially contagious and can trigger severe epidemics through person-to-person contact via droplets in the air.
- From 2010 to 2015, there were 3248 cases reported worldwide, including 584 deaths.
- Historically, plague was responsible for widespread pandemics with high mortality. It was known as the "Black Death" during the fourteenth century, causing more than 50 million deaths in Europe. Nowadays, plague is easily treated with antibiotics and the use of standard precautions to prevent acquiring infection.

Signs and Symptoms

- People infected with plague usually develop acute febrile disease with other non-specific systemic symptoms after an incubation period of one to seven days, such as sudden onset of fever, chills, head and body aches, and weakness, vomiting and nausea.
- There are two main forms of plague infection, depending on the route of infection: bubonic and pneumonic.

  1. Bubonic plague is the most common form of plague and is caused by the bite of an infected flea. Plague bacillus, Y. pestis, enters at the bite and travels through the lymphatic system to the nearest lymph node where it replicates itself. The lymph node then becomes inflamed, tense and painful, and is called a 'bubo'. At advanced stages of the infection the inflamed lymph nodes can turn into open sores filled with pus. Human to human transmission of bubonic plague is rare. Bubonic

plague can advance and spread to the lungs, which is the more severe type of plague called pneumonic plague.

2. Pneumonic plague, or lung-based plague, is the most virulent form of plague. Incubation can be as short as 24 hours. Any person with pneumonic plague may transmit the disease via droplets to other humans. Untreated pneumonic plague, if not diagnosed and treated early, can be fatal. However, recovery rates are high if detected and treated in time (within 24 hours of onset of symptoms).

Treatment

- Untreated pneumonic plague can be rapidly fatal, so early diagnosis and treatment is essential for survival and reduction of complications. Antibiotics and supportive therapy are effective against plague if patients are diagnosed in time. Pneumonic plague can be fatal within 18 to 24 hours of disease onset if left untreated, but common antibiotics for enterobacteria (gram negative rods) can effectively cure the disease if they are delivered early.

Prevention

- Preventive measures include informing people when zoonotic plague is present in their environment and advising them to take precautions against flea bites and not to handle animal carcasses. Generally people should be advised to avoid direct contact with infected body fluids and tissues. When handling potentially infected patients and collecting specimens, standard precautions should apply.

Vaccination

- WHO does not recommend vaccination, expect for high-risk groups (such as laboratory personnel who are constantly exposed to the risk of contamination, and health care workers).

5.13 PLAGUE CONTROL

- Find and stop the source of infection. Identify the most likely source of infection in the area where the human case(s) was exposed, typically looking for clustered areas with large numbers of small animal deaths. Institute appropriate infection, prevention and control procedures. Institute vector control, then rodent control. Killing rodents before vectors will cause the fleas to jump to new hosts, this is to be avoided.
- Protect health workers. Inform and train them on infection prevention and control. Workers in direct contact with pneumonic plague patients must wear standard precautions and receive a chemoprophylaxis

with antibiotics for the duration of seven days or at least as long as they are exposed to infected patients.

Ensure Correct Treatment:

- Verify that patients are being given appropriate antibiotic treatment and that local supplies of antibiotics are adequate.

Isolate Patients with Pneumonic Plague :

- Patients should be isolated so as not to infect others via air droplets. Providing masks for pneumonic patients can reduce spread.

Surveillance:

- Identify and monitor close contacts of pneumonic plague patients and give them a seven-day chemoprophylaxis. Chemoprophylaxis should also be given to household members of bubonic plague patients.
- Obtain specimens which should be carefully collected using appropriate infection, prevention and control procedures and sent to labs for testing.

Disinfection :

- Routine hand-washing is recommended with soap and water or use of alcohol hand rub. Larger areas can be disinfected using 10% of diluted household bleach (made fresh daily).

Ensure Safe Burial Practices :

- Spraying of face/chest area of suspected pneumonic plague deaths should be discouraged. The area should be covered with a disinfectant-soaked cloth or absorbent material.

EXERCISE

1. Define communicable disease. Discuss how it is spread in human beings.
2. Explain various ways of protection from communicable diseases.
3. Explain following diseases with respect to infectious agent, occurrence, reservoir, mode of transmission and control

(i) Typhoid	(ii) Dysentery
(iii) Hepatesis	(iv) Cholera
(v) Influenza	(vi) Common cold
(vii) TB	(viii) Malaria
(ix) Measles	(x) Mumps

4. Explain Chain of disease transmission.
5. Explain control measures for

(i) Plague – Rats	(ii) Malaria- Mosquito
(iii) Flies	

6. Discuss Life cycle of

(i) Mosquito	(ii) Fly

RURAL SANITATION

6.1 INTRODUCTION

- Water, sanitation and waste management are important driving forces for community health in India. A clean environment, open defecation free areas, personal hygiene practices among the individuals, proper solid and liquid waste management, and availability of adequate safe drinking water determine the health of individuals as well as the community. About 88 % of child deaths are estimated to be related to diarrhoeal diseases and according to the World Health Report, 2005, CHERG (Child Health Epidemiology Reference Group) 17% of child mortality was due to diarrhoea.

- Major public health challenges like diarrhoea, cholera, malaria and dengue, polio, Hepatitis A and E, amoebic dysentery and other diseases are also caused due to unsafe drinking water and poor sanitation. Over 1 billion people or 15% of the world's total population practice open defecation. In India 68.84% of the population lives in rural areas. According to the Census of India 2011, 71.59% of rural individual households in Karnataka do not have toilet facilities and are dependent on the use of open spaces for defecation.

- The government of India along with several international organisations, civil society organisations, elected representatives, policy makers and Panchayati Raj Institutions have facilitated several programmes since 1986 to improve Sanitation and the living conditions of rural populations but the implementation and achievement is not up to the mark .

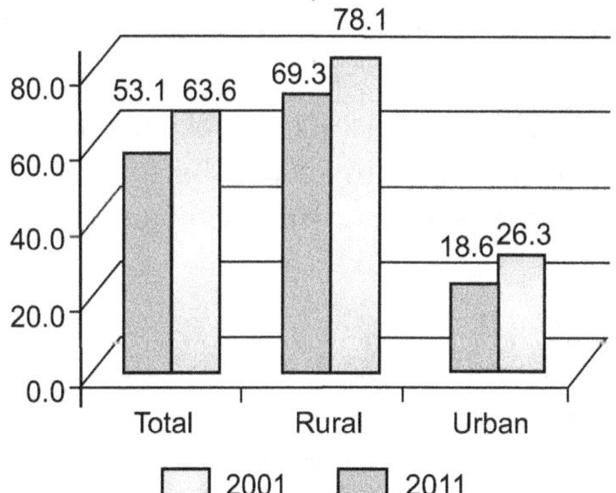

Fig. 6.1 : Percentage of household having no latrine

- We live in a time of rapid change and progress in science and technology with changes in housing structures, lifestyle, food habits, clothing, education, but one of the major challenges that still remains is tackling the problem of open defecation.

Table 6.1 : Social Change in Human Being

Time	Food	Clothing	Education	Housing	Transportation	Relationships	Open Defecation
Stone age	Raw food	Leaves, animal skin etc	Informal	Caves, tree trunks etc.	Bare foot walking, using animals	No relationship	Yes
Modern age	Cooked food	Clothes	Schools, colleges etc.	Proper structure with brick, stone etc.	Motor vehicles	Marriage, parental relationsip etc.	Yes

- In a developing country like India, the diseases related to unsafe drinking water and sanitation in rural areas still have considerable community and public health significance and have impacted socio-economic development.

- Awareness regarding the value of a clean environment, organic farming, nutrient value of faeces and urine, reduction in waterborne diseases through improved sanitation practices and economic development through job creation for masons in their own villages will lead to sustained socio-economic growth.

6.2 HISTORY OF SANITATION

- Open defecation has been practiced since the beginning of time by all creatures.

- Individuals use agricultural fields, wastelands, banks of lakes/rivers, forestlands and open waste places to defecate. Defecation sites are usually far from human settlements.

- Over the years due the influence of factors like modernisation, changed cultural behavior, changes in social life and education, have developed technologies for sanitation ranging from using dry toilets to flush toilets and ecological sanitation.

Dry Latrines :

- Dry toilets are different from water sealed toilets. In this type of a toilet, a seat with a hole is directly connected to a pit. In some communities buckets were used for collecting faeces and were cleaned by persons belonging to lower castes.

Flush Toilets/ Water Sealed Latrines :

- Legislations such as prohibition on manual scavenging and construction of dry latrine gave space for the use of a new technology called flush toilet or water sealed latrines; where-in a connection from the water sealed toilet pan to "Under Ground Drainage (UGD)" and septic tanks came into existence.

Ecological Sanitation:

- Ecological sanitation technology incorporates the principles of recycling faeces, urine and wastewater which are collected separately, treated and used as compost and fertilizer for agricultural crops.

6.2.1 Water and Sanitation

- Surface water in India is scarce and groundwater is deep and difficult to reach. Traditionally, most villagers used water from any source available to them, such as ponds, rivers, springs and wells. Water quantity greatly depended on the season, and water quality was generally poor. As late as 1980, less than a third of the population (31 per cent) had 'full coverage'2 of clean drinking water, and virtually no rural households had access to sanitation facilities. Access to water was a prerequisite to the later introduction of latrines in both rural and urban areas. In 1966-1967, severe drought hit the states of Bihar and Uttar Pradesh in northern India.

- The Government requested emergency help from UNICEF, which responded by airlifting 11 boreholedrilling rigs into the country from the United Kingdom. These rigs could drill far below the earth's surface to tap into groundwater that was otherwise unavailable.

- In addition to meeting the short-term need, the effort showed the potential of what are known as down-the-hole hammer drilling rigs to reach water under India's hardrock areas. The Government of India subsequently made the provision of clean, safe drinking water a cornerstone of its rural development programme and strengthened its collaboration with UNICEF to provide these services. By 1976, almost 300 rigs were in use, with the Government of India and UNICEF each supplying about half the total.

- Early success raised expectations of what the drilling programme could achieve. By the early 1980s, at the beginning of the International Drinking Water and Sanitation Decade, the Government identified an additional 230,000 'problem villages' in need of water. These 'problem villages' were very remote; prone to drought, cholera, or guinea worm disease; or built upon particularly unfavourable sites. To back up the commitment to extend water coverage to more of the rural poor, the Government considerably increased the funds allocated to water and sanitation.

- Besides having increased financial resources and being supported with political will, the effort to expand coverage achieved success because of three other factors: new, locally adapted rigs that could drill boreholes more quickly; the provision of long-term service for the rigs and training support for the operators and engineers; and standardization of drilling specifications. Until 1998, when the responsibility for water well drilling was transferred to state agencies, UNICEF provided spare parts and service on the rigs. UNICEF also provided training over the years to drilling operators and engineers. This support reduced the down time for rigs. Thus, they performed consistently over the long term, drilling an estimated five to eight boreholes a month. Standardizing norms for drilling also helped widen coverage by increasing the number of boreholes drilled. Standards set on the depth and diameter of drilled boreholes provided operators working independently throughout the country with simple, measurable indicators.

- Standards were set for minimum yield for a handpump, minimum surface casing and the surface sealing necessary to protect the borehole from the entry of polluted water. In some cases, conditions did not require the boreholes to be drilled as deeply as specified. However, the standards encompassed the different situations that hydrogeologists and engineers might encounter, thus saving the time and expense that would have been spent in setting specifications for each site.

Current Scenario

- The Indian government launched the much-hyped *Swachh Bharat Mission (Gramin)* or "Clean India Mission (Rural)" an effort to attain nationwide, open defecation-free (ODF) status over five years by facilitating the installation of more than 200,000 subsidized latrines. It may seem encouraging that, by official estimates, there are already about 314 ODF Indian districts (out of total 644) in 11 states (out of 37). But an important question remains: What will it take for people to actually use government-subsidized latrines?

- Answering this question means looking at the latrine infrastructure they're often pretty shabby, small, and largely without a stable water source. But it also means examining the behavioral norms (socio-cultural taboos handed down through generations) that deter people from the very concept of at-home latrines.

Example:

- Some consider having latrines within a home's premises religiously "impure" human excreta and deities cannot coexist under the same roof. The "filth," villagers believe, tarnishes the sanctity of home, potentially inciting divine wrath instead of garnering bliss. The gruesome chore of pit cleaning aggravates risks of impurity. By the same token, at-home latrines prohibit cooking and eating under the same roof. Similarly, trash-bins are rare in rural homes. Household waste is customarily swept out into the open to keep "evil" at bay.

- Latrine pits are often considered the spawning ground of mosquitoes and flies, and thus homeowners fear them.

- Often, the lack of a sustainable water supply, even if there is a latrine, necessitates open defecation in the vicinity of water sources (wells, streams, ponds, tanks, and so on), thus exacerbating risks of health hazards owing to water contamination.

- Some villagers regard open defecation as an opportunity for social networking, where they can meet others in nearby fields to keep up on gossip.

- Working men and women sometimes view open defecation as a convenience. They can leave home very early in the morning and go to the fields, where they defecate as needed. To them, latrine-time interferes with valuable daylight working hours.

- Some studies have even indicated that open defecation is a demonstration of masculinity, strength, and pride).

- Rural societies in some parts of the country out of dire need and desperation have begun crafting social campaigns to transform these kinds of behavioral norms.

- A "No Toilet, No Bride" campaign in Haryana and Punjab states is one example. Launched in 2005 by the state governments, the campaign made great use of social and public media to prompt families with marriage-age girls to demand that potential suitors' families install an at-home latrine as a pre-requisite to marriage. As of August 2017, the effort had led to a 15 percent rise in family investment in at-home latrines and has had four times the impact in areas where there are fewer potential brides. Private sanitation coverage in Haryana has risen by as much as 21 percent for households having marriage-age boys, according to a report in the Journal of Development Economics.

- Meanwhile, nationwide surveys reveal that women are more interested in having at-home latrines than men. For millions of women, having a latrine at home literally stands between dignity and shame, between safety and abuse, between acquiring a urinary tract infection and health (including the health of their children). Women's collectives in some districts have already taken the matter into their own hands and built their own latrines (popularly known as izzat ghar, or "room of dignity").

- Praiseworthy district administrations are also lending a hand by training women with masonry skills or putting them in touch with local or regional self-help groups. In some central Indian districts such as the state of Jharkhand, where open defecation is most common, women have initiated a drive to dig latrine pits called gaddakhodoabhiyaan. The effort is facilitated by district administration and calls for each house in a given community to contribute a digger to build a "leach pit" a type of latrine that doesn't require a sewage pipe connection and soaks waste water in situ. It also reduces the hassle of transporting raw materials in difficult terrain.

- A recent youth-driven social media campaign, "Take Poo to the Loo," is spreading awareness about open defecation and destigmatizing latrines. The campaign developed following a UNICEF's social awareness drive called "The Poo Party," in which Mr. Poo (an animated fecal mascot) aimed to sensitize citizens to the multidimensional health aspects of open defecation (such as diarrhea and dysentery). "Take Poo to the Loo" features a music video written and composed by the famous Indo-Jazz composer Shrikanth Sriram, and famous Indian bands like Indian Ocean and Raghu Dixit have also come out in its support. Quirky, fun, and informative, this campaign song aims to bring youth to the forefront of the movement against open defecation. Making imaginative use of "toilet sounds," it reminds people that, if left unaddressed, open defecation may encroach on personal space and make existence unbearable. Students of prestigious Indian academic institutions, including Indian Institute of Technology Delhi, have begun using the message to help raise awareness about the health and environmental ramifications of open defecation.

Change From Within

- Transforming the behavioral norms of rural populations is a particularly challenging task, and it will only work if rural communities change from within. One approach, as suggested by a recent World Bank study, may be revitalizing the community-led total sanitation (CLTS) program, which is underpinned by three "Ds" (condensed down from five):

1. **Depict:** Use media to show the dignity and self-esteem of households with at-home latrine facilities, and the part latrines can play in elevating cultural enrichment and social standing. Village administrations might even create provisions to officially reward "progressive" households and motivate others to follow their lead.

2. **Demonstrate:** Organize live demonstrations of toilet designs to show how easy they are to operate and maintain.

3. **Divulge:** Make people aware of the potentially negative health outcomes of open defecation and the low cost of constructing latrines at home by conducting workshops and audio-visual training sessions at the elementary school level, and finding ways to articulate social messages via village songs, plays, and folk art, which often connect better than speeches by government officials.

- The local efforts described above have begun the journey of a thousand miles toward improving sanitation coverage in India, but much depends on the extent to which civil society takes up the call, and whether and how the government pitches in to sensitize and support the fight. Half of the challenge lies in increasing the coverage area and ensuring the long-term functional status of the physical infrastructure.

- The movement needs more private financing behind government efforts to increase coverage, and both communities and the government need to share the responsibility of upkeep. The other half of the challenge is uprooting deep-seated beliefs and taboos in the minds of rural populations. Success will require long-term campaigns to spread awareness, the

development of regionally contextualized innovations, focus-group discussions, and women's empowerment. Authorities should also attempt to make rural sanitation governance more transparent by strengthening bonds between villagers and sanitation departments, and by increasing participatory involvement of communities in the mainframe of planning and operation.

6.3 CONCEPT OF RURAL SANITATION

- The concept of sanitation broadly includes liquid and solid waste disposal, personal and food related hygiene and domestic as well as environmental hygiene.

- It would not be wrong to say that it hardly describes the sanitary conditions as they obtain in the villages of India.

- Most of the people still defecate in the open space, most of the villages lack waste disposal and drainage systems and many in the villages are ignorant about the consequences of poor sanitation and unhygienic conditions. As a result, many people suffer and even die of diseases caused by unhealthy practices of personal and environmental hygiene.

- Rural sanitation figures prominently in the National Agenda for governance. At present the extent of sanitation coverage in India is around 16 percent of all rural households. This figure is one of the lowest in the world, at par with countries like Niger and Afghanistan and possibly lower than Bangladesh.

- The absence of safe sanitation contributes significantly to the poor quality of life as reflected by well accepted indicators like Infant mortality and morbidity rates. According to the Union Ministry of Health, around 7,00,000 children die each year due to diarrhoea and other water sanitation-related diseases.

- Rural sanitation is a state subject. The state governments implement the rural sanitation programme under state sector Minimum Need Programme (MNP). The central government supplements their efforts providing financial and technical assistance through the centrally sponsored Rural Sanitation Programme (CRSP).

- Sanitation is used to define a package of health related measures. It is also defined as the means of collecting and disposing of excreta and community liquid wastes in a hygienic way so as not to endanger the health of individual and the community as a whole (WHO, 1986).

- The concept of sanitation was earlier limited to disposal of human excreta by cesspools, open ditches, pit latrines, bucket system etc. To-day, it denotes a comprehensive concept of not only the methods of disposal of human waste but also of liquid and solid wastes including matter originating from food and hygiene. It is viewed as a package.

6.3.1 Planning for Rural Sanitation

- The Environmental Hygiene Committee set up by the Government of India in 1948 recommended that 90 percent of the country's population should be covered with water supply and sanitation facilities within a period of forty years for which the national programme was to be initiated. In the year 1954, sanitation programme was introduced in the health sector.

- The Government launched the national water supply and sanitation programme as part of the First Five Year Plan. The first five-year plan had a provision of Rs. 6 crores for rural water supply and sanitation programmes. It was realised at the end of Second Plan that sanitation was not receiving due importance and it was the lack of health education and community participation which was responsible for this failure.

- A comprehensive strategy for promoting sanitation in rural areas of Indian states was first developed during the International Water Supply and Sanitation Decade (1980-90) with a view to provide the population with protected water supply and basic sanitation facilities over a period of ten years.

- It was envisaged that 25 percent of rural population would be provided with sanitation facilities by the end of the seventh five year plan period (1985-90). At the beginning of the decade, the position of coverage of rural population was 2.8 million i.e. 0.5 percent.

- The year 1985 witnessed the transfer of Rural Sanitation Programme to the Department of Rural Development from the ministry of urban development. Rural sanitation was made a component of 20 point programme and was also included under the Minimum Needs Programme (MNP) in 1987.

- In 1986, a programme was launched to construct one million sanitary latrines to be provided in houses of SC/ST population under Indira Awaas Yojana and to provide 2,50,000 additional latrines to health centres, schools, Panchayat Ghars and Anganwadis under NREP and RLEGP. Rural Sanitation Programme was also added to the sector MNP from 1987-88.

- The data reveals that 0.1 percent of rural population in the country had access to sanitation facilities in 1970. This was increased to 0.5 percent in 1980 and 2.45 percent in 1990s with a population coverage of about 15 million. This rural sanitation programme could not make much head-way even during the decade 1981-91, in spite of its importance.

Advertisements:

- A clearer picture emerged after 1991 census. Only 9.5 percent of rural families and 63.9 percent of urban families of the country (excluding Jammu and Kashmir state) had toilet facilities. Among the major states, the highest achievement in rural sanitation was in Kerala where 44.1 percent of the rural families had toilet facilities.

6.3.2 Central Rural Sanitation Programme

- Central Rural Sanitation Programme (CRSP), a centrally sponsored Rural Sanitation Programme was launched in 1986. Its objective is to improve the quality of life of the rural people and provide privacy and dignity to women. It was designated to provide sanitary latrines to the SCs/STs, landless labourers and people living below poverty line and the resources were shared by the central and state governments on 50: 50 basis.

- The programme was planned with the objective of providing clean, healthy and environmentally acceptable disposal of excreta with a view to create good sanitation and consequent improved health standards. The CRSP is implemented in different states and union territories for improving sanitation facilities through construction of sanitary latrines for individual households.

- The programme provided for cent percent subsidy for construction of latrines for SCs/STs and landless labourers and subsidy as per the rates prevailing in the states for the general public. It also provided for construction of village complex with bathing facilities, hand pumps, latrines, drainage facilities, washing platform etc.

- The criteria and norms under CRSP were modified in February 1991 and the guidelines were revised again in June 1993. The purpose of such revision was to make the programme more holistic to give emphasis on Information, Education and Communication activities, to involve voluntary organisations in a bigger way and the concept of "Sanitary Mart" was also introduced.

- The revised programme aims of generation felt need and people's participation. The subsidy pattern has been changed limiting to 80% for persons below the poverty line for individual household latrines. The unit cost of construction is to be limited to Rs. 2,500 of which 80% could be paid as subsidy to the selected beneficiaries below the poverty line.

- Another salient feature of the revised programme is to develop at least one model village covering facilities like sanitary latrines, conversion of dry latrines garbage pits, soakage pits, drainage, pavement of lanes, sanitary latrines in village institutions, cleanliness in ponds, tanks, clean surrounding around hand pumps and other drinking water- sources.

- But, experiences of CRSP implemented through state governments and CAPART were not encouraging. There was always a wide gap between the number of units sanctioned under CRSP, number of units taken up for construction and the number actually used by the users.

6.3.3 Total Sanitation Campaign (TSC)

- Total Sanitation Campaign (TSC) was initiated on 1st April 1999 under sector reform process. The campaign is community led and people centred. It was launched after restructuring Central Rural Sanitation Programme and is operational in 451 districts with an out IAY of Rs. 4,416 crores in which community contribution is Rs. 812 crore.

- Catchment Area Approach (CAA) has been adopted for monitoring and surveillance by involving various grass roots level educational and technical institutions by utilizing existing resources and strengthening them by providing additional financial resources.

The Components of the TSC are:

1. Construction of household latrines.
2. Construction of sanitary complex for women.
3. Toilets for schools.
4. Toilets for Balwadi/Anganwadi etc.

- Besides these, funds are being provided for Start-Up Activities, Information, Education and Communication and Administrative Charges.

The Main Features of the TSC are as Under:

- Shift from high subsidy to low subsidy regime.
- Greater household involvement and PRI participation.
- Technology options as per choice of beneficiary households.

- Stress on Information, Education and Communication (IEC) as part of the campaign.
- Emphasis on school sanitation, women sanitary complexes.
- Integrating with various rural development programmes.
- Involvement of NGOs and local groups.
- Promoting access to institutional finance and social marketing concept.
- The mission has decided to sanction TSC projects in all districts of the country by 2005-2006 so as to achieve full basic sanitation coverage by 2012. It has been planned to provide all rural schools and Anganwadis with safe drinking water and sanitation by the year 2005-2006. Also, to add vigour to sanitation drive. Government initiated an incentive scheme for fully sanitized and open defecation free Gram Panchayats, Blocks and Districts called the "Nirmal Gram Puraskar" in 2003.

"Nirmal Gram Puraskar" (NGP):

- It is an incentive scheme instituted in October 2003 under the TSC in recognition of the role played by Panchayati Raj Institutions, organisations and individuals in promotion of rural sanitation. As per this scheme, awards are given to Panchayati Raj Institutions at various levels which attain full sanitation coverage in households, schools, Anganwadis with general cleanliness and become open defecation free.

Tenth Plan Strategy:

- The unprecedented sanitation challenge requires new strategies and methods to improve and promote sanitation, which should be accessible to everyone in rural areas. Through the creation of demand and behaviour change Instead of awareness generation to improve physical quality of life in rural areas: sanitation coverage among rural population will be accelerated.

- Toilet facilities especially in all the primary and upper primary schools and integrated sanitary complexes, exclusively for women, would certainly create a kind of ownership among the needy segment of the community. Suitable, cost effective local based affordable multiple designs of individual household toilets should be encouraged. New partners like co-operative milk societies, sugarcane farmers associations and big industries can adopt communities/blocks and promote rural sanitation in their own geographical areas.

6.3.4 Rural Sanitary Marts

- Sanitary marts in India have been supported by UNICEF for nearly a decade with the objective of establishing one-stop shops to meet all sanitary requirements for communities, selling and in some cases producing materials required for the construction of home toilets and sanitary facilities as well as sanitary products.

The Sanitary Marts were Conceived as:

- ".......... retail outlets dealing with not only the materials required for construction of sanitary latrines and other facilities but also those items which are required as a part of the sanitation package".

- The inventory of the typical mart in this latter model included low cost ceramic pans and traps, RCC pit covers, pipes and such other material required for construction of a leach pit latrine as well as readymade cattle trough, food safe, cheap footwear, toilet soap, nail cutter and other items relating to personal hygiene and home sanitation.

- The exact composition of the inventory was to be decided locally at the mart level. The rural sanitary mart was also expected to serve as counseling centre for those interested in building a toilet on their own. The mart would have information on the entire range of technical options including possible variations in super structure and corresponding cost implications.

- A list of masons trained or possessing the skills required to construct such toilets were also available in these marts. Thus, with the aim of promoting 'zero' subsidy and in response to the need to cater for motivated households unable to construct latrines due to the non-availability of information and materials rural sanitary marts were established.

6.3.5 Strategy for School Sanitation

- School sanitation is a vital component of sanitation and this has long been neglected, though there was a feeble attempt made to provide such facilities in Higher Secondary Schools under the recommendations of the Tenth Finance Commission. School Sanitation and Hygiene Education (SSHE), a component of Total Sanitary Campaign (TSC) aims to promote sanitation and hygiene in and through schools to bring about behavioural change that will have a lasting impact.

- The strategies are developed in tune with local needs which are adaptable and acceptable among target groups. These are involvement of child as a change agent to spread the sanitary practices in the proven route of teacher-children-family community emphasis on attitude and behavioural change through hygiene education using the skill approach.

- Child-friendly, especially girl child and disabled-friendly water and sanitation design options, inter-sectoral co-ordination through alliance building with concerned Ministries and Departments, and involvement of community as an equal partner.

- Children are more receptive to new ideas and therefore, the school is the best suitable institution in changing the conditioned habits of people from open defence on to the use of lavatory through motivation and education.

- The experience gained by children through use of toilets in school and sanitation education imparted by teachers would definitely be carried home and passed on to parents, who in most cases do not have formal education.

- Sanitation in India is presently undergoing adjustment aiming at almost total coverage in the beginning of the new millennium. The realisation of these plans will critically depend on affordable and cost effective technologies and implementation systems which are easy to implement together with new administrative system.

The following points are necessary for the success of achieving total sanitation campaign goals of the ministry of Rural Development:

1. We must have a mission to provide sanitary facilities to all dwelling units in rural areas by the year 2010. Since the facility is still to be provided for over hundred million dwelling units, we should target provisioning of sanitary facilities to at least twenty million dwelling units per year. While providing this facility we should ensure provision of adequate water supply.

2. The mission must be executed through village panchayats in conjunction with societal establishments mobilised for this purpose in each of the villages. It will be useful to empower women in all the villages to execute this programme.

3. The ministry of Rural Development can organise state-wise training programme to train the members of sanitation mission in construction and maintenance of modern sanitary facilities. The Environmental Sanitation Institute, Ahmedabad and similar institutions can become the nodal agency for imparting such training can become a public-private partnership programme. Programmes aimed at

employment can be tuned to give such workers good income as well.

4. The Sanitation Mission has to make the entire village community dynamic and provide employment opportunity for certain number of people. Educate the children right from the age of three to make use of sanitary facilities. This should become part of the total sanitation campaign.

The state council for sanitation proposed under urban sanitation sector should also have the mandate for rural sanitation:

1. Subsidy for low cost household toilets should be given to rural Below Poverty Li ne (BPL) families, and it should be at par with subsidy for the urban households. For the success of the scheme, a subsidy of 50 percent of the cost of the unit inclusive e of sub and super structures for the basic twin-pit pour flush system appears to be necessary during the 10th plan.

2. The recommendations made with regard to urban low cost sanitation also apply to the rural segment. Creation and maintenance of a record of locally relevant information regarding various technological options, hydro-geological information, availability of building materials, choices in design and implementation etc. at the block level should be organised through the panchayats, sanitary marts and building centres.

3. For the success of the schemes, and to over-come the huge problem of insanitary practices in the country, a programme of education. Propagation, training designing and development, production and installation, needs to be undertaken. NGOs should be mobilised to support the programme, especially for supervision, monitoring, training and development work. A suitable provision for the participation of non-governmental organisations in the sanitation programme should be made under the head project costs.

- Thus Rural Sanitation Programme envisages promoting "Environmental Sanitation" as a package aiming at addressing the issues to reduce the probability of people's exposure to diseases and providing hygienic environment and taking measures to break the cycle of diseases by improved management of human, animal and domestic wastes.

- Sanitation should become a massive people's programme. This is possible through motivation and awareness education programmes with the concerted efforts of panchayats, voluntary clubs, MahilaMandals and the government machinery. The existing sanitation conditions call for a new strategy of making rural sanitation a people's programme with government participation.

- The awareness campaign must revolve around the concept of total environmental sanitation. Active involvement of PRIs must be ensured in the task of promoting hygiene and environment consciousness in relation to community health and welfare.

6.4 LOW-COST SANITATION

- Unsanitary conditions and contaminated drinking water exact a crippling toll on both the health of the human population and the environment. Approximately 40 percent of the world's population does not have access to improved sanitation. In addition to the indignity suffered by those lacking sanitation facilities, millions of people in the developing world die each year from diseases contracted through direct and indirect contact with pathogenic Bactria found in human excreta. Infectious diseases such as cholera, hepatitis, typhoid and diarrhea are waterborne, and can be contracted from untreated wastewater discharged into water bodies.

- More than half of the world's rivers, lakes, and coastal waters are seriously polluted from wastewater discharge. The cost of inadequate sanitation translates into significant economic, social, and environmental burdens. As the world attempts to realize these goals, we must reassess the lessons learned, evaluate new technologies, identify research gaps, and critically discuss ways forward. Since 2 billion of the 2.6 billion people lacking sanitation live in rural areas, we must complement large-scale urban investments with low-cost, on site technologies that target rural communities. Low cost sanitation options have significantly improved, especially for the reuse of sewage for agriculture or aquaculture.

- The Centrally Sponsored Scheme of Low Cost Sanitation for Liberation of Scavengers started from 1980-81 initially through the Ministry of Home Affairs and later on through the Ministry of Welfare. From 1980-90, it came to be operated through the Ministry of Urban Development and later on through Ministry of Urban Employment and Poverty Alleviation now titled Ministry of Housing and Urban Poverty Alleviation. The main objectives of the Scheme are to convert the existing dry latrines into low cost pour flush latrines and to construct new ones where none exist. The scheme has been continued in the 12th plan period with the intention of converting the remaining latrines serviced by human identified by the Census of India 2011 in urban areas.

- The objective of the Scheme is to convert/ construct low cost sanitation units through sanitary two pit pour flush latrines with superstructures and appropriate variation to suit local conditions (area specific latrines) and construct new latrines where EWS household have no latrines and follow the inhuman practice of defecating in the open in urban areas. This would improve overall sanitation in the towns. The Scheme also encourages adoption of new technologies like bio-digesters and ecosan toilets by the implementing agencies.

6.4.1 Sulabh Sauchalaya – Low Cost Sanitation

- Sulabh flush compost toilet is eco-friendly, technically appropriate, socio-culturally acceptable and economically affordable. It is an indigenous technology and the toilet can easily be constructed by local labour and materials. It provides health benefits by safe disposal of human excreta on-site. It consists of a pan with a steep slope of 25°-28° and an especially designed trap with 20 mm waterseal requiring only 1 to 1.5 litres of water for flushing, thus helping conserve water.

- It does not need scavengers to clean the pits. There are two pits of varying size and capacity depending on the number of users. The capacity of each pit is normally designed for 3 years' usage. Both pits are used alternately. When one pit is full, the incoming excreta is diverted into the second pit. In about two years, the sludge gets digested and is almost dry and pathogen free, thus safe for handling as manure.

- Digested sludge is odourless and is a good manure and soil-conditioner. It can be dug out easily and used for agricultural purposes. The cost of emptying the pit can be met partially from the cost of manure made available. Sulabh toilet can also be constructed on the upper floors of buildings. It has a high potential for upgradation, and can later be easily connected to sewers when introduced in the area. Sulabh has so far constructed over a million individual household toilets in different parts of the country.

Operation and Maintenance

Operation and maintenance of a Sulabh flush compost toilet is very easy and simple:

- Before use, wet the pan by pouring only a little quantity of water.

- After defecation, pour 1.5 to 2 litres of water in the pan for flushing.

- Pour about half litre of water in the pan after urination.

- The pan should be cleaned once a day with a brush or a broom and with soap powder periodically.

- One of the pits is to be used at a time by plugging the drain for the other pit.

- Kitchen, bathroom waste water or rain water should not be allowed to enter the pits.

- Other solid wastes like kitchen waste, rags, cotton, sweepings etc. should not be thrown in the pan, this could block the toilet.

- When the first pit in use is full, the flow should be diverted to the second pit and the filled up pit should be desludged after 1.5 to 2-year rest period. The first pit can then be put to reuse, when the second pit fills up.

6.4.2 Prevention of Pollution

- To check pollution of drinking water sources, the pits in fine soils (effective size 0.2mm or less) should be located at a minimum distance of 3 metres from open wells and shallow hand pumps provided ground water table throughout the year is 2 metre or more below the bottom of the pit; if water table is higher, the distance should be increased to 10 metres.

- In coarser soils (effective size more than 0.2mm), the same safe distances can be maintained by providing 500mm thick sand envelope of 0.2mm sand all round the pit and sealing the pit bottom by some impervious material like puddled clay, polythene sheet, lean cementconcrete or cement stabilised soil.

- Normally bacteria do not move beyond 3 metres horizontally in homogeneous soil and vertically they do not permeate more than 1 metre, however there can be marginal deviations depending upon the types and compaction of the soil. It may be noted that chances of ground water occur due to higher hydraulic load. Since in this system hydraulic load is only 1.5 to 2 litres per use, there is no such chance of ground water pollution.

Manure from Human Excreta

- One of the major difficulties for the use of human excreta as manure is the presence of bacterial and other pathogens. Human excreta contain a full spectrum of pathogens causing various infections. It should be free from pathogens before being used as manure.

- Another problem is psychological/chamber (250mm x 500mm internal size) should be constructed at the place from where the pipe is bifurcated to connect the two pits. The pipes of drains should have a minimum gradient of 1:15.

Cost of Sulabh Flush Compost Toilet

- The cost of Sulabh flush composting toilets varies widely to suit people of every economic stratum. The cost ranges from US$ 10 to US$ 1000 per unit. It depends upon materials of construction of pits and seat as well as of the superstructure. The pits can be constructed with bricks or any locally available materials like stones, wood logs, burnt clay rings, concrete rings or even used coaltar drums.

- Similarly, the quality of superstructure ranges from simple gunny bag sheets, or thatch to well finished tiles with R.C.C. roof, doors, wash basin, etc. Cost varies also due to size and capacity of the pits, varying from 2 years to 20 years capacity for each pit. Keeping the basic design unchanged, Sulabh has a number of such toilet models for demonstration.

- Manure from Human Excreta One of the major difficulties for the use of human excreta as manure is the presence of bacterial and other pathogens. Human excreta contains a full spectrum of pathogens causing various infections. It should be free from pathogens before being used as manure. Another problem is psychological/ religious taboos associated with it.

- The studies carried out by the Sulabh have revealed that the content of a Sulabh toilet pit is almost free from pathogens when taken out after two years of resting period. To make it completely pathogen free, digested sludge is sun dried for 2 to 3 weeks. During drying of sludge big lumps are formed making it difficult to mix in soil homogeneously.

- Sulabh developed a technology to granulate such dried lumps into small size graded granules which look like processed tea leaves. Before granulating, it is processed in a ball mill to break it into small pieces. Then it is passed through the mass mixer where the

moisture content of manure is regulated by adding water.

- Such manure has a good percentage of plant nutrients. Besides, it increases humus and water holding capacity of the soil. The Institute has carried out experiments to monitor its manurial effects on different vegetables and flowering plants. In all the cases tested, the effect of manure on the growth of plants was very encouraging.

6.5 WATER AND SANITATION

- Every eight seconds a child dies of a water-related disease. Every year more than five million human beings die from illnesses linked to unsafe drinking water, unclean domestic environments and improper excreta disposal.

- At any given time perhaps one-half of all peoples in the developing world are suffering from one or more of the six main diseases associated with water supply and sanitation (diarrhoea, ascaris, dracunculiasis, hookworm, schistosomiasis and trachoma).

- In addition, the health burden includes the annual expenditure of over ten million person-years of time and effort by women and female children carrying water from distant, often polluted sources.

- Nearly a quarter of humanity still remains today without proper access to water and sanitation.

- During the International Drinking Water Supply and Sanitation Decade (1981-1990), some 1600 million people were served with safe water and about 750 million with adequate excreta disposal facilities. However, because of population growth of 800 million people in developing countries, by 1990 there remained a total of 1015 million people without safe water and 1764 million without adequate sanitation.

- Overall progress in reaching the unserved has been poor since 1990. Approximately one billion people around the world still lack safe water and more than two billion do not have adequate excreta disposal facilities. Rapid population growth and lagging rates of coverage expansion has left more people without access to basic sanitation today than in 1990.

- Another problem with coverage goals is the magnitude of resources needed to achieve them. At the Global Consultation of Safe Water and Sanitation for the 1990s, held in New Delhi in 1990, it was stated that universal coverage by the year 2000 would require US$ 50 billion per year, a five-fold increase in current investment levels.

- In 1992, WHO concluded its monitoring of the Decade with the estimate that a total of US$ 133.9 billion had been invested in water supply and sanitation during the period 1981-1990, of which 55% was spent on water and 45% on sanitation. Urban areas received 74% of the total and rural areas only 26%. Contrary to widespread perceptions, almost two-thirds of all funds were provided by national sources and only a third by external organizations.

- WHO estimates that it costs an average of US$ 105 per person to provide water supplies in urban areas and US$ 50 in rural areas, while sanitation costs an average of US$ 145 in urban areas and US$ 30 in rural areas.

- Water supply and sanitation can be viewed as a process having three interactive elements. The most fundamental of these elements is the availability of safe drinking water and sanitary means of excreta disposal. This means 20 to 40 litres of water per person per day located within a reasonable distance from the household. Safe water implies protection of water sources as well as proper transport and storage within the home. It also means facilities for bathing and for washing clothes and kitchen utensils which are clean and well-drained. Sanitary excreta disposal is the isolation and control of faeces from both adults and children so that they do not come into contact with water sources, food or people. To break the transmission chain of faecally related diseases, good standards of personal and domestic hygiene, which begin with handwashing after defecation, are essential.

- A second element in the water and sanitation development process is the use and care of water and sanitation facilities. People must use these facilities properly to obtain the health benefits inherent in them. This means knowing how to protect and store water safely, how to maintain personal and domestic cleanliness, how to care for excreta disposal facilities and how to avoid or minimize unsanitary environmental conditions. Knowledge transfer, behaviour change and personal responsibility are the key factors.

- The third of the interactive elements is the institutional support from the communities, developing agencies and government policies that provide a framework for water and sanitation improvements. Experience has shown that community-based efforts, whether in a small village or a large metropolis, are most effective in identifying and meeting peoples' needs. Governments, especially at the regional and national levels, are more effective as facilitators of the development process than providers of water and sanitation improvements.

- Water contaminated by human, chemical or industrial wastes can cause a variety of communicable diseases through ingestion or physical contact:

- Water-borne diseases: caused by the ingestion of water contaminated by human or animal faeces or urine containing pathogenic bacteria or viruses; include cholera, typhoid, amoebic and bacillary dysentery and other diarrhoeal diseases.

- Water-washed diseases: caused by poor personal hygiene and skin or eye contact with contaminated water; include scabies, trachoma and flea, lice and tick-borne diseases.

  - ➤ **Water-Based Diseases:** caused by parasites found in intermediate organisms living in water; include dracunculiasis, schistosomiasis and other helminths.

  - ➤ **Water-Related Diseases:** caused by insect vectors which breed in water; include dengue, filariasis, malaria, onchocerciasis, trypanosomiasis and yellow fever.

- No single type of intervention has greater overall impact upon the national development and public health than does the provision of safe drinking water and the proper disposal of human excreta. The direct effects of improved water and sanitation services upon health are most clearly seen in the case of water-related diseases, which arise from the ingestion of pathogens in contaminated water or food and from insects or other vectors associated with water. Improved water and sanitation can reduce morbidity and mortality rates of some of the most serious of these diseases by 20% to 80%.

6.5.1 Drinking Water Supply Programme and Policies at a Glance

Table 6.2

Year	Event
1949	The Environment Hygiene Committee (1949) (Bhor Committee) recommended the provision of safe water supply to cover 90 per cent of India's population in a timeframe of 40 years.
1950	The Constitution of India specifies water as a State subject.
1969	National Rural Drinking Water Supply Programme was launched with technical support from UNICEF and Rs.254.90 crore was spent during this phase with 1.2 million bore wells dug and 17,000 piped water supply schemes provided.

Conti...

1972-73	Introduction of the Accelerated Rural Water Supply Programme (ARWSP) by the Government of India to assist States and Union Territories to accelerate the pace of coverage of drinking water supply.	2005	The Government of India launches the Bharat Nirman Program, with emphasis on providing drinking water within a period of five years to 55,069 uncovered habitations, habitations affected by poor water quality and slipped back habitations based on 2003 survey. Revised sub Mission launched as component of ARWSP for focused funding of quality affected habitations.
1981	India as a party to the International Drinking Water Supply and Sanitation Decade (1981-1990) declaration sets up a national level Apex Committee to define policies to achieve the goal of providing safe water to all villages.	2007	Pattern of funding under Swajaldhara changed: 50:50 Centre-State shares.
1986	The National Drinking Water Mission (NDWM) launched to accelerate the process of coverage of the country with drinking water	2009	National Rural Drinking Water Programme launched from 1/4/2009 by modifying the earlier Accelerated Rural Water Supply Programme and subsuming earlier sub Missions, Miscellaneous Schemes and mainstreaming Swajaldhara principles.
1987	First National Water Policy drafted by Ministry of Water Resources giving first priority to drinking water supply.	2010	Department of Drinking Water Supply renamed as Department of Drinking Water and Sanitation
1991	The National Drinking Water Mission (NDWM) renamed as Rajiv Gandhi National Drinking Water Mission (RGNDWM).	2011	Department of Drinking Water and Sanitation upgraded as to the Ministry of Drinking Water and Sanitation
1994	The 73rd Constitution Amendment makes provision for assigning the responsibility of providing drinking water to the Panchayati Raj Institutions.	2012	Twelfth five - year plan focusing on piped water supply with 55 LPCD, earmarking of 5% funds for coverage of quality affected as well as 60 JE/AES affected districts
1999	Formation of separate Department of Drinking Water Supply in the Ministry of Rural Development, Govt. of India. For ensuring sustainability of the systems, steps are initiated to institutionalize community participation in the implementation of rural drinking water supply schemes through sector reform. Sector Reform ushers in a paradigm shift from the 'Government-oriented supply-driven approach' to the 'Peopleoriented demand driven approach'. The role of the government reoriented from that of service provider to facilitator. Total Sanitation Campaign (TSC) as a part of reform principles initiated in 1999 to ensure sanitation facilities in rural areas with the specific goal of eradicating the practice of open defecation. TSC gives strong emphasis on Information, Education and Communication, Capacity Building and Hygiene Education for effective behavioural change with involvement of PRIs, CBOs, and NGOs	2013	Launch of special Programme to address the rural water supply and sanitation issues of four low income States with collaboration of World Bank.
		2014	Focus on Innovationsin rural drinking water.
		2016	Focus on reforms in NRDWP to make it more outcome. Performance linked financing of the programme while keeping in mind that States with limited revenues and average performance are not left behind as well.
2002	Scaling up of sector reform initiated in the form of Swajaldhara Programme. The National Water Policy revised; priority given to serving villages that did not have adequate sources of safe water and to improve the level of service for villages classified as only partially covered. India commits to the Millennium Development Goals to halve the proportion of people without sustainable access to safe drinking water and basic sanitation by 2015, from 1990 levels.		

6.5.2 Water Quality Criteria

- For any water body to function adequately in satisfying the desired use, it must have corresponding degree of purity. Drinking water should be of highest purity. As the magnitude of demand for water is fast approaching the available supply, the concept of management of the quality of water is becoming as important as its quantity.

- Each water use has specific quality need. Therefore, to set the standard for the desire quality of a water body, it is essential to identify the uses of water in that water body. In India, the Central Pollution Control Board (CPCB) has developed a concept of designated best use.

- According to this, out of the several uses of water of a particular body, the use which demands highest quality is termed its designated best use. Five designated best uses have been identified. This classification helps the water quality managers and planners to set water quality targets and design suitable restoration programs for various water bodies.

Table 6.3 : Water Quality Criteria

Designated-Best-Use	Class of Water	Criteria
Drinking Water Source without conventional treatment but after disinfection	A	Total Coliforms Organism MPN/100ml shall be 50 or less pH between 6.5 and 8.5 Dissolved Oxygen 6mg/l or more Biochemical Oxygen Demand 5 days 20C 2mg/l or less
Outdoor bathing (Organised)	B	Total Coliforms Organism MPN/100ml shall be 500 or less pH between 6.5 and 8.5 Dissolved Oxygen 5mg/l or more Biochemical Oxygen Demand 5 days 20C 3mg/l or less
Drinking water source after conventional treatment and disinfection	C	Total Coliforms Organism MPN/100ml shall be 5000 or less pH between 6 to 9 Dissolved Oxygen 4mg/l or more Biochemical Oxygen Demand 5 days 20C 3mg/l or less
Propagation of Wild life and Fisheries	D	pH between 6.5 to 8.5 Dissolved Oxygen 4mg/l or more Free Ammonia (as N) 1.2 mg/l or less
Irrigation, Industrial Cooling, Controlled Waste disposal	E	pH betwwn 6.0 to 8.5 Electrical Conductivity at 25C micro mhos/cm Max.2250 Sodium absorption Ratio Max. 26 Boron Max. 2mg/l
	Below-E	Not Meeting A, B, C, D and E Criteria

6.5.3 Water and Sanitation in India's Census-2012

- The Census-2012 has given a very dismal picture water and sanitation facilities in India. Improper planning and casual implementation of schemes coupled with rampant corruption and irregularities in the concerned departments as well as lack of awareness among the public particularly in the rural areas have made India to be left behind in this most basic and important aspect of public health. The statistics on drinking water and sanitation of Assam also indicate a very disturbing picture.

- The Census-2011 report on water and sanitation says that out of 246,692,667 (191,963,935 in 2001) surveyed households 43.5% (36.7% in 2001) in India have tap water, 11% (18.2% in 2001) has wells, 42% (41.2% in 2001) hand pump/tube well and 3.5% (3.9% in 2001) has other sources of drinking water. The Census-2011 added two new queries on treated and untreated drinking water and on covered or uncovered source of drinking water.

- Here the report says only 32% of Indians use treated drinking water while 11.6% do not use treated drinking water. Similarly only 1.6% households use drinking water from covered sources while 9.4% do not have that. In Assam out of 6,367,295 households only 10.5% (9.2% in 2001) has tap water, 18.9% (26.7% in 2001) has wells and majority 59.4% (46.9% in 2001) use tube wells for drinking water followed by 11.3 % (14.6%) having drinking water from other sources.

- This is almost ten percentage drop of use of wells and more than 10 percent increase of the use of tube wells for drinking water in Assam. Likewise the only three percent fall of the use of drinking water from other sources like ponds and rivers is also a matter of concern and indicates the failure of concerned departments in policy making and implementation.

- In the rural sector the statistics of drinking water condition of the Census-2011 both the national and state figures are of mixed results. While 30.8% (24.3% in 2001) household in rural India has tap water sources for drinking water Assam has only 6.8% (5.4% in 2001). The national figure of wells for drinking water in the rural sector is 13.3% (22.2% in 2001) Assam has 19% (29% in 2001). 51.9% (48.9% in 2001) of rural household in India have drinking water from tube wells while the percentage in Assam is 61.5% (51.4%).

- On other sources of drinking water in rural India the findings are 4% (4.5% in 2001) and in rural Assam is 12.6% (16.2% in 2001). On treated drinking water the percentage in rural India is 17.9% and that of rural Assam is 5.8%. On untreated water the national figure in the rural sector is 13% and in Assam it is only 1%. Only 1.5% rural households in India collect drinking water from covered sources while in Assam the percentage is dismal 1.1%. Similarly on uncovered sources of drinking water the national figure of rural India is 11.8% and in Assam it is 18%.

- In the urban sector the all India Census-2011 findings of drinking water are like this:-70.6% (68.7% in 2001) using tap water, only 6.2% (7.7% in 2001) using wells, 20.8% (11.8% in 2001) using hand pumps/ tube wells and 2.55 (2.3% in 2001) using other sources. In Assam statistics are very unimpressive:-30.2% (31.4% in 2001) using tap water which is 1.02% less than the last census, 17.8% (24.6% in 2001) using wells, 48% (35.9% in 2001) using tube wells and 4% (5.1% in 2001) using from other sources. Similarly on the use of treated water in the urban sector Assam's figure (29.4%) is far less than the national figure (62%). 4.5% of urban households in Assam have drinking water from covered sources while 13.1% do not have such.

- The distance of availability of water from the households is one important matter of concern. The 2011 Census reveals encouraging picture of Assam from the national level. While 46.6% (39% in 2001) Indian households have availability of drinking water inside their premises in Assam the figure is well above 54.8% (37.9% in 2001). The availability of drinking water near the households in India is 35.8% (44.3% in 2001).

- In Assam it is also less than national figure, from 39.7% in 2001 to 26.7% in 2011. However the state is ahead on the availability of drinking water away from the household. Here Assam's figure is 18.5% (22.5% in 2001) and that of India is 17.6% (16.7% in 2001). On this same category Assam also has impressive figures in the rural sector than the all India figures with 50.4% (33.6% in 2001) of households having drinking water source available within the premises while the national figure is 35% (28.7% in 2001).

- There is a fall in households in rural Assam of availability of drinking water near the premises from 41.9% in 2001 to 29.3% this time. In India the figure is 42.9% (51.4% in 2001). The availability of drinking water away from the households in the rural sector is 22.1% (19.5% in 2001) at the national level while in Assam it is 20.4% (24.5% in 2001).

- Assam too has advanced in availability of drinking water sources within the households in the urban sector than the all India level with 78.8% (63.2% in 2001) while national figure is 71.2% (65.4% in 2001). Drinking water sources near urban households in India is 20.7% (25.2% in 2001) and in Assam is 12.8% (26.3% in 2001) while away from the households the national figure is 8.1% (9.4% in 2001) and the that of the state is 8.4% (10.5% in 2001).

- On 6th March, the WHO/UNICEF's Joint Monitoring Programme on sanitation for Millennium Development Goal released its report on India which indicated that 59% (626 million) Indians still does not have access to toilets and they use open defecation. The Census-2011 gives another disturbing account of India on sanitation which says 53.1% (63.6% in 2001) households in India does not have a toilet. In the rural sector the percentage is 69.3% (78.1% in 2001) and in the urban areas it is 18.6% (26.3% in 2001).

- In Assam overall 35.1% (35.4% in 2001) households have no toilet. That means in the last ten years the governmental schemes could reach only 0.3% of the households. In rural Assam the picture is as same as in 2001 with 40.4% households with no toilet in 2011 Census. In the urban areas the state has 6.3% (5.4% in 2001) households with no toilet.

- Assam also has over all only 28.5% (15.9% in 2001) households have toilets with water closet (all India figure is 36.4% as against 18% in 2001), 34.7% (43.9% in 2001) households have pit latrines (all India 9.4% as against 11.5%) and 1.8% (4.3% in 2001) households have other types of toilet (all India 1.1% as against 6.9% in 2001). In the rural sector Assam has 20.6% (8.6% in 2001) households having toilets with water closet (all India 19.4% as against 7.1% in 2001), 37.2% (46.9% in 2001) households have pit latrines (all India 10.5% as against 10.3% in 2001) and 1.8% (4% in 2001) households have other toilets (all India 0.8% as against 4.5% in 2001).

- Similarly in the urban sector the state has an impressive development with 71% (58.9% in 2001) households having toilet with water closet (all India 72.6% as against 46% in 2001). In the pit latrine category Assam has urban households 21.01% (26.4% in 2001) where the national figure is 7.1% (14.6% in 2001). The state has 1.7% (9.3%) urban households with other toilets (all India 1.7% as against 13% in 2001 Census).

- The increase of allocation of funds for rural drinking water in this year's union budget from Rs 11,000 Crores to Rs 14,000 Crores and from Rs 1500 Crores to Rs 3500 Crores for the rural sanitation programme is the immediate step taken by the union government to address this problem. However policy making and implementation and public awareness hold the key to improve the water and sanitation standards of India.

6.5.4 Sustainable Sanitation

- Sustainable sanitation recognizes that in order to be sustainable, a sanitation approach must be socially acceptable and economically viable. In this way, sustainable sanitation is a loop- based approach that differs fundamentally from the current linear concepts of waste water management, and that does not only recognize technology, but also social, environmental and economic aspects.

- Sustainable sanitation is an approach that considers sanitation holistically. It recognizes that human excreta and wastewater are not waste product, but valuable resources. This view is based on the fact that wastewater and excreta contain significant amount of energy plant nutrients and also water that can be recycled and reused, thus protecting natural resources.

- The main objective of a sanitation system is to protect and promote human health by providing a clean environment and breaking the cycle of disease. In order to be sustainable, a sanitation system has to be not only economically viable, socially acceptable, and technically and institutionally appropriate, it should also protect the environment and the natural resources.

- Today, the need for sustainability means that resource saving and protection of the environment are vital and there is a need for innovation and rethinking. This cannot be achieved by conventional methods. Also, in our emerging consumer and chemical societies it will not be enough that residents pay for sanitation and water services they have to be partners to make sanitation sustainable.

- Sustainable sanitation is a simple approach: the most basic principle is that is considers wastewater and excreta not as a waste, but as resources, that sanitation has to be socially acceptable and should be as economically viable as possible. There is no one- fit-all approach much rather, the most adequate solution has to be found from case to case, considering climate and water availability, agricultural practices, socio-cultural preferences, affordability, safety and technical prerequisites – just to name a few.

When improving an existing and/or designing a new sanitation system, sustainability criteria related to the following aspects should be considered:

- **Health and Hygiene :** Includes the risk of exposure to pathogens and hazardous substances that could affect public health at all points of the sanitation system from the toilet via the collection and treatment system to the point of reuse or disposal and downstream populations. This topic also covers aspects such as hygiene, nutrition and improvement of livelihood achieved by the application of a certain sanitation system, as well as downstream effects.

- **Environment and Natural Resources :** Involves the required energy, water and other natural resources for construction, operation and maintenance of the system, as well as the potential emissions to the environment resulting from its use. It also includes the degree of recycling and reuse practiced and the effects of these (e.g. reusing wastewater; returning nutrients and organic material to agriculture), and the protection of other non-renewable resources, e.g. through the production of renewable energies (such as biogas).

- **Technology and Operation :** Incorporates the functionality and the ease with which the entire system including the collection, transport, treatment and reuse and/or final disposal can be constructed, operated and monitored by the local community and/or the technical teams of the local utilities. Furthermore, the robustness of the system, its vulnerability towards power cuts, water shortages, floods, earthquakes etc. and the flexibility and adaptability of its technical elements to the existing infrastructure and to demographic and socio-economic developments are important aspects.

- **Financial and Economic Issues :** Relate to the capacity of households and communities to pay for sanitation, including the construction, operation, maintenance and necessary reinvestments in the system. Besides the evaluation of these direct costs also direct benefits e.g. from recycled products (soil conditioner, fertiliser, energy and reclaimed water) and external costs and benefits have to be taken into account. Such external costs are e.g. environmental pollution and health hazards, while benefits include increased agricultural productivity and subsistence economy, employment creation, improved health and reduced environmental risks.

- **Sociocultural and Institutional Aspects:** The criteria in this category refer to the socio-cultural acceptance and appropriateness of the system, convenience, system perceptions, gender issues and impacts on human dignity, the contribution to food security, compliance with the legal framework and stable and efficient institutional settings.

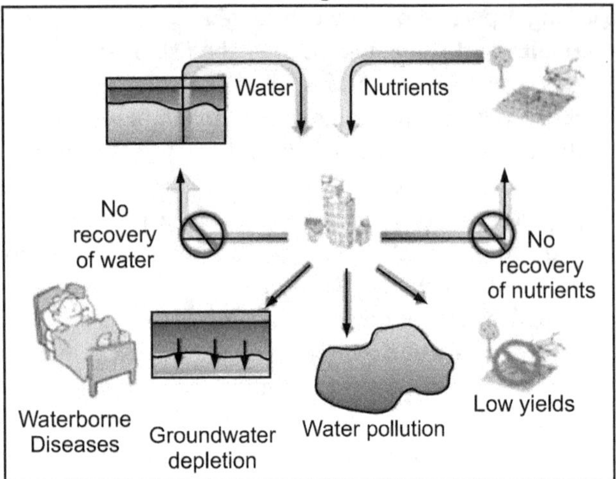

Fig. 6.2 : Sustainable Sanitation

Conventional approaches to wastewater management that regard wastewater as a waste, and often are dysfunctional, have serious drawbacks.

6.6 LOW COST SANITATION METHODS

Various low cost sanitation methods developed and tried. They are described here:

1. **Open Defecation**

   - Where there are no latrines people resort to defecation in the open. This may be indiscriminate or in special places for defecation generally accepted by the community, such as defecation fields, rubbish and manure heaps, or under trees. Open defecation encourages flies, which spread faeces-related diseases. In moist ground the larvae of intestinal worms develop, and faeces and larvae may be carried by people and animals.

Fig. 6.3 : Open defecation

   - Surface water run-off from places where people have defecated results in water pollution. In view of the health hazards created and the degradation of the environment, open defecation should not be tolerated in villages and other built-up areas. There are better options available that confine excrete in such a way that the cycle of reinfection from excrete-related diseases is broken.

2. **Shallow Pit**

   - People working on farms may dig a small hole each time they defecate and then cover the faeces with soil. This is sometimes known as the "cat" method. Pits about 300 mm deep may be used for several weeks. Excavated soil is heaped beside the pit and some is put over the faeces after each use.

   - Decomposition in shallow pits is rapid because of the large bacterial population in the topsoil, but flies breed in large numbers and hookworm larvae spread around the holes. Hookworm larvae can migrate upwards from excrete buried less than I m deep, to penetrate the soles of the feet of subsequent users.

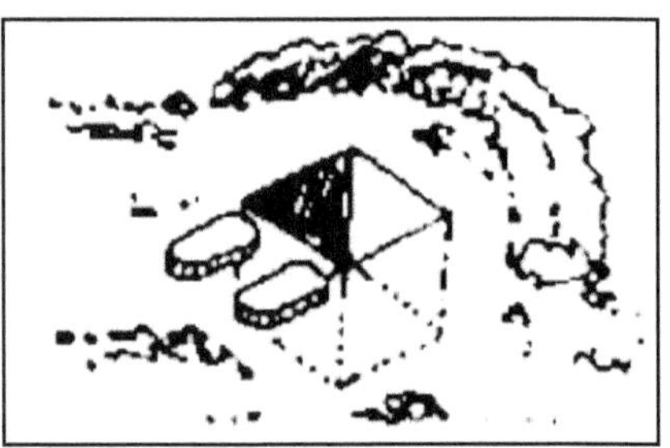

Fig. 6.4 : Shallow pit

3. **Trench/Pit Latrine**

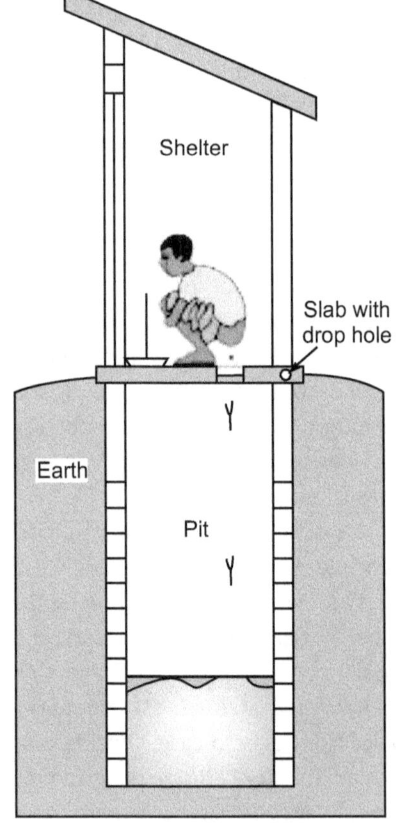

Fig. 6.5 : Trench/pit Latrine

- Trench latrine consists of a shallow pit 60 to 90 cm deep and 90 to 120 cm. Wide, either circular, rectangular and square, a wooden plank having a hole in the middle and covered from three sides without either by a tin or thatched. In a trench latrine the users have to put some earth and grass-leaves after defecation. The site of the trench latrine has to be changed after every six months when it is full.

Comment:

- Trench latrine, although better than open defecation, does not serve the purpose.

- It is very difficult to change its place every six months.

- More over it does not eliminate fly breeding and odour.

- Sometimes bacteria produced in the night soil used to float on the surface of the latrine. Therefore, the people could not adopt this system in general.

4. Bore-Hole Latrine

- Bore-latrine was developed under the joint collaboration of the All India Institute of Hygiene and Public Health, Calcutta, and the Rockefeller Foundation in pre independence days. It consists of a circular hole, usually 40 cms in diameter, bored vertically into the ground by means of an earth auger to a depth of 6-8 meters. The latrine floor and its super structure are also provided. This is used in the African, Western Pacific and South American countries.

Comment:

- The greatest difficulty in a borehole latrine is the collapse or caving-in of the pit walls particularly in alluvial soils.

- It requires special equipments for its construction and the possibility of water pollution is also very high in this system.

- Fly breeding is another serious problem in this type of latrine

- Due to these reasons the borehole latrine could not get acceptance on a mass scale. (Bhaskaran; 1966)

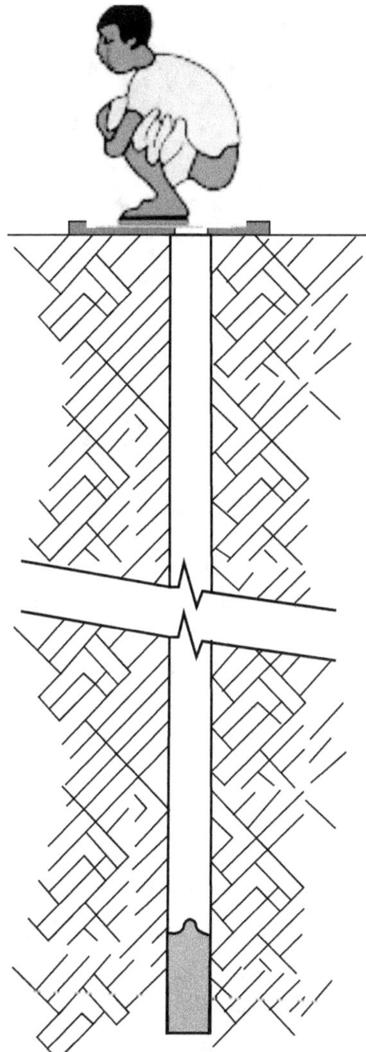

Fig. 6.6 : Bore hole latrine

5. Dug-Well Latrine

- It is about 75 cm in diameter and 3-6 metres deep in hard soil. The well is lined to prevent caving-in of the soil and is brimmed with concrete around its entry point. The squatting plate is placed over the pit for defecation with a suitable super structure for the privacy of the users.

Comment:

- The dug-will latrine is fairly expensive and

- It is difficult to construct it at places where the water table is high.

- The function of the dug-well is the same as that of borehole. If it penetrates ground water, it carries with it the risk of contamination.

Fig. 6.7 : Dug well Latrine

- Because of these limitations, the dug-well latrine cannot be used in most of the parts of India.

6. Ulta Matka Privy

- A large earthenware pot (matka) is buried upside down in a pit, at least 0.75 metre below the ground level, which acts a lining. The Matka is about 1.5 metre high. A hole is made to fix a pipe joining, the wc for discharging the waste. A layer of horse dung is laid at the bottom in the beginning to accelerate the process of decomposition.

- Salt water is also flushed once a week to help liquification of the faecal matter. The maximum leaching takes place from the open end at the bottom and a little from the porous side of the Matka.

Comment:

- The gases produced in the pit are absorbed by the soil. When the Matka gets filled up; it is kept close for about two years and the digested material can be used as fertilizer. During this period an alternative Matkahas to be provided

- Although Matka privy has been tried in rural areas, it has not gained popular acceptance and adoption on account of its very temporary nature.

7. Goupri

- Gopuri is a form of compost latrine provided with two tanks, which are constructed over the ground level, instead of being dug in the ground. A movable seat with pan is fixed on the tank in use. The filled up tank is covered with gray earth, ashes, leaves and domestic waste materials.

- Gopuri latrine has permeable bottom and, like other compost latrines, the night soil filling it up is transformed into compost after a certain period of time. It has a vent pipe, which keeps it reasonably odourless (Bhaskaran; 1966)

Comment:

- Although there is no handling of faeces or urine, fly breeding can be the problem with the Gopuri type. The main disadvantage, however, is the location of its receptacles above the ground level which makes it unsuitable for most of the households in India (Winblad; 1980)

8. Sopa Sandas Latrine

- It is type of compost latrine, which was first introduced in the State of Maharashtra. It consists of

    (i) RCC or stone slab with concrete or mosaic WC Pan.

    (ii) Steep sloping pipe with a tin flap at the upper end.

- Rectangular pit of 120 x 90 x 90 cm. divided into two parts with partial honey-comb brick lining.

- Y-pipe to connect both pits.

- A vent pipe to carry away odours from the pit. It has been observed that the tin flap prevents bad odour from entering the toilet and also prevents the passing of flies, etc. The flap, however, wears out after use and has to replace from time to time. The pit is covered with a sheet of tin when full and eventually the human excreta is transformed into compost. It is shoveled from the pit in course of time and spread in the fields to augment the fertility of the soil.

9. Hagebu Latrine

- This is also a modified form of pit latrine, which is dug upto 90 cms in the first instance with a diameter of 52.5 cm. The diameter is gradually increased to 1120 cm. at the base, while the depth is extended upto 480 to 510 cm. A latrine seat is placed over the pit and a suitable superstructure is also built (Bhaskaran; 1966)

Comment:

- Here also space is a problem, as once the pit is filled up, another fresh pit hole has to be dug nearby. The question of its popular acceptance, therefore, does not arise.

10. Barapalli Type of Latrine

- For the environment of villages, Dr. Edwin Abbot developed a pit latrine in the Barapalli village of the State of Orrissa, India. This type of latrine is 90 to 150 cm.deep with a diameter of 75 cm. About 200 sq. ft. of land is needed for the installation of this latrine. A container of tin or an earthen vessel is provided to store water for flushing and also for washing after defecation. A broom is also provided for cleaning the pan. (Bhaskaran; 1966)

11. Prai Type Latrine

- After a number of trials and intensive research spread over a long period, the Planning and Research Action Institute, Lucknow, India, has designed hand flush water-seal latrine with a pan and a trap. Two types were developed- a direct type in which the latrine is placed over the pit and the other type in which the latrine is away from the pit and connected with a pipe.

- Two or three litres of water are usually sufficient to make by its side, which is connected to the latrine. The contents of the first pit may be used as manure after it is left for a sufficient period.

Comment:

- The one pit system is not very filled up. In most of the cases house owners dropped the idea of getting constructed the other one. Secondly, the cost of the second pit would go up with the passage of time.

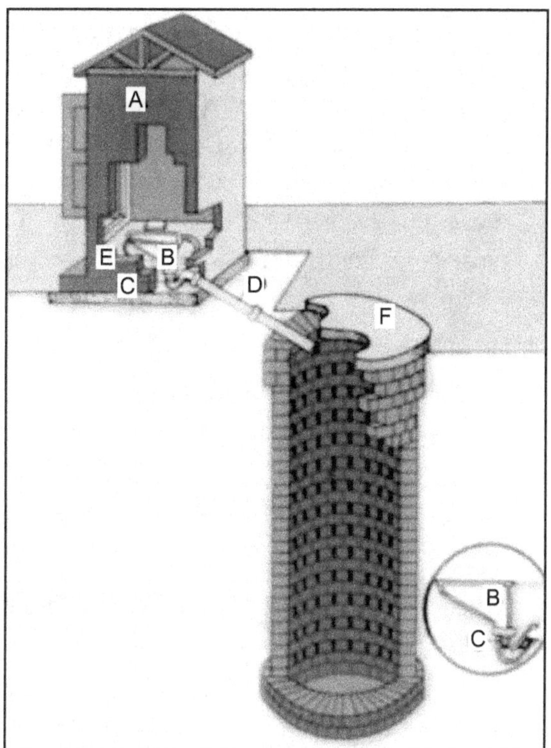

Fig. 6.8 : Prai type latrine

- The PRAI type latrine was constructed mostly during the later part of the 40's and in the early 50's, but it remained confined mostly to rural areas.

12. Ventilated Improved Pit (VIP)

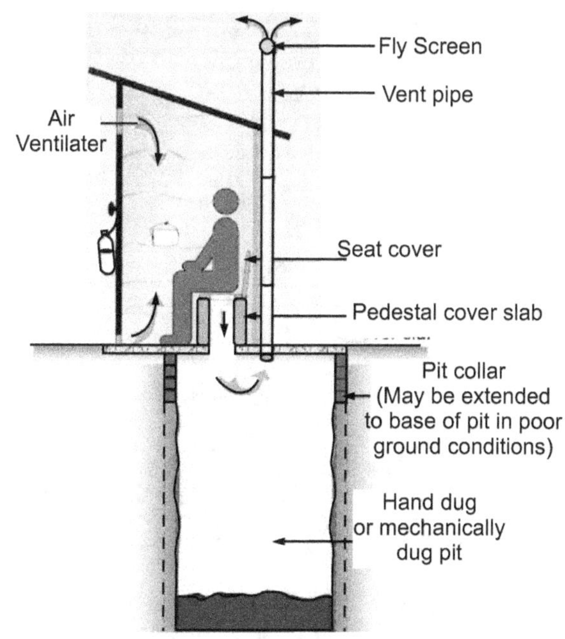

Fig. 6.9 : Ventilated Improved

- This is based on an improved design of the pit latrine developed in Rhodesia. Aerobic action takes place in the through a cycle provided by air-suction through the toilet seat and up the vent pipe, which is warmed by the sun and thereby draws up the air. It is claimed that the odour emanating from the pit is expelled through the vent pipe.

Comment:

- As flies cannot enter the pit through the flue pipe, chances of fly breeding in the pit is reduced.

- Other deficiencies, however, continue to exist. The latrine cannot be constructed at places where the groundwater table is high.

- When the pit gets filled up, a new latrine has to be constructed.

- However, the VIP latrine is a considerable improvement over the traditional pit latrine.

13. Ventilated Improved Double Pit Latrine (VIDP)

- VIDP latrines differ from the VIP in only respect, viz. that it has two alternating pits. When one is filled up, it should rest for atleast one year before it is emptied to ensure pathogen destruction. The operation and maintenance is the same as that of the VIP. Constructing a separate wall in the VIP pit provides two pits, or by once the pit is full, is precluded.

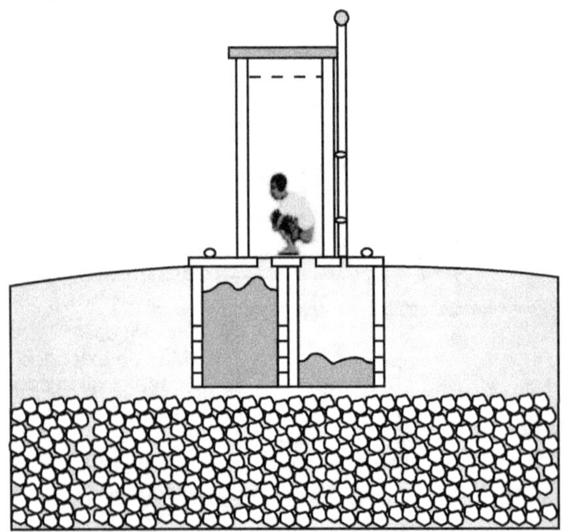

Fig. 6.10 : Ventilated Improved Double Pit Latrine (VIDP)

Comment:

- As the VIP and VIDP are designed for use without water, they require good maintenance, which consists principally of keeping the squatting plate and superstructure clean. Otherwise, the risk of fly and mosquito breeding is increased.

- The ventilation system has also to be designed properly to minimize fly and other nuisance.

14. Reid's Odourless Earth Closet (ROEC)

- An alternate design for the VIP latrine is the ROEC. In this latrine, the pit is completely offset and the excrement is introduced into the pit via a chute. A vent pipe is provided as in the VIP latrine.

Comment:

- A serious disadvantage of the ROEC, however, is that the chute is easily fouled with the excrement and this may cause fly breeding. The chute is, therefore, to be cleaned regularly by brush.

15. RCA Latrine

- An appropriate design of latrine for rural areas has been designed under the RCA project. Particularly the Poonamallee Centre, near Madras, did considerable work in this direction. A composite design was prepared which embodied the salutary features of the earlier designs.

16. Chemical Toilet

- The chemical closet consists of a metal tank containing a solution of caustic soda. A seat with cover is placed over the tank, which is ventilated by a flue rising through the house roof. The excreta deposited in the tank is liquefied and sterilized by the chemical, which also destroys all the pathogens. After several months of operation, this spent chemical and liquefied matter are drained or removed.

Comment:

- This system is quite satisfactory, safe and hygienic. But, the cost of construction and maintenance of chemical toilet is prohibitive, which is why this system could not become popular.

- This system cannot, therefore, be recommended for a large-scale use.

17. Aqua Privy

- The aqua privy consists of a tank filled with water into which plunges a drop pipe hanging from the latrine floor. The excreta and the urine fall through the drop pipe into the tank where they under go anaerobic decomposition as in septic tank. The digested sludge, which reduced to about a quarter of the volume of the deposited excreta, accumulates in the tank and has to be removed at intervals.

Comment:

- In aqua privy, there is no provision of soak pits for the discharge of the effluents. This is therefore, not hygienic.

- Secondly, whenever the water level falls below the drop-pipe, the smell comes out of the tank and the entire surrounding is filled with stink.

- This system could not, therefore, be adopted on a large scale.

18. Vietnam Compost Toilet (The Double Vault Latrines)

- The double vault latrines consists of two receptacles, each with a volume of 300 litres. The receptacles are covered with a squatting slab, which has two holes, foot rests and a channel for urine. Faeces are deposited in one of the receptacles, which can be used for about three to six months by a household of 5-10 persons. Urine is drained away and collected in a jar behind the latrine.

- The input in the receptacles is, thus, only faeces, ashes and toilet paper. The contents are fairly dry and the decomposition process is the first container is two-thirds full; it is filled with dried and powered earth, ash or paper and then sealed.

- The other one is used in its place. When the second vault is nearly full, the first is open and emptied. The decomposed excrement, how odourless, provides a good fertilizer. Vietnamese health authorities claim that after 45 days in a sealed container, all bacteria and pathogens get killed.

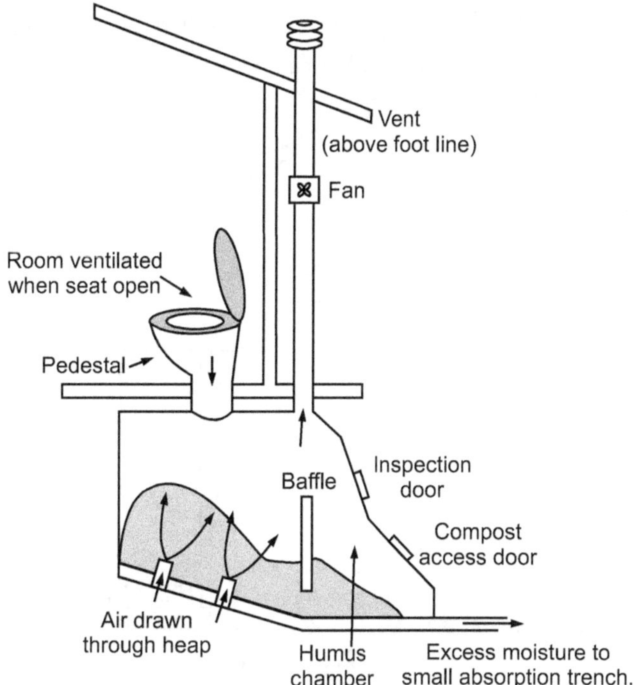

Fig. 6.11 : Double Vault Composting Latrine

Comment:

- From health point of view, this system should be acceptable but it cannot be recommended in India where water is used after ablutions.

- Moreover, it is built above the ground; the latrine cannot be constructed inside the house and has to be located.

Common Problems with Composting Toilets are:

- Odours from the compost pile caused by ineffective composting or not enough ventilation

- Clogging in compost chamber caused by not regularly removing humus

- The wrong system in wrong location. Wet composting toilets are best located away from a dwelling, while dry composting toilets are more suitable under houses.

6.7 THE POUR FLUSH LATRINE

- These latrines have a special pan cast in a cover slab above the pit, which provides a 'water seal' to contain bad smells. The pit can be directly below the hole or 'offset' so that, for example, the latrine can be inside the home and the pit outside. Offset pits require a pipe to be fitted between pan and pit and need more water to flush. It's also possible to have two pits so that, when one is full, the other can be used.

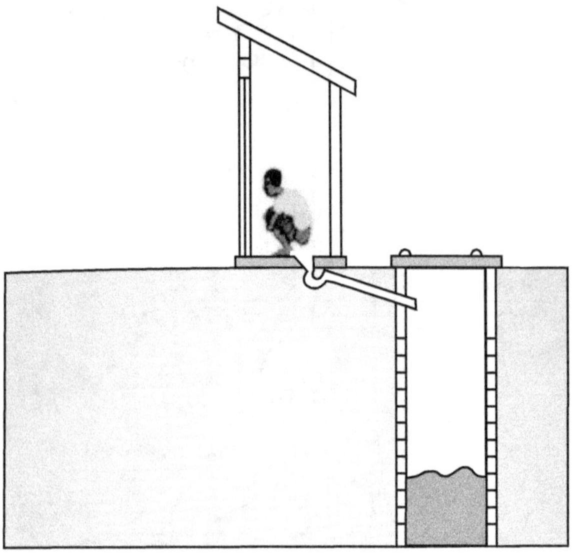

Fig. 6.12 : Pour flush latrine

- Septic tanks are already discussed in previous modules.

6.8 SCHEMES AND POLICIES

6.8.1 National Urban Sanitation Policy

- Sanitation is defined as safe management of human excreta, including its safe confinement treatment, disposal and associated hygiene-related practices. While this policy pertains to management of human excreta and associated public health and environmental impacts, it is recognized that integral solutions need to take account of other elements of environmental sanitation, i.e. solid waste management; generation of industrial and other specialized / hazardous wastes; drainage; as also the management of drinking water supply.

- According to Census 2001, 27.8% of Indians, i.e. 286 million people or 55 million households live in urban areas projections indicate that the urban population would have grown to 331 million people by 2007 and to 368 million by 2012. 12.04 million (7.87 %) Urban households do not have access to latrines and defecate in the open. 5.48 million (8.13%) Urban households use community latrines and 13.4 million households (19.49%) use shared latrines. 12.47 million (18.5%) households do not have access to a drainage network. 26.83 million (39.8%) households are connected to open drains.

- The status in respect of the urban poor is even worse. The percentage of notified and non-notified slums without latrines is 17 percent and 51 percent respectively. In respect of septic latrines the availability is 66 percent and 35 percent. In respect of underground sewerage, the availability is 30 percent and 15 percent respectively. More than 37% of the total human excreta generated in urban India, is unsafely disposed.

- This imposes significant public health and environmental costs to urban areas that contribute more than 60% of the country's GDP. Impacts of poor sanitation are especially significant for the urban poor (22% of total urban population), women, children and the elderly. The loss due to diseases caused by poor sanitation for children under 14 years alone in urban areas amounts to Rs. 500 Crore at 2001 prices (Planning Commission-United Nations International Children Emergency Fund (UNICEF), 2006).

- Inadequate discharge of untreated domestic / municipal wastewater has resulted in contamination of 75 percent of all surface water across India. The Millennium Development Goals (MDGs) enjoin upon the signatory nations to extend access to improved sanitation to at least half the urban population by 2015, and 100% access by 2025. This implies extending coverage to households without improved sanitation, and providing proper sanitation facilities in public places to make cities open defecation free.

6.8.2 National Water Policy

- A scarce natural resource, water is fundamental to life, livelihood, food security and sustainable development. India has more than 18 % of the world's population, but has only 4% of world's renewable water resources and 2.4% of world's land area. There are further limits on utilizable quantities of water owing to uneven distribution over time and space.

- In addition, there are challenges of frequent floods and droughts in one or the other part of the country. With a growing population and rising needs of a fast developing nation as well as the given indications of the impact of climate change, availability of utilizable water will be under further strain in future with the possibility of deepening water conflicts among different user groups.

- Low consciousness about the scarcity of water and its life sustaining and economic value results in its mismanagement, wastage, and inefficient use, as also pollution and reduction of flows below minimum ecological needs. In addition, there are inequities in distribution and lack of a unified perspective in planning, management and use of water resources. The objective of the National Water Policy is to take cognizance of the existing situation, to propose a framework for creation of a system of laws and institutions and for a plan of action with a unified national perspective.

The present scenario of water resources and their management in India has given rise to several concerns, important amongst them are:

- Large parts of India have already become water stressed. Rapid growth in demand for water due to population growth, urbanization and changing lifestyle pose serious challenges to water security.

- Issues related to water governance have not been addressed adequately. Mismanagement of water resources has led to a critical situation in many parts of the country.

- There is wide temporal and spatial variation in availability of water, which may increase substantially due to a combination of climate change, causing deepening of water crisis and incidences of water related disasters, i.e., floods, increased erosion and increased frequency of droughts, etc.

- Climate change may also increase the sea levels. This may lead to salinity intrusion in ground water aquifers / surface waters and increased coastal inundation in coastal regions, adversely impacting habitations, agriculture and industry in such regions.

- Access to safe water for drinking and other domestic needs still continues to be a problem in many areas. Skewed availability of water between different regions and different people in the same region and also the

intermittent and unreliable water supply system has the potential of causing social unrest.

6.8.3 National Rural Health Mission

- The National Rural Health Mission (NRHM) was launched by the Hon'ble Prime Minister on 12th April 2005, to provide accessible, affordable and quality health care to the rural population, especially the vulnerable groups. The Union Cabinet vide its decision dated 1st May 2013, has approved the launch of National Urban Health Mission (NUHM) as a Sub-mission of an over-arching National Health Mission (NHM), with National Rural Health Mission (NRHM) being the other Sub-mission of National Health Mission.

- NRHM seeks to provide equitable, affordable and quality health care to the rural population, especially the vulnerable groups. Under the NRHM, the Empowered Action Group (EAG) States as well as North Eastern States, Jammu and Kashmir and Himachal Pradesh have been given special focus.

- The thrust of the mission is on establishing a fully functional, community owned, decentralized health delivery system with inter-sectoral convergence at all levels, to ensure simultaneous action on a wide range of determinants of health such as water, sanitation, education, nutrition, social and gender equality. Institutional integration within the fragmented health sector was expected to provide a focus on outcomes, measured against Indian Public Health Standards for all health facilities.

6.8.4 National Urban Health Mission

- The National Urban Health Mission (NUHM) as a sub-mission of National Health Mission (NHM) has been approved by the Cabinet on 1st May 2013.

- NUHM envisages to meet health care needs of the urban population with the focus on urban poor, by making available to them essential primary health care services and reducing their out of pocket expenses for treatment. This will be achieved by strengthening the existing health care service delivery system, targeting the people living in slums and converging with various schemes relating to wider determinants of health like drinking water, sanitation, school education, etc. implemented by the Ministries of Urban Development, Housing and Urban Poverty Alleviation, Human Resource Development and Women and Child Development.

6.8.5 NUHM would Endeavour to Achieve its Goal Through

- Need based city specific urban health care system to meet the diverse health care needs of the urban poor and other vulnerable sections.

- Institutional mechanism and management systems to meet the health-related challenges of a rapidly growing urban population.

- Partnership with community and local bodies for a more proactive involvement in planning, implementation, and monitoring of health activities.

- Availability of resources for providing essential primary health care to urban poor.

- Partnerships with NGOs, for profit and not for profit health service providers and other stakeholders.

- NUHM would cover all State capitals, district headquarters and cities/towns with a population of more than 50000. It would primarily focus on slum dwellers and other marginalized groups like rickshaw pullers, street vendors, railway and bus station coolies, homeless people, street children, construction site workers.

- The centre-state funding pattern will be 75:25 for all the States except North-Eastern states including Sikkim and other special category states of Jammu and Kashmir, Himachal Pradesh and Uttarakhand, for whom the centre-state funding pattern will be 90:10.The Programme.

- Implementation Plans (PIPs) sent by the by the states are apprised and approved by the Ministry.

6.8.6 Central Rural Sanitation Programme

- Individual Health and hygiene is largely dependent on adequate availability of drinking water and proper sanitation. There is, therefore, a direct relationship between water, sanitation and health. Consumption of unsafe drinking water, improper disposal of human excreta, improper environmental sanitation and lack of personal and food hygiene have been major causes of many diseases in developing countries.

- India is no exception to this. Prevailing High Infant Mortality Rate is also largely attributed to poor sanitation. It was in this context that the Central Rural Sanitation Programme (CRSP) was launched in 1986 primarily with the objective of improving the quality of life of the rural people and also to provide privacy and dignity to women.

- The concept of sanitation was earlier limited to disposal of human excreta by cesspools, open ditches, pit latrines, bucket system etc. Today it connotes a comprehensive concept, which includes liquid and solid waste disposal, food hygiene, and personal, domestic as well as environmental hygiene.

- Proper sanitation is important not only from the general health point of view but it has a vital role to play in our individual and social life too. Sanitation is one of the basic determinants of quality of life and human development index. Good sanitary practices prevent contamination of water and soil and thereby prevent diseases. The concept of sanitation was, therefore, expanded to include personal hygiene, home sanitation, safe water, garbage disposal, excreta disposal and waste water disposal.

- A comprehensive Baseline Survey on Knowledge, Attitudes and Practices in rural water supply and sanitation was conducted during 1996-97 under the aegis of the Indian Institute of Mass Communication, which showed that 55% of those with private latrines were self-motivated. Only 2% of the respondents claimed the existence of subsidy as the major motivating factor, while 54% claimed to have gone in for sanitary latrines due to convenience and privacy. The study also showed that 51% of the respondents were willing to spend upto Rs.1000/- to acquire sanitary toilets.

- Swachh Bharat or Swachh Bharat Abhiyan (Hindi: स्वच्छभारतअभियान, English: Clean Indian Mission) is a national level campaign by the Government of India covering 4041 statutory towns to clean the streets, roads and infrastructure of the country. This campaign was officially launched on 2 October 2014 at Rajghat, New Delhi, where Prime Minister Narendra Modi himself wielded broom and cleaned a road. The campaign is India's biggest ever cleanliness drive and 3 million government employees and schools and colleges students of India participated in this event.

- The mission was started by Narendra Modi, the Prime Minister of India, nominating nine famous personalities for this campaign, and they take up the challenge and nominate nine more people and so on(like the branching of a tree). It has been carried forward since then with famous people from all walks of life joining it.

- Swachh Bharat Abhiyan was announced by Prime Minister of India Narendra Modi on Indian Independence Day and launched on 2 Oct 2014, Gandhi Jayanti. On this day, Modi addressed the citizens of India in a public gathering held at Rajghat, New Delhi, India and asked everyone to join this campaign. Later on this day, Modi himself swept a parking area at Mandir Marg Police Station followed by pavement in Valmiki Basti, a colony of sanitation workers, at Mandir Marg, near Connaught Place, New Delhi. On 2 October, Anil Ambani, an Indian industrialist and a participant in this event, told in a statement.

- I am honoured to be invited by our respected Prime Minister Shri Narendrabhai Modi to join the "Swachh Bharat Abhiyan"... I dedicate myself to this movement and will invite nine other leading Indians to join me in the "Clean India" campaign.

- Indian President Pranab Mukherjee asked every Indian to spend 100 hours annually in this drive. This campaign is supported by the Indian Army, Border Security Force, Indian Air Force and India.

6.8.7 Objectives of Swachh Bharat Abhiyan

- Construction of individual, cluster and community toilets.

- To eliminate or reduce open defecation. Open defecation is one of the main causes of deaths of thousands of children each year.

- Not only latrine construction, the Swachh Bharat Mission will also make an initiative of establishing an accountable mechanism of monitoring latrine use.

- Public awareness will also be provided about the drawbacks of open defecation and promotion of latrine use.

- Proper, dedicated ground staff will be recruited to bring about behavioural change and promotion of latrine use.

- For proper sanitation use, the mission will aim at changing people's attitudes, mindsets and behaviours.

- Villages to be kept clean with Solid and Liquid Waste Management.

- Solid and liquid waste management through gram panchayats.

- To lay water pipelines in all villages, ensuring water supply to all households by 2019.

- To make India Open Defecation Free (ODF) India by 2019, by providing access to toilet facilities to all.

- To provide toilets, separately for Boys and Girls in all schools by 15.8.2015.

- To provide toilets to all Anganwadis.

6.8.8 Swachh Bharat Mission: Gramin Areas

- The Nirmal Bharat Abhiyan has been restructured into the Swachh Bharat Mission (Gramin). The mission aims to make India an open defecation free country in Five Years. Under the mission, One lakh thirty four thousand crore rupees will be spent for construction of about 11 crore 11 lakh toilets in the country.

- Technology will be used on a large scale to convert waste into wealth in rural India in the forms of bio-fertilizer and different forms of energy. The mission is to be executed on war footing with the involvement of every gram panchayat, panchayat samiti and Zila Parishad in the country, besides roping in large sections of rural population and school teachers and students in this endeavor.

- As part of the mission, for rural households, the provision for unit cost of individual household latrine has been increased from Rs 10,000 to Rs 12,000 so as to provide for water availability, including for storing, hand-washing and cleaning of toilets. Central share for such latrines will be Rs 9,000 while state share will be Rs 3,000. For North Eastern states, Jammu and Kashmir and special category states, the Central share will be 10,800 and the state share Rs 1,200. Additional contributions from other sources will be permitted.

6.8.9 Objectives of the Swachh Bharat Mission (Gramin)

- Bring about an improvement in the general quality of life in the rural areas.

- Accelerate sanitation coverage in rural areas to achieve the vision of Swachh Bharat by 2019 with all Gram Panchayats in the country attaining Nirmal status.

- Motivate communities and Panchayati Raj Institutions promoting sustainable sanitation facilities through awareness creation and health education.

- Encourage cost effective and appropriate technologies for ecologically safe and sustainable sanitation.

- Develop community managed environmental sanitation systems focusing on solid and liquid waste management for overall cleanliness in the rural areas.

EXERCISE

1. Discuss history of sanitation.

2. Explain main features of Total Sanitation Campgain.

3. Discuss silent features of Swachh Bharat Mission.

4. Explain VIP and VIDP Latrines.

5. Give overview of schemes and policies by Govt of India on rural Sanitation.

6. Explain water Quality Criteria

7. What is sustainable sanitation? Explain in detail.

8. Discuss Composting type latrine.

MODEL QUESTION PAPERS

PAPER I

Time : 3 Hours   **Max. Marks : 60**

Instructions to the Candidates :

1. Each question carries 12 marks.

2. Attempt **any five** questions of the following.

3. Illustrate your answers with neat sketches, diagram etc., wherever necessary.

4. If some data is noticed to be missing, you may appropriately assume it and should mention it clearly.

---

1. **(a)** Explain significance of Screen and grit chamber Classify screens. **(6)**

    (b) Explain working of trickling filter with sketch. Which equation is used in design of HRTF? Explain the same with meaning of each and every term. **(6)**

2. **(a)** Design a septic tank for 100 users. Assume all necessary data. **(6)**

    (b) Write detailed note on DO sang curve **(6)**

3. **(a)** Differentiate between Industrial and municipal wastewater. **(6)**

    (b) Explain with examples Waste Volume reduction techniques. **(6)**

4. **(a)** List our various techniques used for dissolved solids removal from industrial waste. Explain air stripping method in detail. **(6)**

    (b) What is CETP? Explain need for CETP. **(6)**

    Also discuss O and M problems in CETP.

5. **(a)** Define Communicable Disease. Explain Transmission modes of communicable disease. **(6)**

    (b) Explain methods for vector and rodent control. **(6)**

6. **(a)** Discuss status of rural sanitation in India **(6)**

    (b) Enlist various low cost excreta disposal methods. Explain VIP and VIDP. **(6)**

PAPER II

Time : 3 Hours   **Max. Marks : 60**

Instructions to the Candidates :

1. Each question carries 12 marks.

2. Attempt **any five** questions of the following.

3. Illustrate your answers with neat sketches, diagram etc., wherever necessary.

4. If some data is noticed to be missing, you may appropriately assume it and should mention it clearly.

1. **(a)** Derive Equation for Sludge age.   (6)

   (b) Explain in detail activated sludge process.   (6)

2. **(a)** Enlist the factors governing the reoxygenation and deoxygenation of stream.   (6)

   (b) Mention various methods of wastewater disposal. Discuss their merits and demerits. Explain the conditions favourable for their adoption.   (6)

3. **(a)** Write a difference between domestic waste water and industrial waste water.   (6)

   (b) Explain strength of reduction.   (6)

4. **(a)** Explain Air stripping with schematic diagram.   (6)

   (b) Write notes on :   (6)

   (i) Adsorption

   (ii) Physical adsorption

5. **(a)** Explain various ways of protection from communicable diseases.   (6)

   (b) Explain control measures for   (6)

   (i) Plague – Rats

   (ii) Malaria- Mosquito

   (iii) Flies

6. **(a)** Give overview of schemes and policies by Govt. of India on rural Sanitation.   (6)

   (b) What is sustainable sanitation? Explain in detail.   (6)